FUNCTIONAL SIGNS

FUNCTIONAL SIGNS
A New Approach from Simple to Complex

Harry Bornstein, Ph.D.
Professor Emeritus
Gallaudet College
and
I. King Jordan, Ph.D.
Professor of Psychology
Gallaudet College
Illustrated by **Ralph Miller, Sr.**

[redacted]

University Park Press
Baltimore

UNIVERSITY PARK PRESS
International Publishers in Medicine and Human Services
300 North Charles Street
Baltimore, Maryland 21201

Manufactured in the United States of America by The Maple Press Company.

Library of Congress Cataloging in Publication Data
Bornstein, Harry.
Functional Signs

Includes index.
1. Sign language 2. Deaf—Means of communication.
I. Jordan, I. King. II. Title.
HV2474.B65 1983 419 83-16825
ISBN 0-8391-1839-2

Contents

Preface

Some will look at this book and think "What, another sign language vocabulary book?" We want to assure you that this is not the case. This is not simply another sign vocabulary book. Here is how it differs: A core vocabulary of 330 functional signs is described in terms of how important each sign characteristic (i.e., handshape, location, or movement) is for understanding. The implications of this may be quite profound.

Traditionally, teachers of sign vocabulary have thought it necessary to give equal emphasis to all aspects of a sign. Frequently, the result has been to label as incorrect the production of a sign in which any of the characteristics differ significantly from the citation form. Teachers of special children frequently report tremendous frustration in not being able to successfully elicit a reasonable version of the citation form of a sign from their students. Nevertheless, they have continued to try to elicit such signs because they believed it necessary for the children to be understood, and, we imagine, succeeded only in increasing their (and the children's!) frustration. Or else they reluctantly accepted a version of a sign which was not likely to be understood by those who had not been involved in the instructional process with that individual. Our research shows that these practices need not be followed. While it is true that some of the characteristics for some signs are absolutely essential for understanding, other characteristics may not be necessary at all. We call those characteristics that are necessary for understanding, *fragile* characteristics. Fragile, because, without their being made correctly, the sign is not understood or is confused with an altogether different sign. We call those characteristics that are *not* necessary for understanding, *robust* characteristics. Robust, because, without them, understanding is still strong.

Perhaps an example will help make this clear. The sign for *father* is made by placing the thumb of the "5" hand at the forehead and wiggling the fingers. The wiggling motion is robust. That is, if the hand remains perfectly still, the sign will still be understood to mean *father.* The location on the forehead is fragile. That is, if the sign is made in any other location, it will probably not be read as *father.*

Through careful study of each of these characteristics of every sign, we are able to present a numerical value of understandability when a given characteristic is missing. Based upon these numerical values, we have developed a series of suggestions as to how you might better teach special children and hearing adults who might need to learn signs. We hope that these suggestions are of value to you.

Several people have been of significant assistance in the preparation of this text. Thanks to Macalyne Fristoe and Lyle L. Lloyd for providing us with the list of words. Many thanks to our signing model, Thelma Gonzales Schroeder. Technical assistance provided at Gallaudet College by Linda Carman of the Computer Center, Tom Allen of the Office of Demographic Studies, and Sandy White Maley and Tom Klagholz of the TV studio was of tremendous help. Finally, our most sincere appreciation to several graduate students who have been involved in virtually every stage of the research, Al Hoekstra, Dot Johnson, Jeanne Karlecke, Kristin Nowell, Pam Rush, and Pam Ryan.

Introduction

For some years now manual signs have increasingly been used in settings and by people outside of the deaf social community. Signs are now commonly used in some educational settings for both the hearing impaired and for those with no hearing impairment, in particular, those who have shown delayed spoken language development and who may be severely handicapped, such as the profoundly retarded and the autistic. Further, a large number of parents and such professionals as teachers, speech therapists, and interpreters have learned to use signs to communicate with those persons who can profit from their use. It is for all of these users that we believe that the information in this book will be helpful.

Notwithstanding this surge of use of manual signs with severely handicapped persons, the amount of success, i.e., the number of signs learned, is usually quite limited. This is not surprising because signs are almost always introduced late in a child's life, after a history of failure in speech or language development. Still, however limited, the acquisition of any number of signs is almost always a significant advance over previous failures. Consequently, a number of investigators, rather than being discouraged by these results, have begun to investigate how signs can be better taught to handicapped persons.

We do not propose to review once again the literature on why signs are used as an augmentative system of communication. Many have done that very well indeed (e.g., Reichle et al., 1981). It seems obvious to us that signs are useful because they are portable and do not require complete or even partial vocalization to convey meaning. It is also obvious that there are a very large number of adults, both parents and professionals, who are willing to learn to use signs with handicapped persons. What is less obvious is what cognitive and motor ability levels are needed by the handicapped to enable them to use signs with any facility. Rather than attempt to measure these levels directly, most investigators have tried instead to determine what attributes or characteristics of signs are related to their being better learned by handicapped persons. When we have determined which are the easiest signs to learn, then we would be in a better position to ascertain minimum skill levels, and to specify reasonable curricula.

To date, the principal attributes investigated are:

- the functional utility of the sign to the learner
- the representational value of the sign, i.e., how well one can guess what the sign means (transparency) or, when told what

a sign is "supposed to represent," the concept appears reasonable (translucency).

- the physical and motoric characteristics of the sign.

As might be expected those signs that are functionally useful to the signer are indeed learned most quickly. Somewhat unexpectedly, the findings on representational value are still somewhat equivocal (Kohl, 1982; Luftig and Lloyd, 1981). It is quite possible that what is transparent or translucent to college students and therapists is beyond the cognitive skills of most of the handicapped. More work needs to be done in this area. Finally, there is some evidence to support the more rapid learning of the following sign characteristics:

- one-handed signs are more easily learned that two-handed signs
- two-handed symmetrical signs more than nonsymmetrical signs
- signs that involve touching another part of the body more than nontouching signs
- signs clearly visible to the signer rather than those peripherally visible to the signer.

Our work deals with signs in quite a different way. We suspected that all of the visual information inherent in a sign may not be needed for that sign to be recognized and understood. If this proved to be the case, we might be able to simplify further the learning task for the handicapped. Also we believe that this approach would be consistent with and could be used together with the information reported above.

BASIC LOGIC OF THIS WORK

Manual signs can be described in terms of movement, handshape, and location with reference to the face and body.[1] Each of these characteristics or features of a sign represents a cluster of visual information that *may or may not* need to be present for a reader to be able to recognize the sign. If a given cluster of information *need not be present,* then it might be possible to use a simpler form of that sign with those children and/or adults who have difficulty learning the usual version of a sign. A sign can be regarded as simpler if the movement usually executed need not be made, the usual handshape not be formed, and/or the usual site of location ignored. By using simpler forms it may be possible to reduce the motoric and/or perceptual demands upon the learner, thereby enabling him or her to read and/or execute more signs and communicate better with family, friends, and teachers. We call simpler sign forms that are still highly understandable, *robust.*

---

[1] Other characteristics such as palm orientation and facial expression are also useful descriptors, but were not included in this work because they were not easily amenable to experimental or classroom control.

On the other hand, if a cluster of information *must be present* for a sign to be recognized and understood, then that cluster can be thought of as critical visual information. We call such signs, *fragile.* Teachers of the American Sign Language and of Signed English as well as interpreters should be prepared to articulate these critical features clearly so that they are more easily understood by deaf persons. Further, hearing persons new to signs may enhance their own reading of signs more rapidly if teaching procedures emphasize attending to the critical feature(s) of a sign.

It is also important to know why a sign feature is critical visual information. There appear to be two basic reasons: First, the feature may be necessary for the reader to be able to make any kind of reasonable judgment as to what the sign is. Second, that feature may be necessary to distinguish one sign from a different, similarly executed sign. A convenient and perfectly logical example of the latter type can be seen with the different days of the week. Handshape in the form of a different manual letter is the critical feature for distinguishing between these signs. Otherwise, a viewer would be confused as to which day is meant. However, we have found that most confusions are not perfectly logical, except after the fact. These *confusion possibilities* should also be known both to those who communicate with the hearing impaired and to those who teach the severely handicapped. Severely handicapped children might be much less frustrated if their teachers were able to anticipate possible confusions and reduce undesirable misunderstandings.

HOW WE OBTAINED THIS INFORMATION

Although you can use the information in this book without knowing how it was obtained, we think that this knowledge will give you a better feel for the technique. The procedures were really quite simple and straightforward.

Fristoe and Lloyd (1979) examined 20 manuals used for handicapped children throughout the United States. They selected the 330 most frequently appearing words. With but a few modifications, we selected a sign for each of these words. Subsequently, videotapes were made of a congenitally deaf native signer who signed each sign in three or four simpler versions. Eleven tapes were made, each containing about 30 sign versions. In random sequence, each sign was shown a) with shapeless mittens on the hands, b) without background location, i.e., hands only, without head or body, and c) in a fixed position. For more than 70% of these signs that do not move significantly from one location to another, the fixed position was in the center of movement. For the remaining signs that do move from location to location, both initial and final locations were used.

Subjects were permitted seven seconds to write down the English gloss of the presented sign.

The videotapes were presented to deaf college students at Gallaudet College and the National Technical Institute for the Deaf during the 1980–81 and 1981–82 academic years. Each student wrote down the English gloss for each sign presentation. Eight hundred and seventy free response protocols were obtained (Bornstein and Jordan, 1981).

For each of the 330 signs, the percentage of subjects who correctly understood each simplified form of the sign was determined (Bornstein and Jordan, 1982a). Signs correctly recognized by 80% or more of the subjects are characterized as robust signs. Those correctly recognized by 20% or fewer are thought to be fragile. Sometimes, the subjects thought the simplified form stood for an entirely different sign. In short, they confused the sign with another. Those signs chosen or confused by 10% or more of the subjects are reported in this book.

Finally, we want to bring to your attention the fact that the students who viewed these tapes had no context or situational clues to help them recognize the simpler sign forms. We believe that if situational clues had been present then it is likely that even more students would have been able to recognize the simpler forms. Hence, the robust figures presented in this book are probably underestimates of what is likely to happen in school or home settings. Put in another way, most of these simpler forms are likely to be even more robust in real life than we claim herein.

THE BASIC INFORMATION

Let us begin with a sample page that illustrates what we mean when we say "simpler form" (see sample on page xiii). On the upper, outer edge of the page, we show the complete or usual form of the sign that represents "yes" in both American Sign Language and Signed English (Bornstein et al., 1983). Fragile features are shown in red. A word description is provided as well.

Looking down the column closest to the inner edge of the page you can see the same sign formed without reference to face or body, then without movement, and finally, without the correct handshape. Accompanying each sign is a number ranging from 0 to 10.[2] This indicates how understandable the simpler form is, with 10 being understandable by almost everyone, whereas 0 represents a simpler sign form that is correctly recognized by almost no one. Simpler forms are always ordered with the most understandable form on top of the page to least understandable on the bottom of the page. To the side of each simpler form is the sign or signs that were confused with "yes" when that kind of simpler form was shown.

Now let us explain how this book has been organized so that you can make the most effective use of it. It is divided into ten sections according to how well each sign was understood in simplified form. Within each section, signs are ordered according to their average under-

---

[2] The scale represents nothing more than dropping the last digit of percent recognition. Thus 100 equals 10, 70 is changed to 7, etc. The actual percentages obtained are given in the appendix of this book, Tables A-1 to A-4.

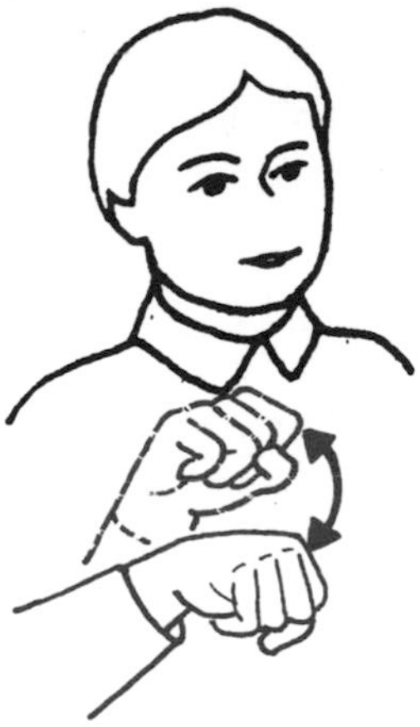

YES

Handshape: S
Location: In front of body
Movement: Shake up and down

understandable

10

confused with

2

"S" **4**

Saturday **1**

0

bye-bye **7**

standability over the three characteristics, from most to least understandable. Table 1 shows the ten sections.

Table 1. Sign content by section

Section number	Sign characteristics	Number of signs
1	All robust	21
2	Two robust/zero fragile	54
3	Two robust/one fragile	30
4	One robust/zero fragile	42
5	One robust/one fragile	38
6	One robust/two fragile	12
7	Zero robust/zero fragile	27
8	Zero robust/one fragile	56
9	Zero robust/two fragile	32
10	All fragile	18

HOW TO USE ROBUSTNESS

As we stated earlier, we consider a sign to be robust when it is recognized by 80% or more of the readers who view it in a simplified form. Section 1 includes 21 signs, all of which are robust when any of the three characteristics of the signs are missing. The sign words are *morning, out, time, baby, day, talk, butterfly, bread, glass, dance, bear, stand, drink, tree, good-bye, help, have, sad, our, quiet,* and *kick*. Other things being equal, we think that these are the signs that should be taught first to a handicapped learner. Readers should be able to understand even the clumsiest version of these signs. Therefore, teaching these signs should facilitate early, quick success in learning with obvious advantages.

Section 2 contains an additional 54 sign words that are robust on two characteristics and fragile on none. The sign words are *happy, in, put, close, afternoon, climb, later, want, hurt, boat, clean, sick, paint, night, key, fish, sweep, table, love, afraid, lion, can't, when, blue, walk, will, movie, butter, stop, me, show, cup, which, see, make, bring, hospital, hurry, money, number, fall, why, I, please, shoe, buy, book, house, before, good, jump, soda pop,* and *bowl*. These are the next most likely candidates for early instruction.

On a global level, therefore, we suggest that you judge whether motor coordination is most likely to interfere with or impede a child's ability to a) form the proper handshape, b) execute the correct movement, and/or c) place the hand(s) in proper relation to face or body. If you judge, for example, that a child will have difficulty making handshapes properly, then you can select the most robust signs when handshapes have been simplified. Robustness on handshape can be found in Table A-1 in the Appendix. Table A-1 orders the signs from most to least robust. Tables A-2, A-3, and A-4 provide similar information for location and movement. After you have found the signs you are interested in, return to the appropriate page in the text to cross-check each sign form for its confusion signs so that you can be fully aware of all potential instructional problems.

HOW TO DEAL WITH FRAGILITY

It is not until Section 3 that we encounter our first signs that contain a fragile characteristic, i.e., a simpler form that is understood by 20% or fewer of the readers. By definition, therefore, this characteristic must be presented in a reasonably accurate manner if the sign is to be recognized by most people. As noted earlier we have drawn such critical features in red as an extra signal to you. Red alerts you to adjust your teaching strategies. Those of you who teach the severely handicapped should first ascertain if the child is physically capable of executing the proper movement, handshape, or placement. If not, you can delay the introduction of that sign for as long as possible or consult the next section for further suggestions. However, if the child is physically capable, then you should stress or emphasize the fragile characteristic when teaching the sign. Most probably, more trials will be required before the child can produce an acceptable version of the sign characteristic.

If your overall teaching strategy is to allow the child to make a mistake and then correct him/her, we suggest that you correct the fragile characteristic first and again, last, before asking the child to repeat that sign. However, you may be able to devise a better strategy depending upon the sign you are working with. Those of you who prefer not to allow the child to form the sign incorrectly by molding his or her hands could emphasize a fragile characteristic by touching, pointing, tracing-in-air, or by any other emphatic physical action. Here again, the specific sign may suggest the best way to emphasize the fragile characteristic.

Parents, interpreters, teachers, and other professional workers with the deaf should be taught that their signs will be more easily understood if fragile characteristics are executed accurately. Conversely, this is probably the very same feature they should maximally attend to if they wish to improve their own reading of the sign.

Since we have arranged this book in order of decreasing robustness (and increasing fragility), it is clear that after 144 signs (Sections 1, 2, 4, and 7), the language instructor of the severely handicapped will be teaching signs that are fragile on one or more dimensions. What happens then? Until now we have not explicitly considered the physical and motoric capabilities of the student other than to assume that they are equal to the task. But some children may not be up to the task and there is some evidence that suggests that motoric considerations do relate to the learning of signs. Consequently, we have further described the signs in this book in seven ways that reflect how a sign is formed or executed. These include the number of hands involved in the signs, whether they execute symmetrical or asymmetrical movements, whether they touch each other or other parts of the body, etc. These seven characteristics are summarized in Table 2. As an aside, these summary statistics offer a very interesting description of these highly functional signs. In any event, what value is this information to you? As noted earlier there is some empirical research that suggests that the following kinds of signs are learned more rapidly (Kohl, 1981; Lloyd and Doherty, 1982):

- signs that "require" one or both hands to touch each other or other parts of the body
- two-handed signs that form symmetrical movements
- signs that are clearly visible to the signer

In addition, Kiernan (1982) has demonstrated that retarded children are able to imitate and learn simpler handshapes more rapidly than complex handshapes in the British Sign Language.

Therefore, when you must work with signs that are fragile, you should consider choosing first those that are two-handed, touching, symmetrical, and clearly visible to the signer. Further, where a sign need not necessarily touch another part of the body, you might wish to have it touch for instructional purposes.

To enable you to use this information more efficiently, we have used substantially the same descriptive language given in Table 2, where appropriate, in the word description of the citation form of each of the 330 signs.

Table 2. Frequency distributions of seven characteristics of the 330 signs

Sign characteristic	N	% Total
1 Number of hands		
Two	183	55
One	147	45
2 Visibility to signer		
Clear	284	86
Peripheral	46	14
3 Touch (other parts of body)		
Usual	209	63
Not necessarily	31	9
Rare	90	27
4 Movement symmetry		
Symmetrical	101	31
Asymmetrical	229	69
5 Number of movements		
Two or more	171	52
One movement	157	48
None	2	1
6 Location change		
Change from one location to another	95	29
No change in location	235	71
7 Handshape change		
Change in handshape	22	7
No change in handshape	308	94

In passing, we would like to note that signs can be described in an alternative fashion, i.e., by prehension and unilateral/bilateral movement patterns (Dennis et al., 1982; Dunn, 1982). These authors believe that it is possible to diagnose and subsequently train those with delayed limited motor development so that they can better execute a given sign or signs. As yet we do not know how much training is required for a given prehension and/or motor pattern with an autistic or severely retarded child. Nor do we know how much transfer there will be to the formation of manual signs by that same kind of child. Empirical information on these matters would be most desirable.

Finally, what happens if training fails to help the child develop the necessary motor coordination required to execute a reasonable number of robust signs? Then it seems to us that you should seriously consider other nonvocal systems as a more suitable communication tool. Manual signs require some minimal degree of manual dexterity and this is a fact that simply cannot be ignored.

THE DEAF-BLIND

There is one special group to which we wish to draw your attention—the deaf-blind. Here we have a group of students who may need to use some combination of tactile and visual information in recognizing a sign. On the face of it, handshape and movement may be the most useful characteristic because they can be sensed tactilely. However, recent information suggests that about two-thirds of deaf-blind children have some "usable" vision (Stein et al., 1982). Some of those who retain some degree of vision may be able to see large movements and/or some locations. If so, a combination of fragile signs on movement and/or location and robust on handshape may be effective. Where vision provides little or no information, then handshape and movement may be the features to which the child may best attend. Look for robust location signs. In all probability, you will have to select signs to fit the child's individual capabilities.

THE REMAINDER OF THE SIGN LANGUAGE

By now you know that this book contains only 330 signs. What do you do if you want to use a sign that is not in this book? We do have some suggestions for you. Our analyses of the understandability of simpler forms of these 330 signs has revealed, *on the average,* that two-handed, symmetrical signs that are clearly visible to the signer are somewhat more robust than one-handed or asymmetrical signs that are peripherally visible to the signer (Bornstein and Jordan, 1982b).

If we *assume* that the findings for our 330 signs can be generalized to other signs, then symmetry and visibility should be your basis for expecting robustness in other signs. You will remember also that two-handed, symmetrical signs are learned more quickly. Finally, regarding specific handshapes and locations, our research shows that by and large robust signs are made in the area in front of the body with a "5" or "open B" handshape.

Lacking specific information, you can use these generalizations to guide you in sign selection. You must remember, however, that similarity or confusion with other signs will *surely* affect these generalizations. Always be alert to the probability of another sign being confused with any given simpler sign form and *alter* your instructional strategy accordingly.

ABOUT THE SIGNS IN THIS BOOK

Those of you who are very experienced in signing may have some questions about how the signs in this book were selected and how they should be used. We hope that we have been able to anticipate them.

The 330 signs in this book cover only a very basic, functional vocabulary. Since Signed English and the American Sign Language (ASL), at this language level, are made up largely of the same signs, this corpus is by and large common to both Signed English and ASL. The signs, of course, are not always used in the same order, inflected in the same ways, and don't always have the same meanings. However, these distinguishing characteristics are really quite tangential to the matter of recognizing and executing simpler sign forms.

There is another class of signs in this corpus where the "correct" English gloss is entirely dependent upon context. In this sense they are analogous to English homophones. The signs are *brown-beer, bird-chicken, handkerchief-cold, hungry-wish, my-mine, nice-clean, boring-dry, sleep-pillow-bed, prefer-favorite, wonder-consider, cute-candy-sugar, will-future*, and the verb *"to be"-sure-true.*

There are a small number of signs that can be regarded as relatively recent in origin. These are the several forms of the verb "*to be*" (i.e., *is, am,* and *are*), the pronouns *he* and *she, juice,* and the signs for two vehicles, *truck* and *bus.*

We would like to note that many of the confusions found are quite fascinating and may not be explainable in full in linguistic terms. For example, the simplified forms for *mother, father,* and *polite* in the no location condition are visually identical. Yet viewers do not confuse one with the others equally. Also, *rabbit* is a frequent confusion for simpler forms of *horse* but not vice-versa.

Finally, the Index lists both the target functional vocabulary (in bold type) *and* those signs that were confused with the simpler forms. The drawings of these latter signs can be used for reference purposes if you should so desire.

FINAL WORDS

As the reader may have guessed by now, we value empirically derived information highly. We believe that it is far and away the most dependable way to develop improved methods of communication with the types of students we have referred to throughout this introduction. Yet we have made any number of suggestions about applications of this information without any empirical evidence to support these applications. Obviously, we are uncomfortable in doing this. Our justifications are threefold: 1) the need is great; 2) the logic is intuitively compelling; 3) we will obtain additional evidence on the efficacy of these suggestions and we will modify them as appropriate.

REFERENCES

Bornstein, H., & Jordan, I.K. Studies on the intelligibility of signs. Paper presented at AAMD convention, 1980.

Bornstein, H., & Jordan, I.K. Further studies on the intelligibility of signs. Paper presented at AAMD convention, 1981.

Bornstein, H., & Jordan, I.K. Simpler signs. Symposium presented at AAMD convention, 1982a.

Bornstein, H., & Jordan, I.K. The relationship between sign characteristics and understandability of simpler sign forms. Paper presented at IASSMD, 1982b.

Bornstein, H., Saulnier, K.L., & Hamilton, L.B. *The Comprehensive Signed English Dictionary*. Washington, D.C.: Gallaudet College Press, 1983.

Dennis, R., Reichle, J., Williams, W., & Vogelsburg, R.J. Motoric factors influencing the selection of vocabulary for sign production programs. *Journal of Association for Severely Handicapped*, 1982, *7*, 20-32.

Dunn, M.L. *Pre-Sign Language Motor Skills*. Tuscon, Arizona: Communication Skill Builders, 1982.

Fristoe, M., & Lloyd, L.L. Signs used in manual communication training with persons having severe communication impairment. *AAESPH Review*, 1979, *4*, 364-373.

Kiernan, C.C. The imitation and learning of hand postures (dez) by normal and mentally handicapped children. Paper presented at IASSMD, 1982.

Kohl, F.L. Effects of motoric requirements on the acquisition of manual sign responses by severely handicapped students. *American Journal of Mental Deficiency*, 1982, *85*, 396-403.

Lloyd, L.L., & Doherty, J.E. The influence of production made on recall of signs in normal adult subjects. A paper presented at the 1982 Gatlinburg Conference on Research in Mental Retardation/Developmental Disabilities.

Luftig, R.L., & Lloyd, L.L. Manual sign translucency and referential concreteness in the learning of signs. *Sign Language Studies*, 1981, *30*, 49-60.

Reichle, J., Williams, W., & Ryan, S. Selecting signs for the formulation of an augmentative communication modality. *Journal of Association for Severely Handicapped*, 1981, *6*, 48-56.

Stein, L.K., Palmer, P., & Weinberg, B. Characteristics of a young deafblind population. *American Annals of the Deaf*, 1982, *127*, 828-837.

TRIPLE ROBUST

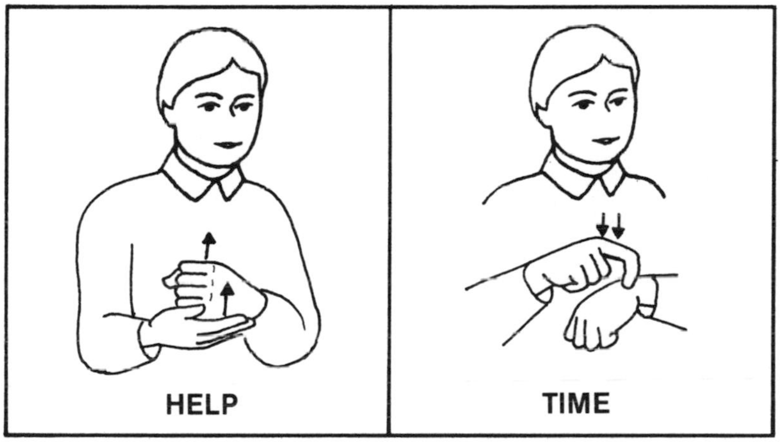

MORNING

Handshape: Open B, both
Location: In front of body
Movement: Up Slightly

understandable

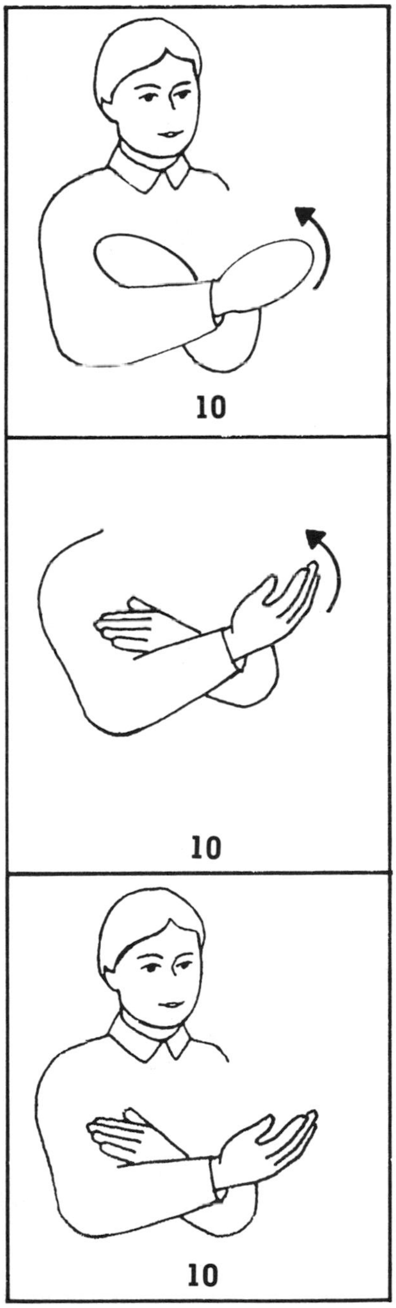

OUT

Handshape: Flat O; C
Location: In front of body
Movement: Pull hand out

understandable

10

10

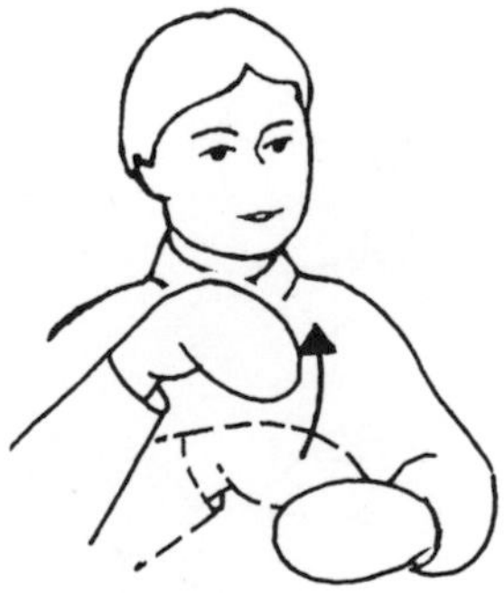

9

TIME

Handshape: Bent 1; S
Location: In front of body
Movement: Tap wrist twice

understandable

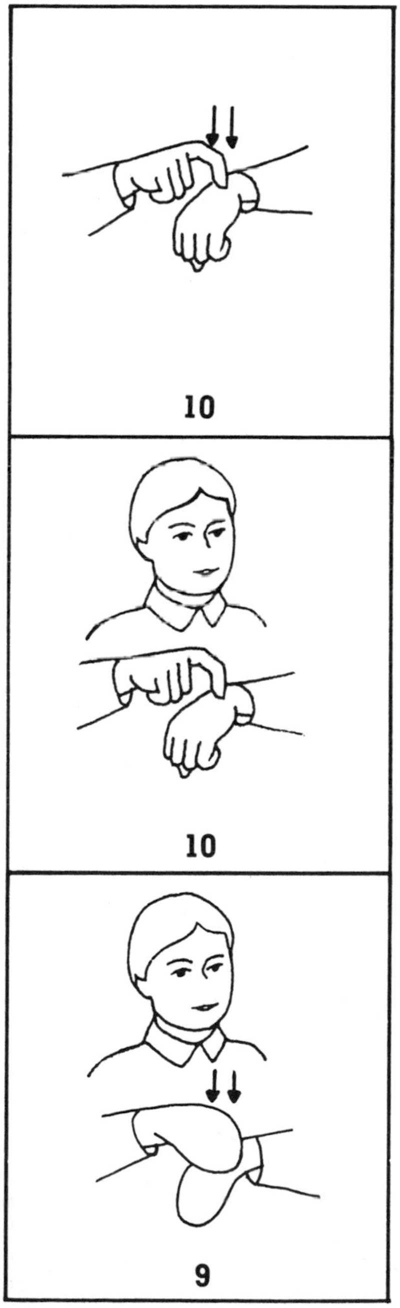

BABY

Handshape: Open B, both
Location: In front of body
Movement: Rock back and forth

understandable

10

10

8

DAY

Handshape: 1; B
Location: In front of body
Movement: Lower hand to elbow

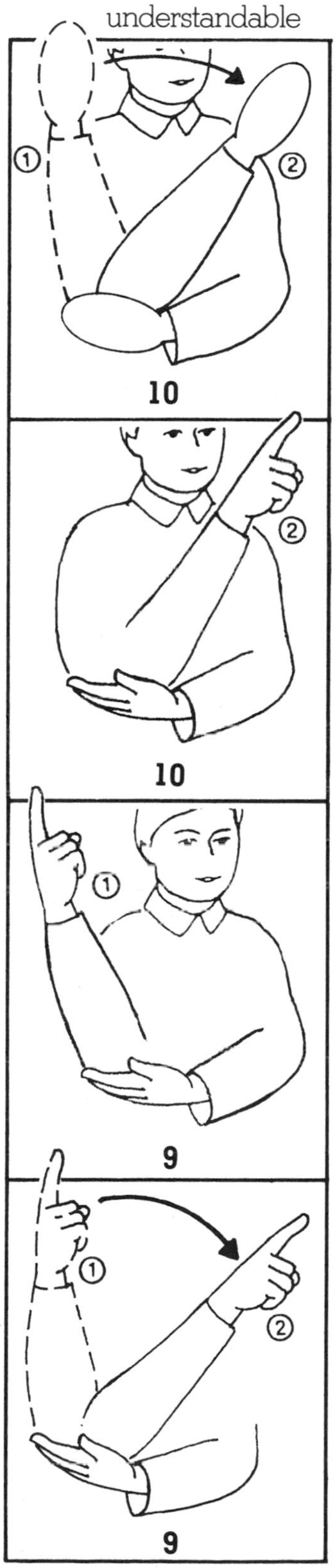

TALK

Handshape: 1, both
Location: In front of mouth
Movement: Back and forth alternately

understandable

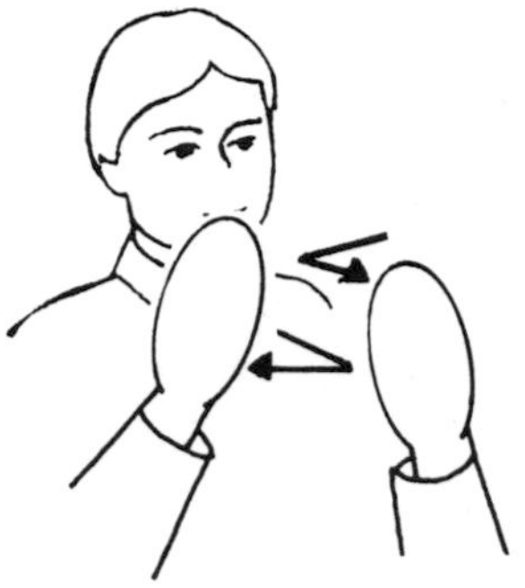

10

9

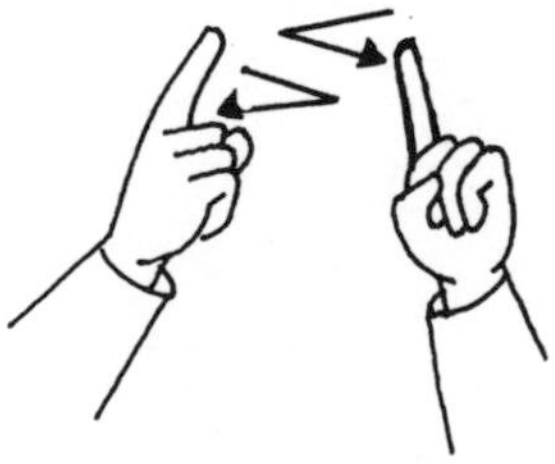

9

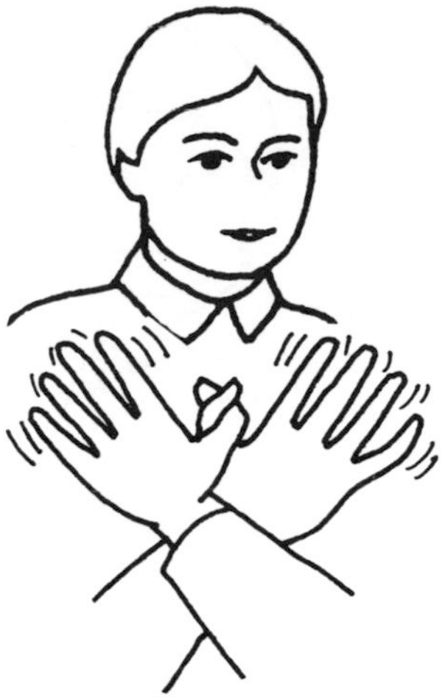

BUTTERFLY

Handshape: 5, both
Location: In front of body
Movement: Wiggle Fingers

understandable

10

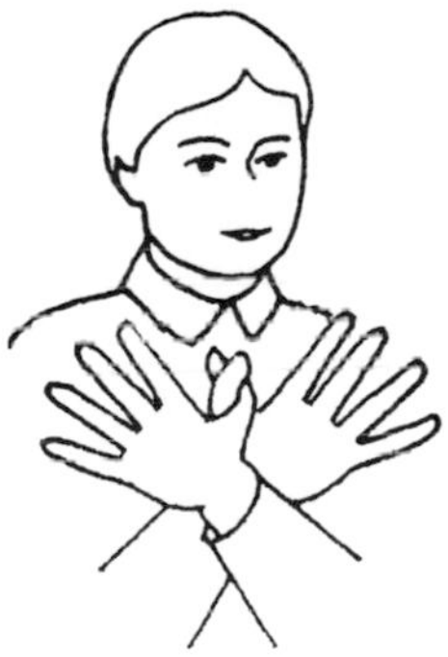

9

9

BREAD

Handshape: Open B, both
Location: In front of body
Movement: Scratch down twice

understandable

9

9

9

GLASS

Handshape: C; open B
Location: In front of body
Movement: Lift slightly

understandable

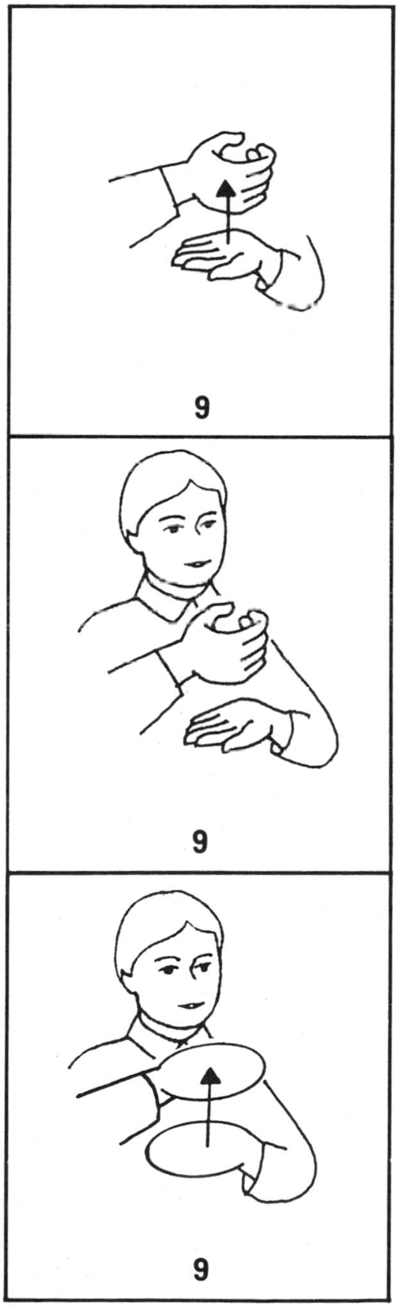

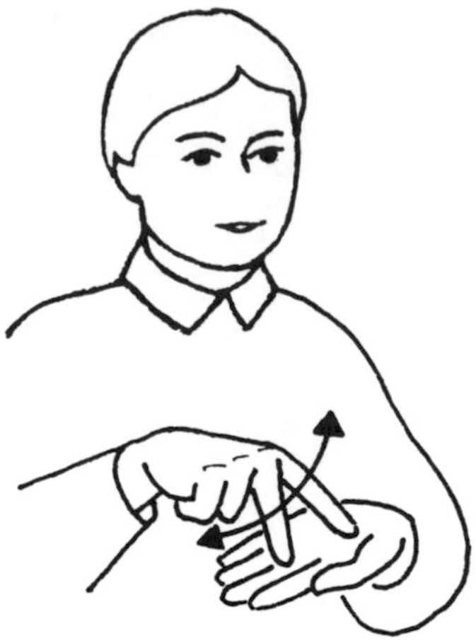

DANCE

Handshape: Inverted V; open B
Location: In front of body
Movement: Back and forth

understandable

10

9

confused with

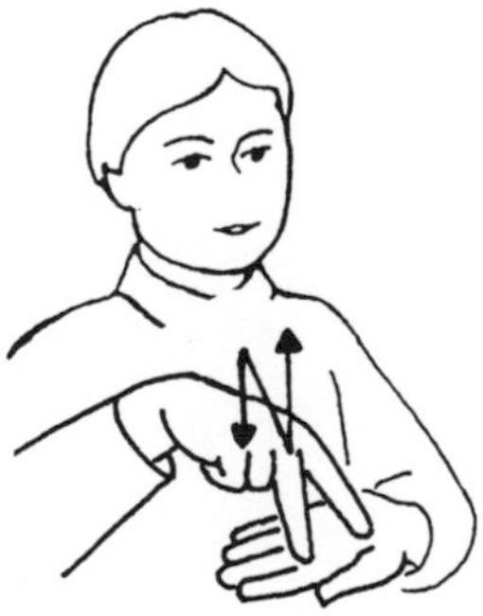

jump 1

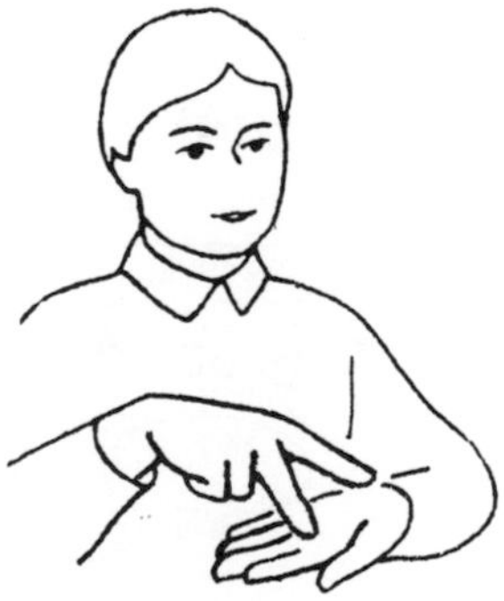

8

understandable

9

9

9

BEAR

Handshape: Bent 5, both
Location: Chest
Movement: Scratch twice

STAND

Handshape: Inverted V; open B
Location: In front of body
Movement: None

understandable

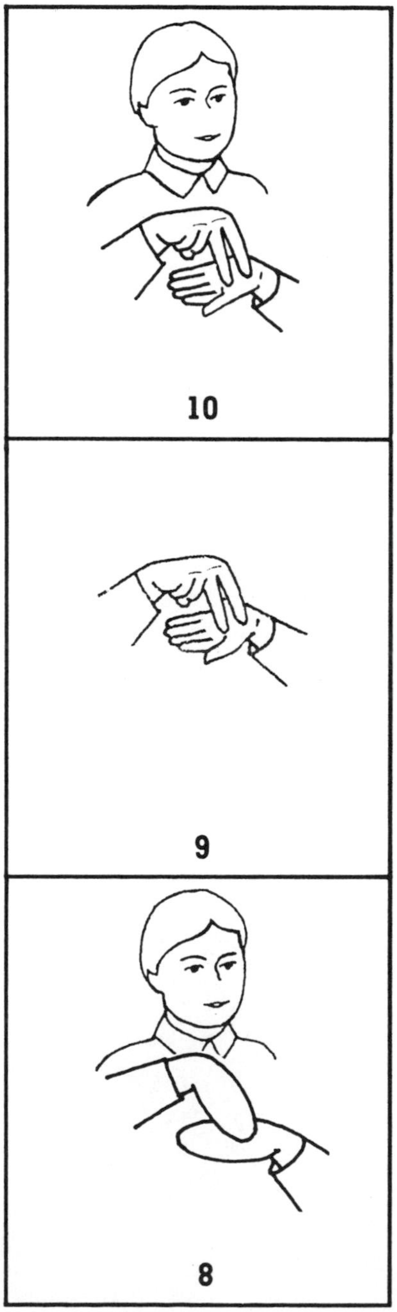

DRINK

Handshape: C
Location: Mouth
Movement: Like drinking

understandable

10

9

8

TREE

Handshape: 5; open B
Location: In front of body
Movement: Vibrate slightly

understandable

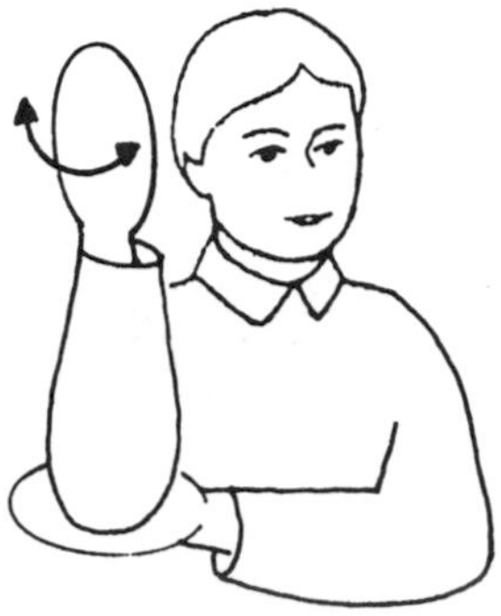

9

9

9

understandable

GOOD-BYE

Handshape: Open B
Location: In front of body
Movement: Wave twice

HELP

Handshape: S; open B
Location: In front of body
Movement: Lift Slightly

understandable

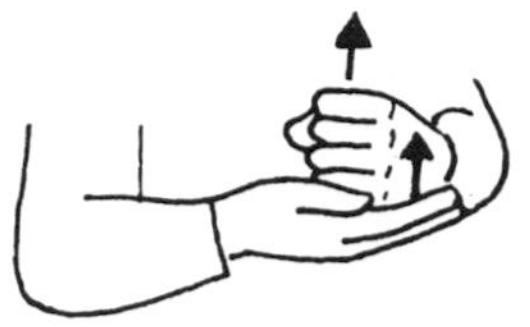

10

9

8

HAVE

Handshape: Open B, both
Location: Chest
Movement: Touch Chest

understandable

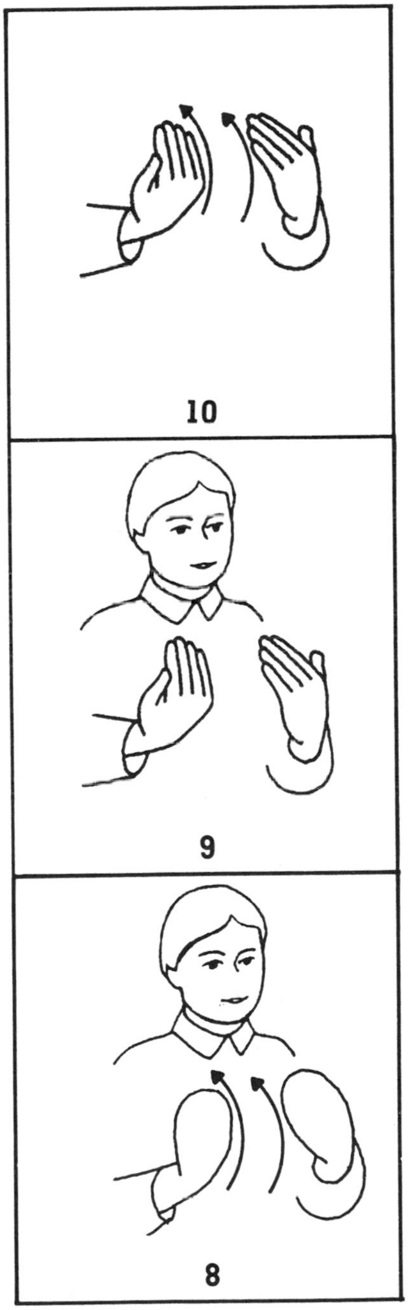

SAD

Handshape: 5, both
Location: Face
Movement: Down Slowly

understandable

9

8

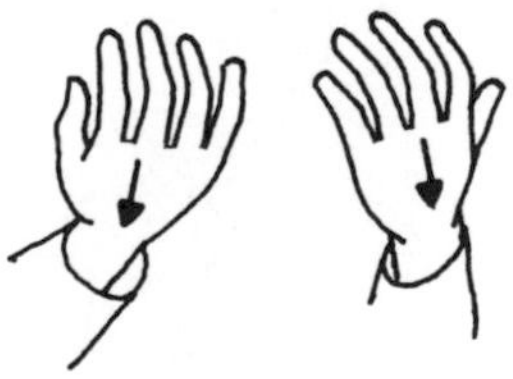

8

understandable

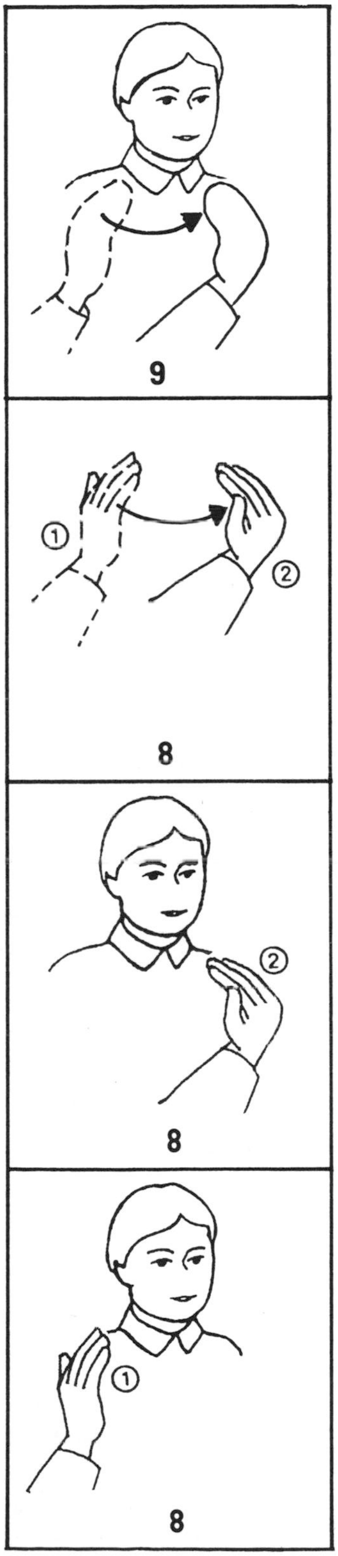

OUR

Handshape: B
Location: Chest
Movement: Arc in front of chest

QUIET

Handshape: B, both
Location: Mouth
Movement: Down and apart

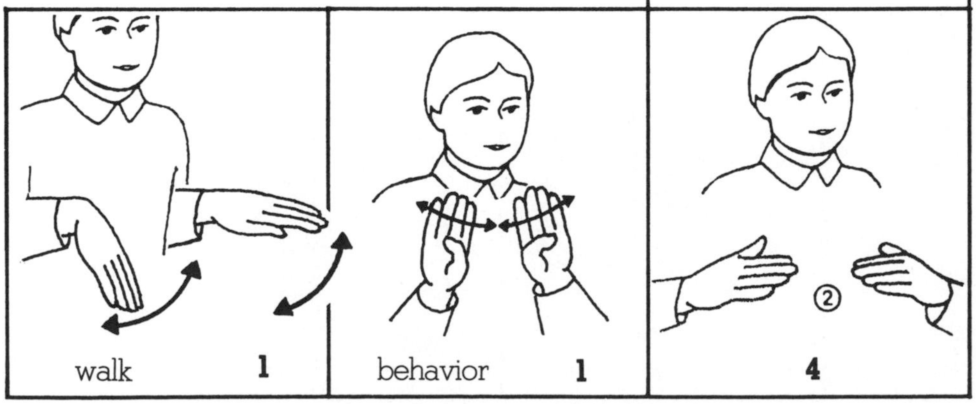

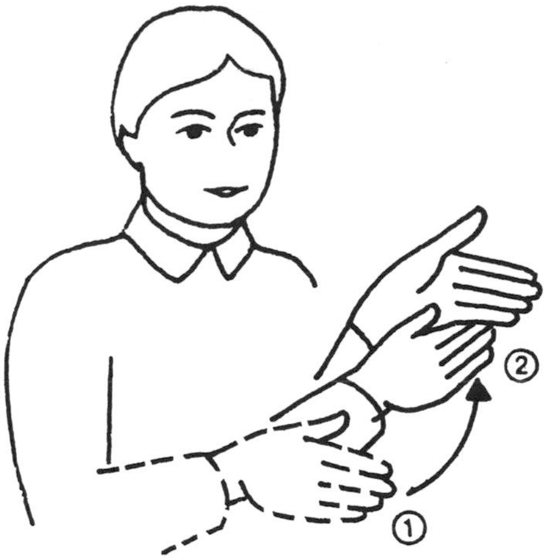

KICK

Handshape: B, both
Location: In front of body
Movement: Chop up

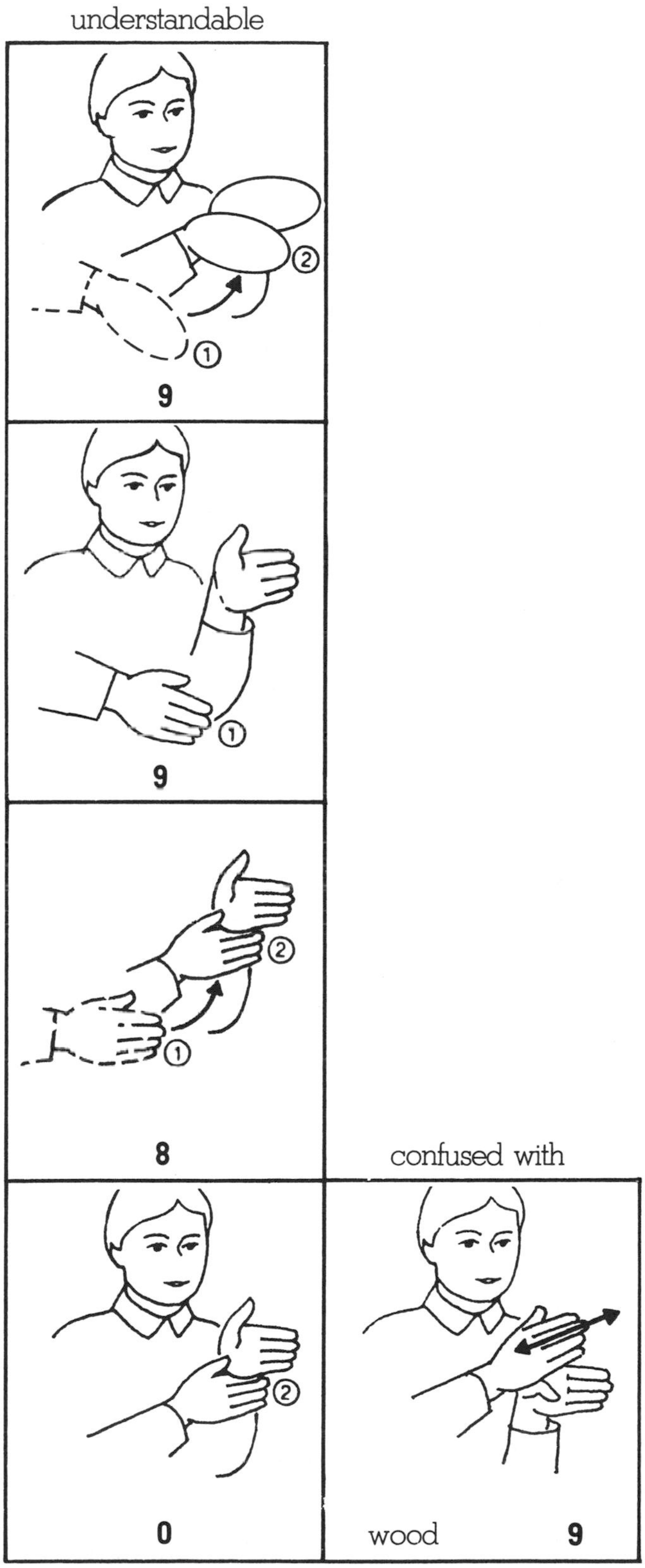

DOUBLE ROBUST

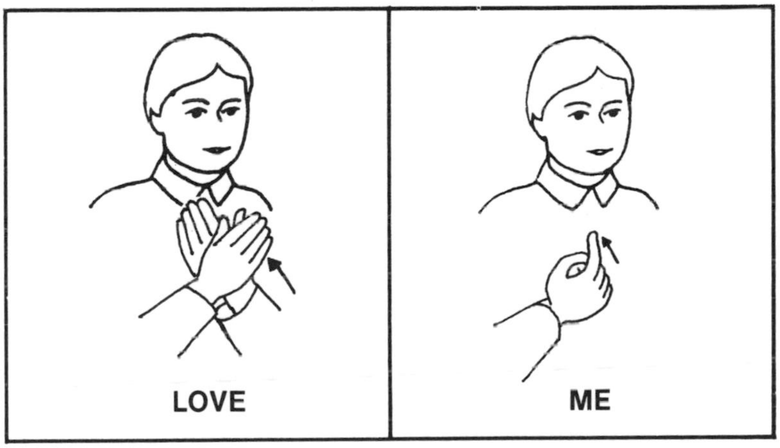

ZERO FRAGILE

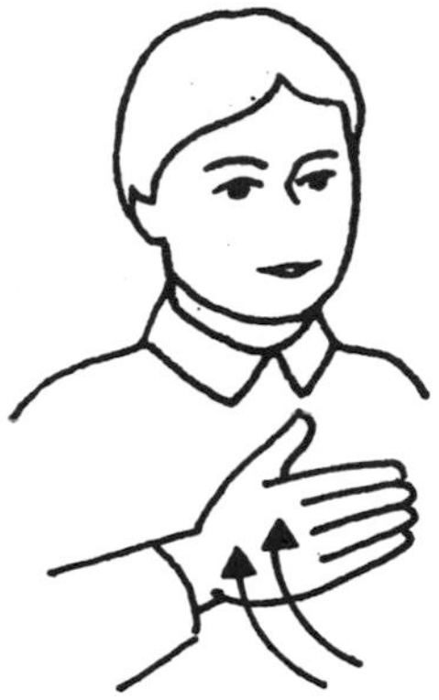

HAPPY

Handshape: Open B, both
Location: Chest
Movement: Up and out twice

understandable

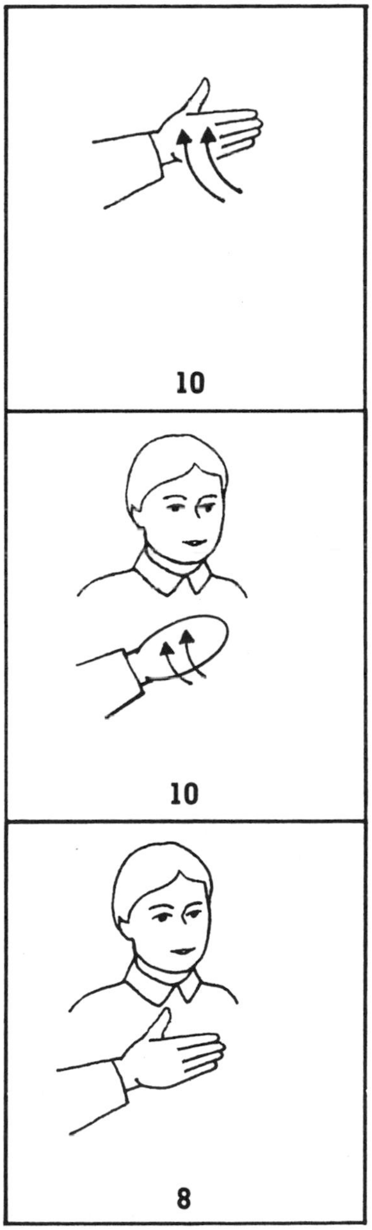

IN

Handshape: Flat O; C
Location: In front of body
Movement: Place O hand into C

understandable

10

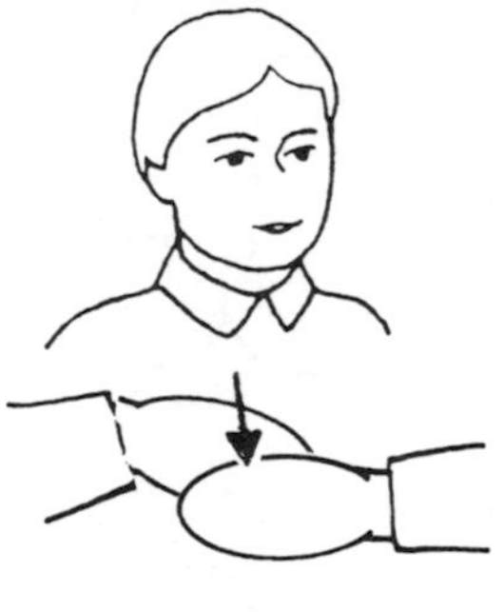

9

8

understandable

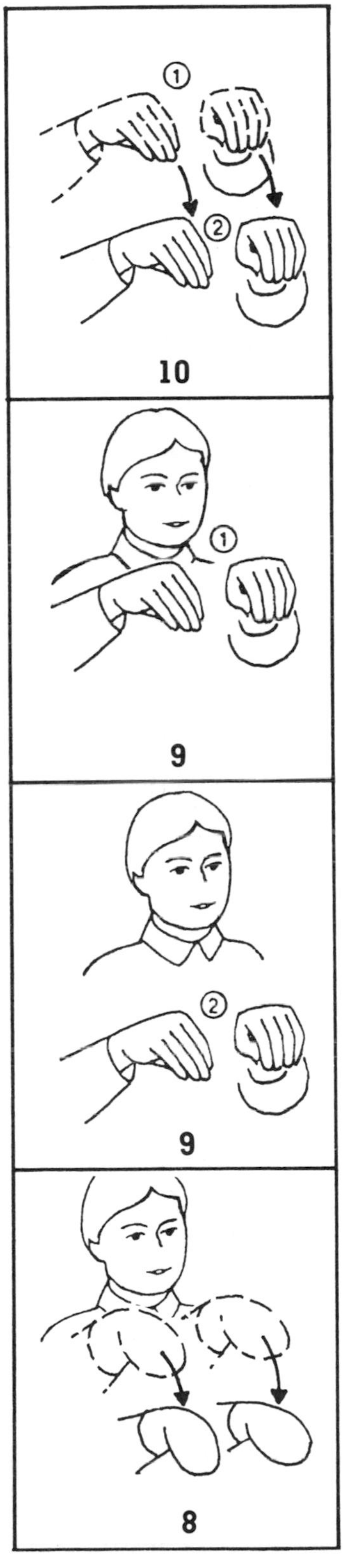

PUT

Handshape: Flat O; both
Location: In front of body
Movement: Arc forward

CLOSE

Handshape: B; both
Location: In front of body
Movement: Bring index fingers together

understandable

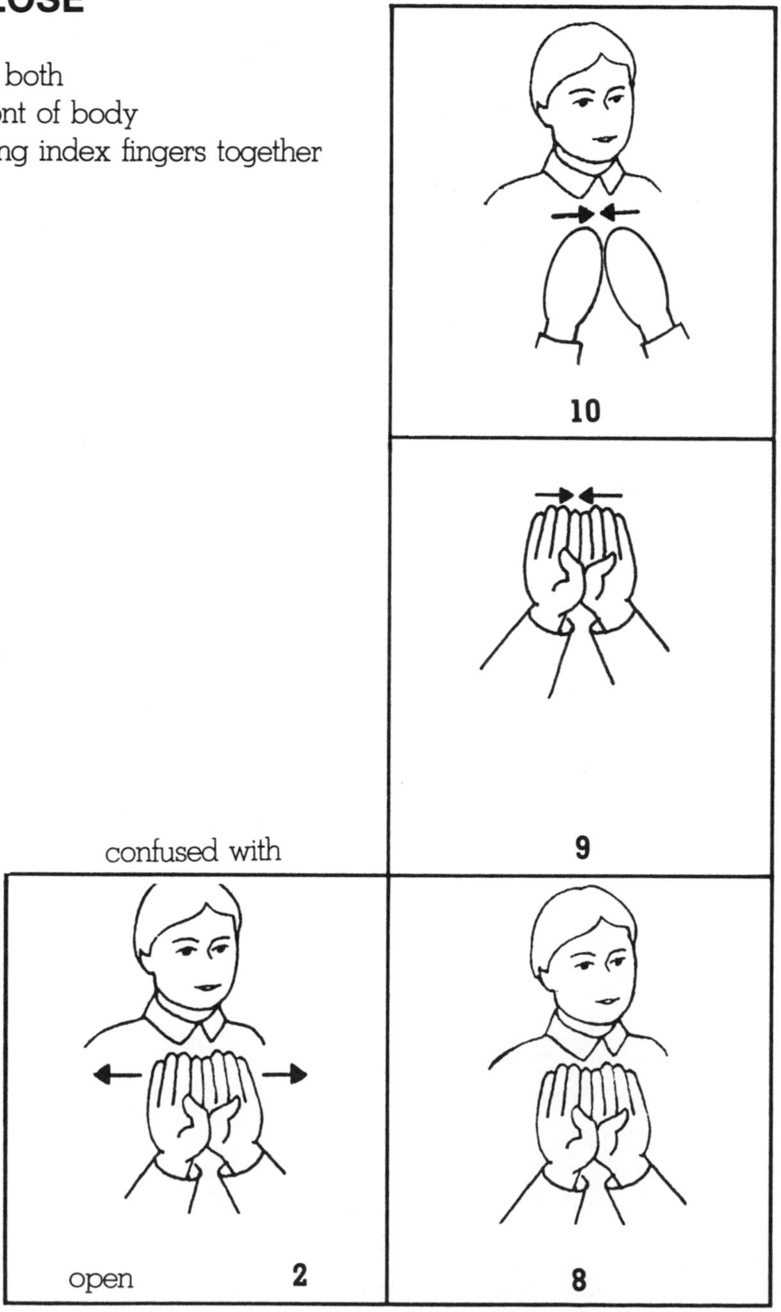

AFTERNOON

Handshape: Open B, both
Location: In front of body
Movement: Lower hand slightly

understandable

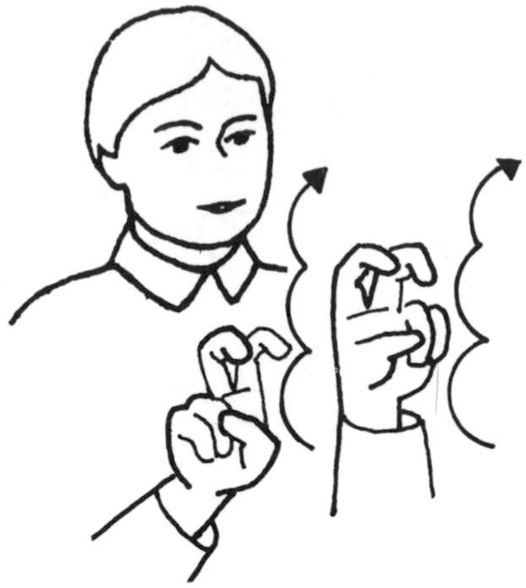

CLIMB

Handshape: Bent V
Location: In front of body
Movement: Alternately move hands up

understandable

10

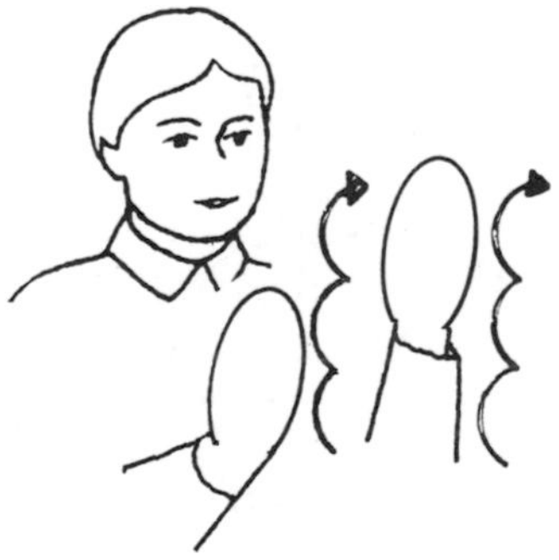

9

8

LATER

Handshape: L
Location: In front of body
Movement: Lower finger

understandable

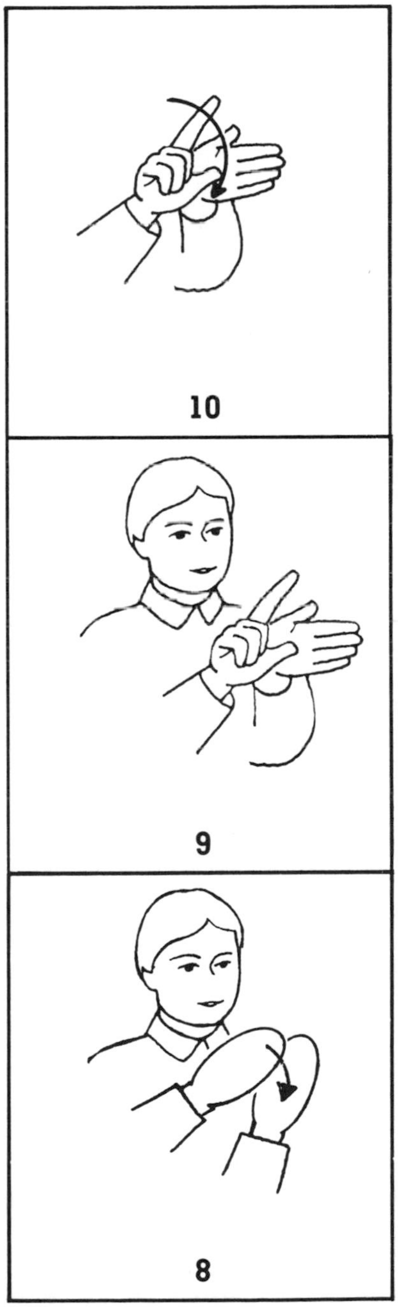

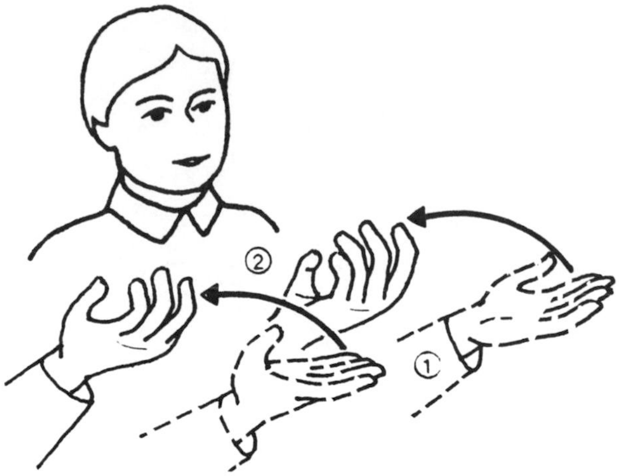

WANT

Handshape: Bent 5, both
Location: In front of body
Movement: In toward body

understandable

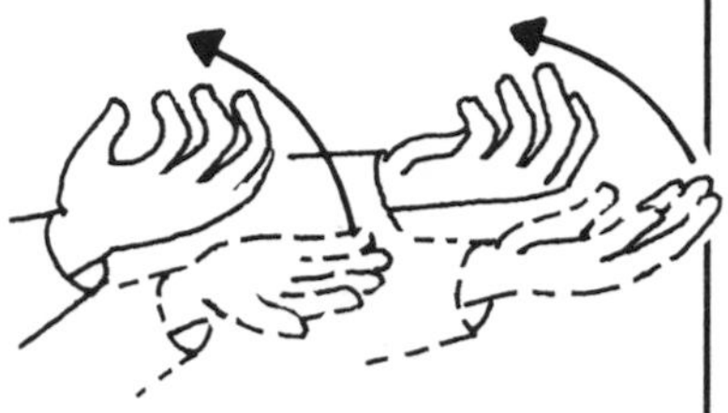

10

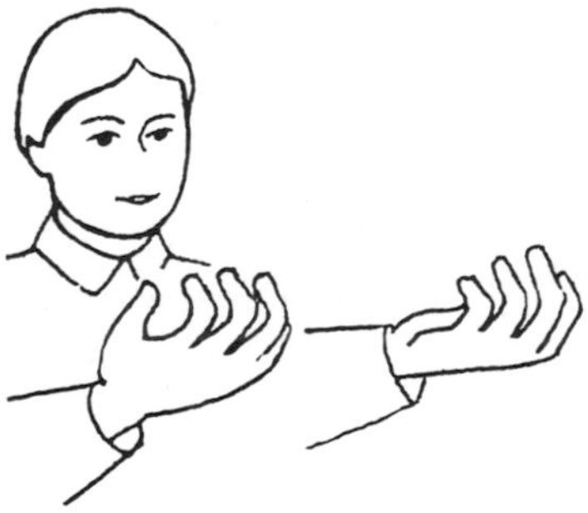

9

confused with

bring 1

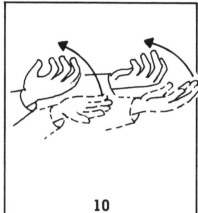

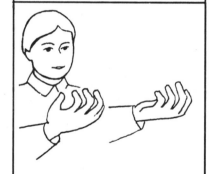

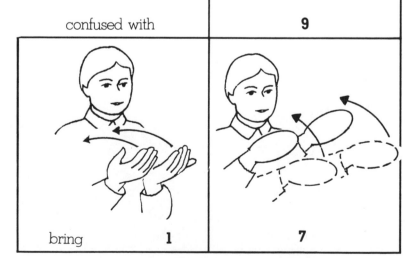

7

HURT

Handshape: 1, both
Location: In front of body
Movement: Back and forth slightly

understandable

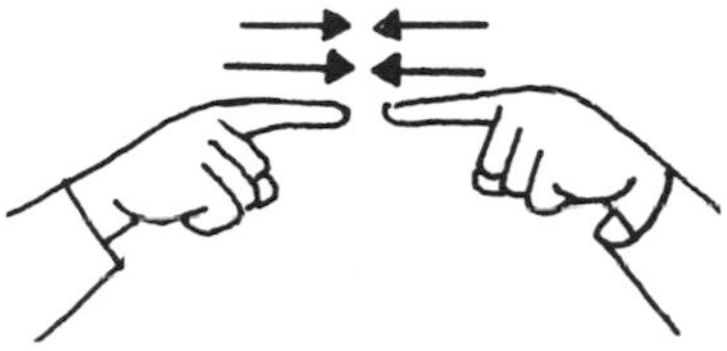

10

9

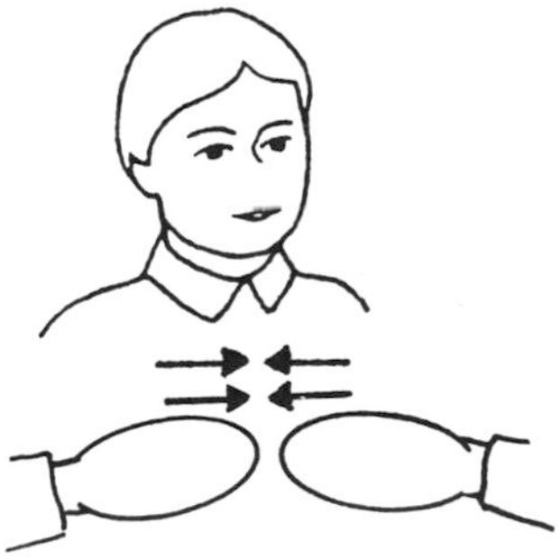

7

confused with

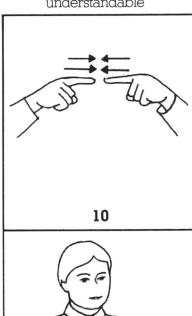

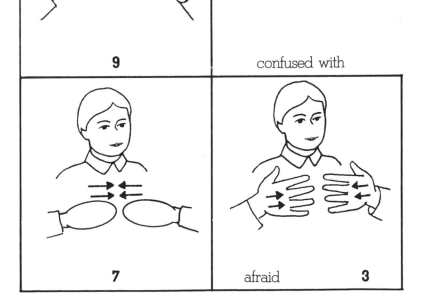

afraid 3

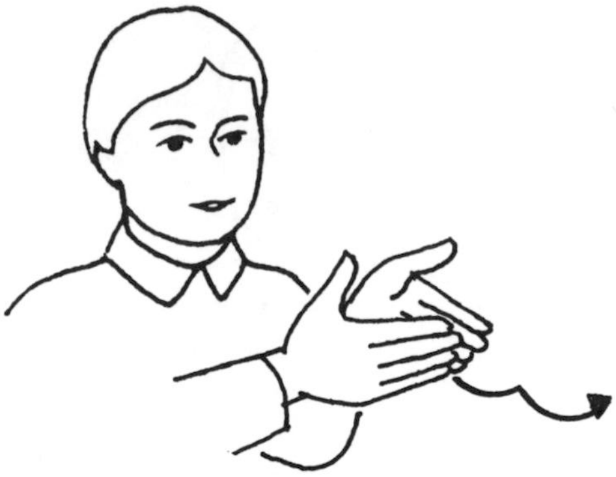

BOAT

Handshape: B, both
Location: In front of body
Movement: Forward Twice

understandable

10

10

confused with

book **2**

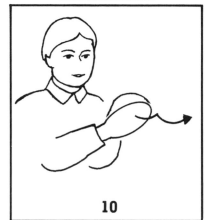

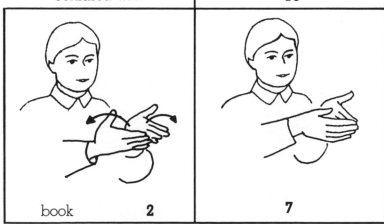

7

CLEAN

Handshape: Open B, both
Location: In front of body
Movement: Brush across palm

understandable

10

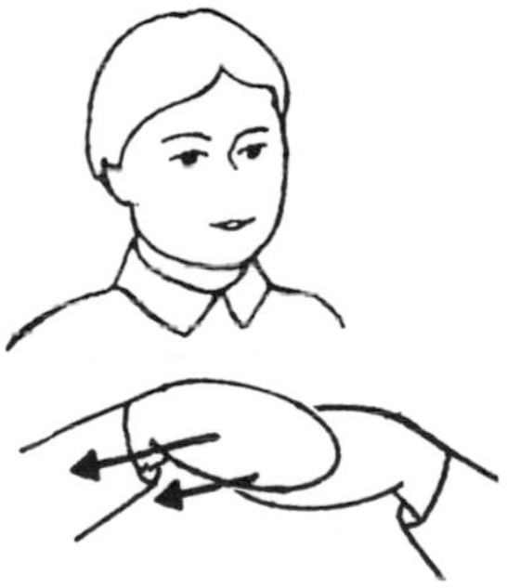

10

confused with

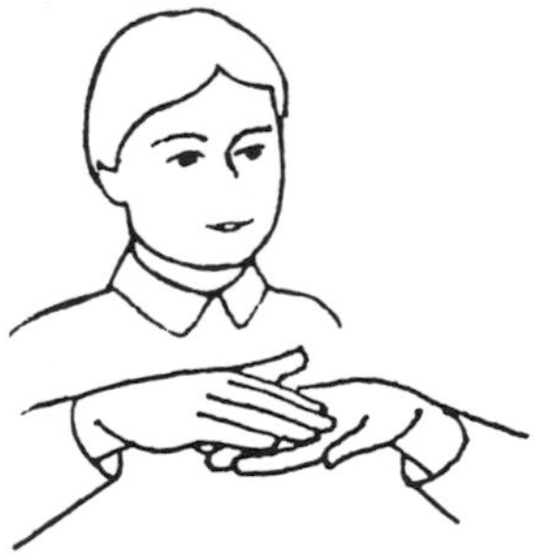

7

excuse **2**

SICK

Handshape: 5, bent middle finger
Location: Forehead
Movement: Tap

understandable

10

9

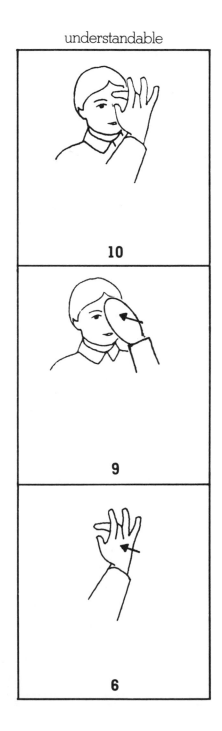

6

PAINT

Handshape: Open B, both
Location: In front of body
Movement: Brush up and down

understandable

10

9

6

confused with

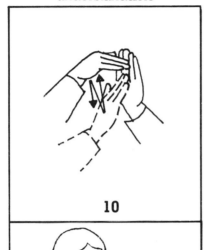

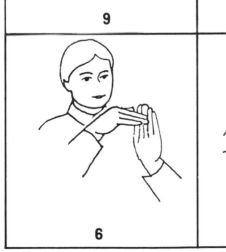

reject **1**

NIGHT

Handshape: Open B, both
Location: In front of body
Movement: Down over wrist

understandable

confused with

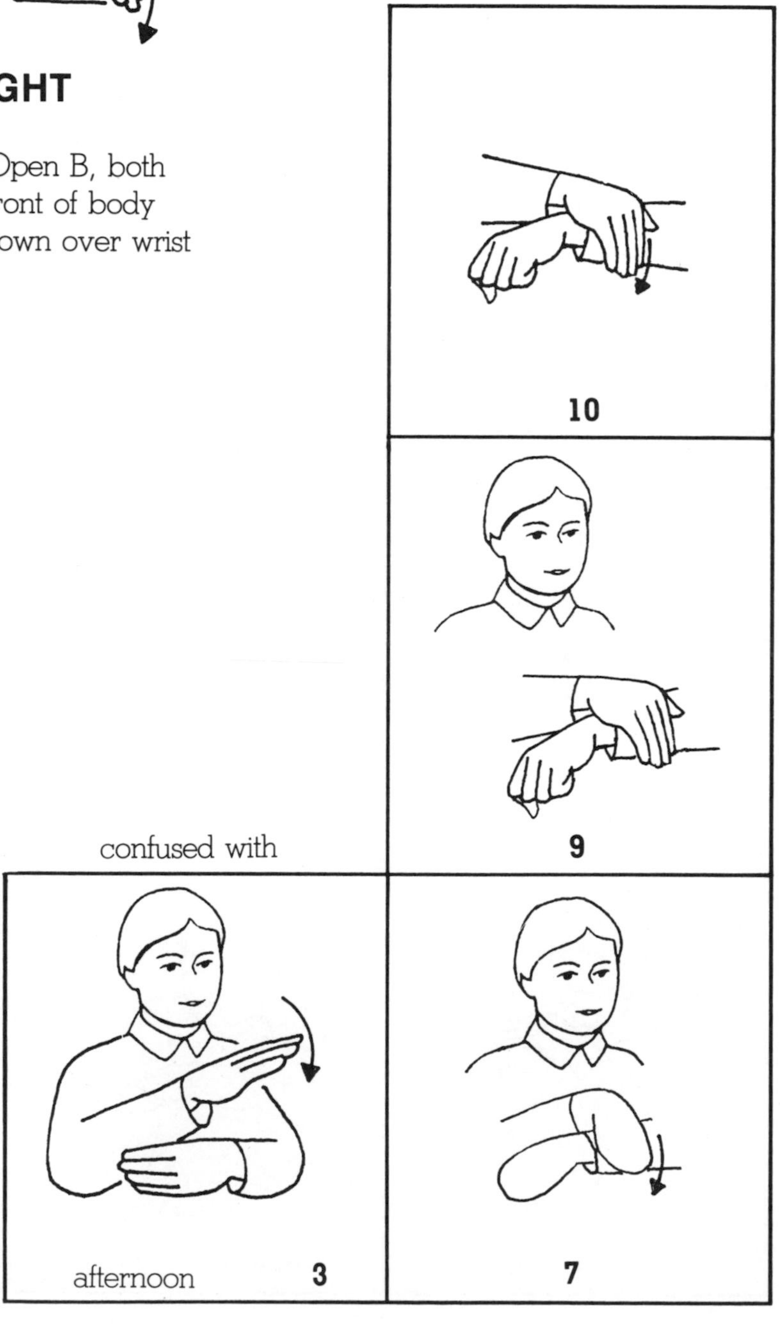

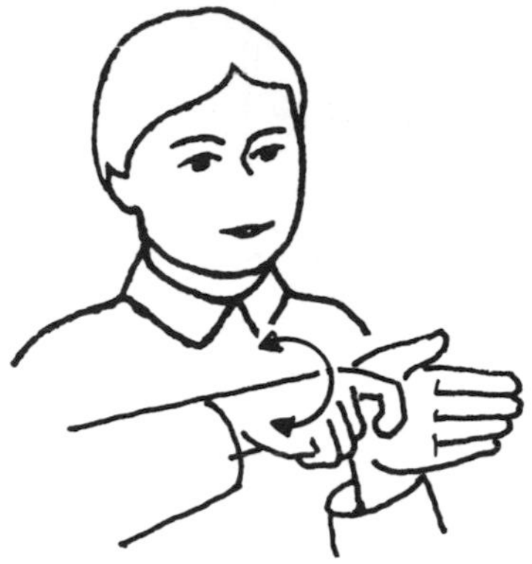

KEY

Handshape: X
Location: In front of body
Movement: Twist Twice

understandable

9

9

7

confused with

screw 1

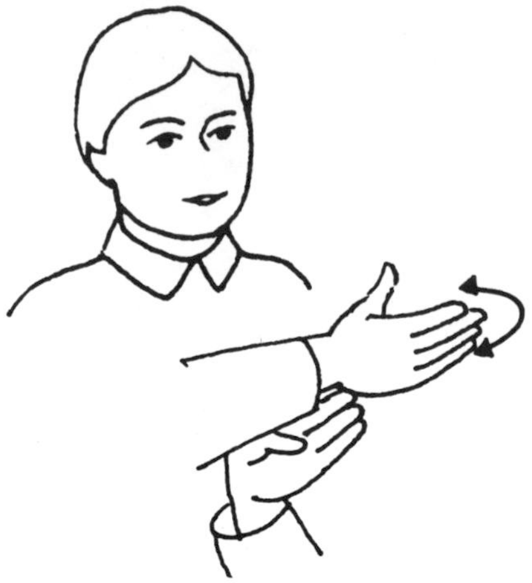

FISH

Handshape: Open B, both
Location: In front of body
Movement: Flutter and forward

understandable

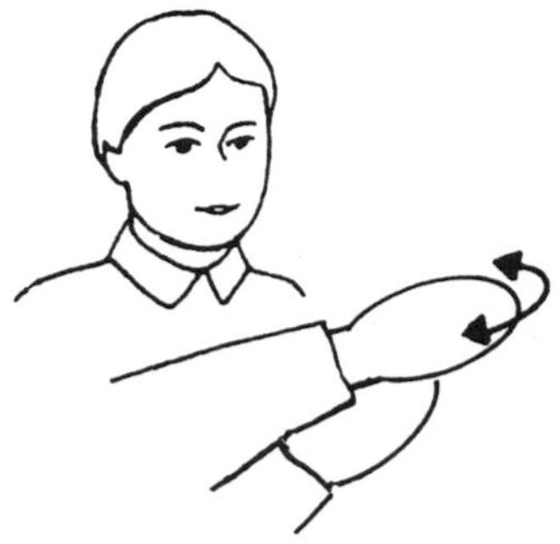

10

10

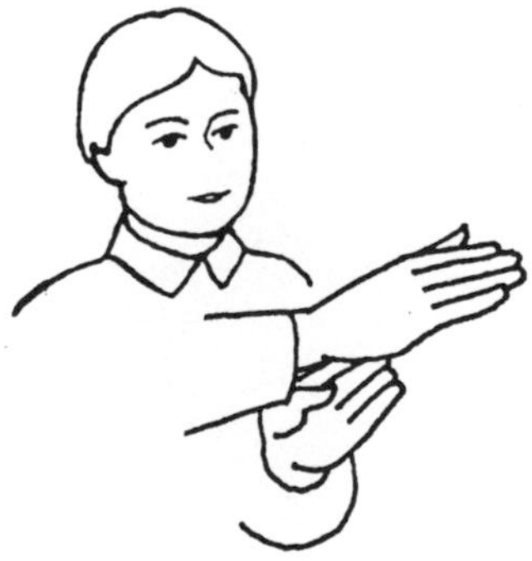

6

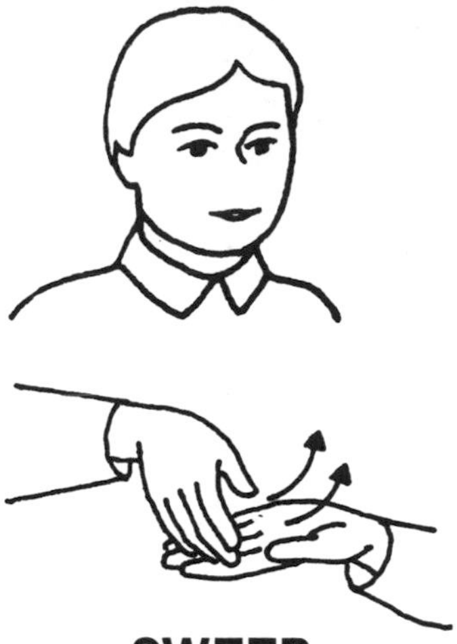

SWEEP

Handshape: Open B, both
Location: In front of body
Movement: Sweep palm

understandable

9

9

8

TABLE

Handshape: Open B, both
Location: In front of body
Movement: Tap forearm twice

understandable

9

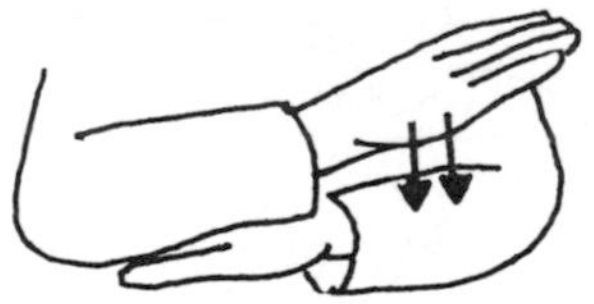

8

7

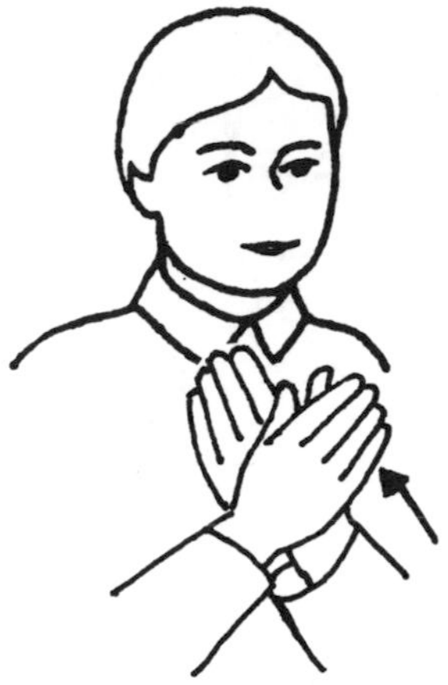

LOVE

Handshape: S, both
Location: Chest
Movement: Hug self

understandable

9

9

6

confused with

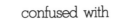

rest 3

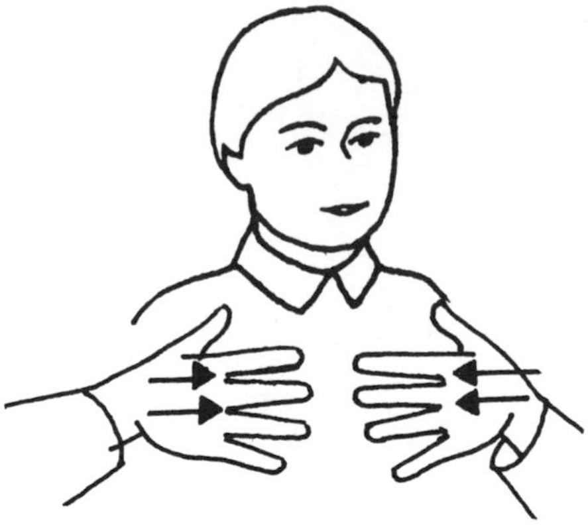

AFRAID

Handshape: 5, both
Location: Chest
Movement: Out toward shoulders

understandable

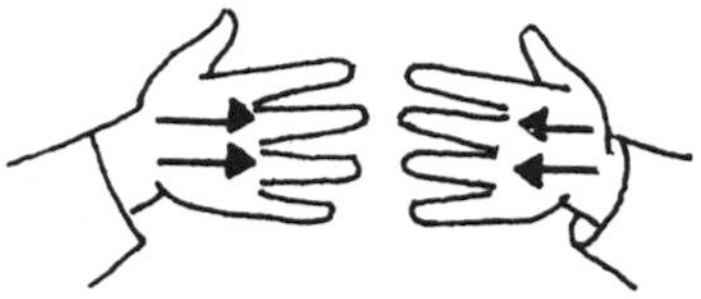

10

10

confused with

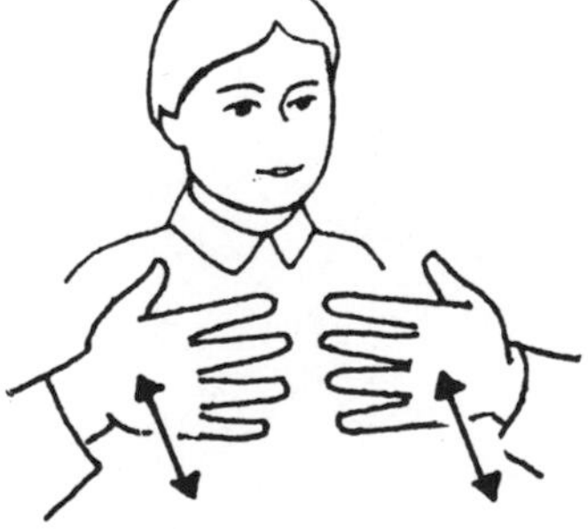

breath **3**

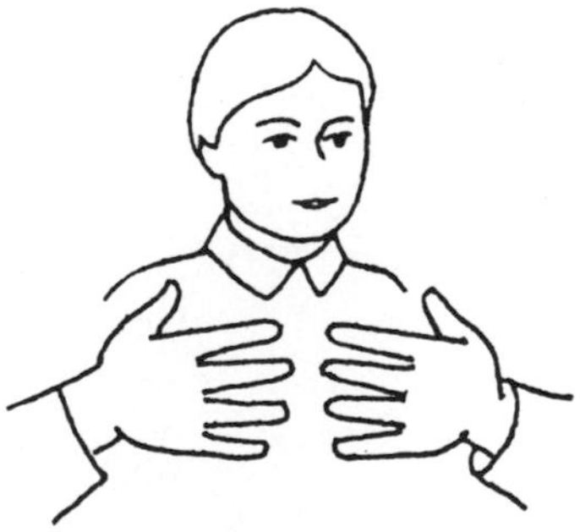

5

LION

Handshape: Bent 5
Location: Head
Movement: Back over hair

understandable

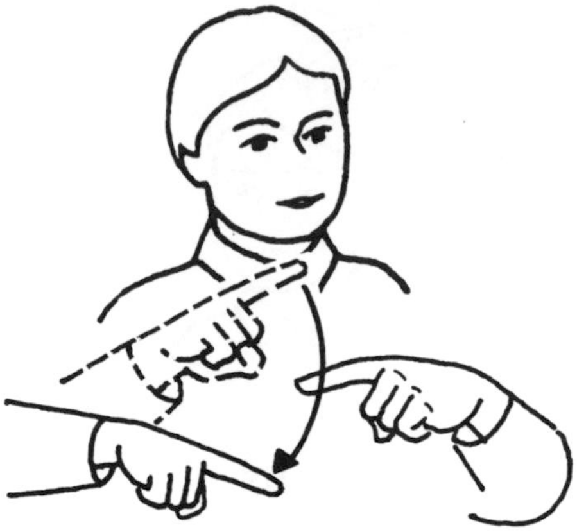

CAN'T

Handshape: 1, both
Location: In front of body
Movement: Strike tip of finger

understandable

10

9

7

confused with

than **1**

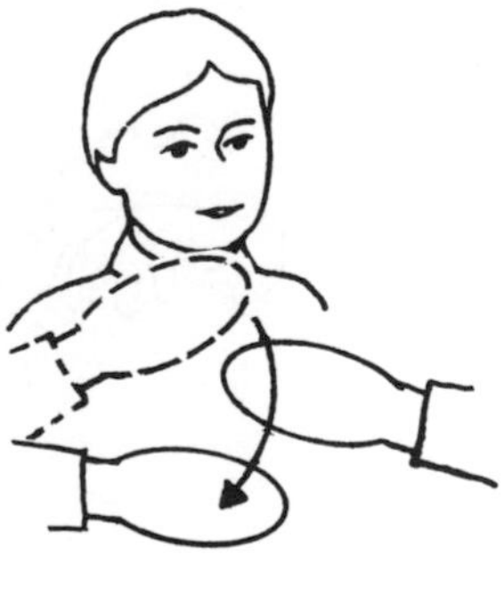

7

WHEN

Handshape: 1, both
Location: In front of body
Movement: Circle tip of finger once

understandable

10

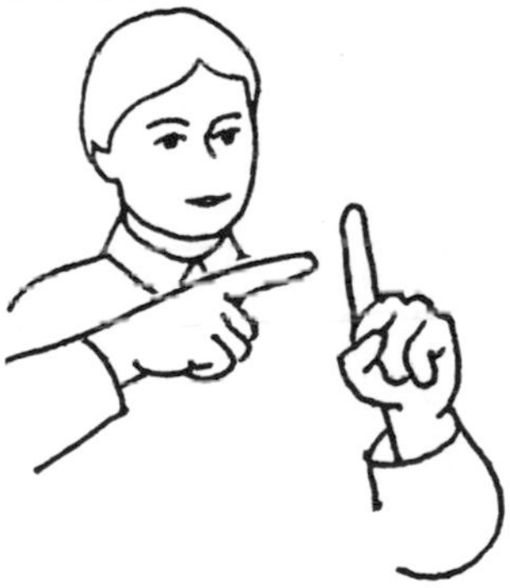

9

confused with

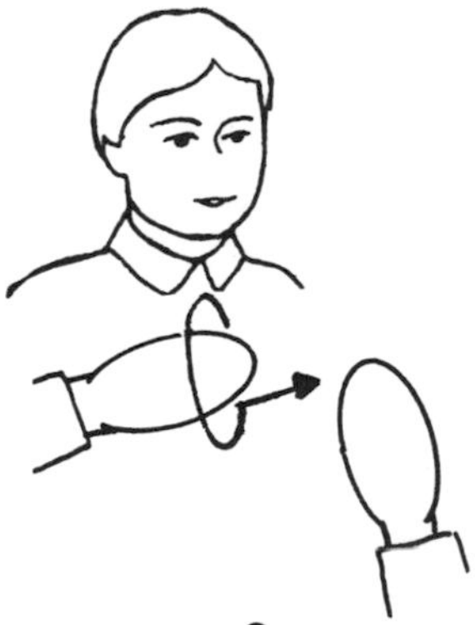

6

middle 3

BLUE

Handshape: B
Location: In front of body
Movement: Shake Twice

understandable

9

8

confused with

mirror 1

past 1

6

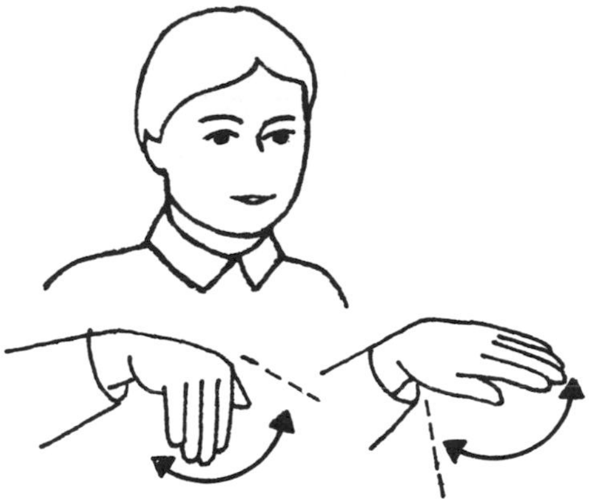

WALK

Handshape: Open B, both
Location: In front of body
Movement: Alternately forward and back

understandable

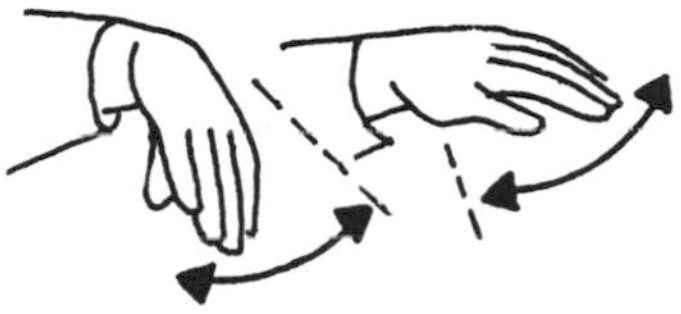

10

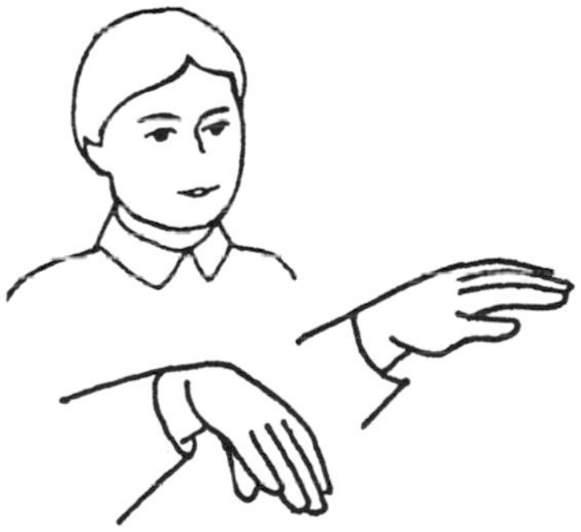

9

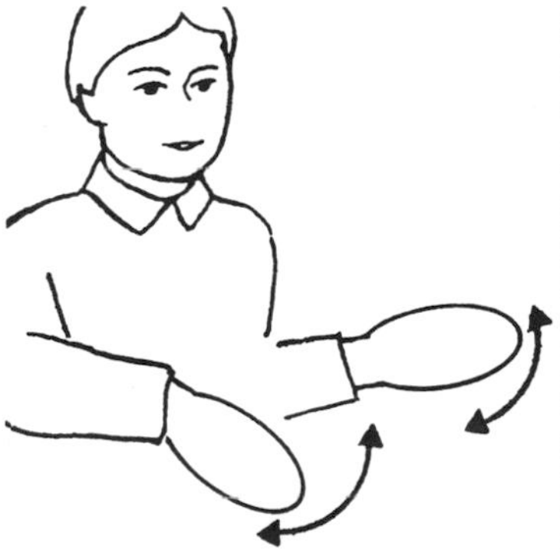

5

confused with

move **1**

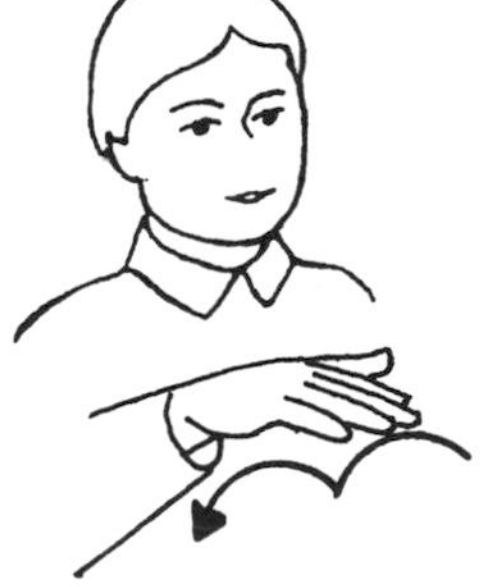

children **1**

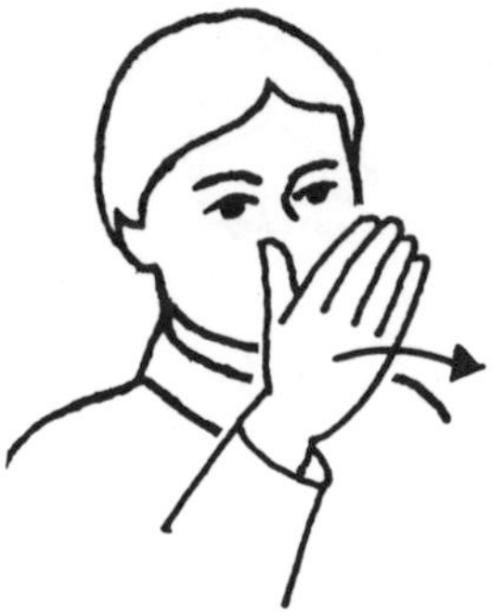

WILL

Handshape: Open B
Location: Cheek
Movement: Forward

confused with	understandable
 future **1**	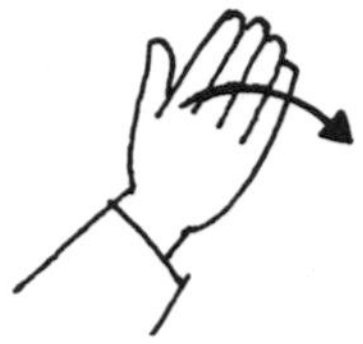 **9**
 future **1**	 **9**
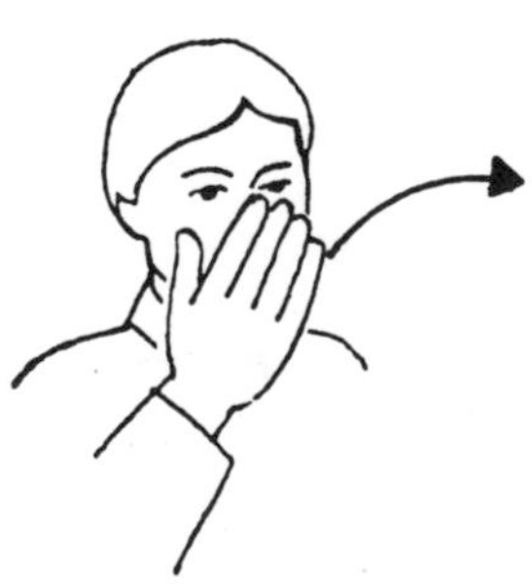 future **2**	 **7**

MOVIE

Handshape: 5, both
Location: In front of body
Movement: Shake slightly

understandable | confused with

9

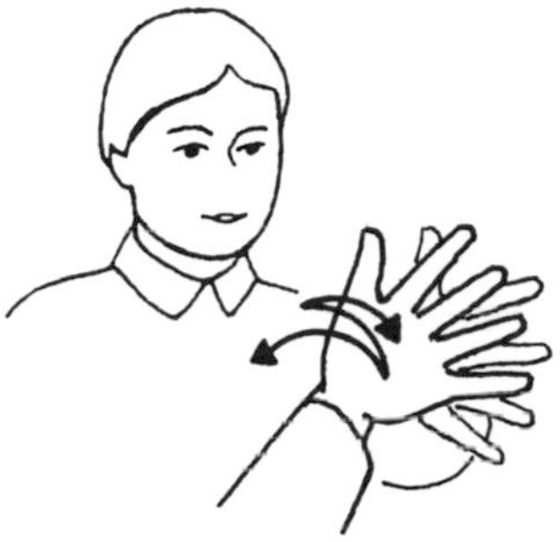

cheese 1

8

cheese 1

7

BUTTER

Handshape: H
Location: In front of body
Movement: Brush palm twice

understandable

10

8

confused with

soap **2**

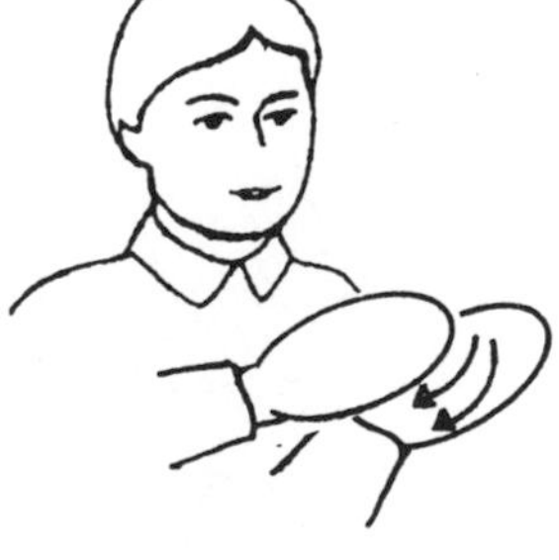

6

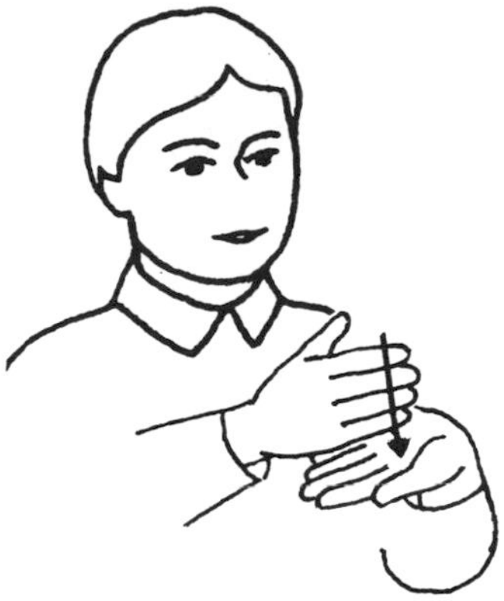

STOP

Handshape: Open B, both
Location: In front of body
Movement: Strike palm

understandable

9

9

confused with

5

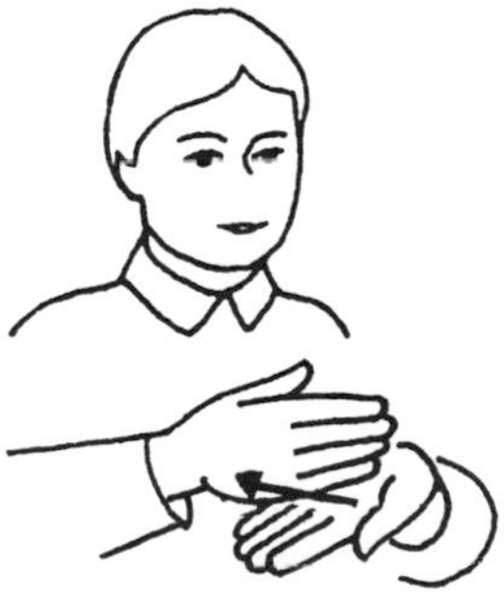

some **1**

ME

Handshape: 1
Location: Chest
Movement: Touch chest

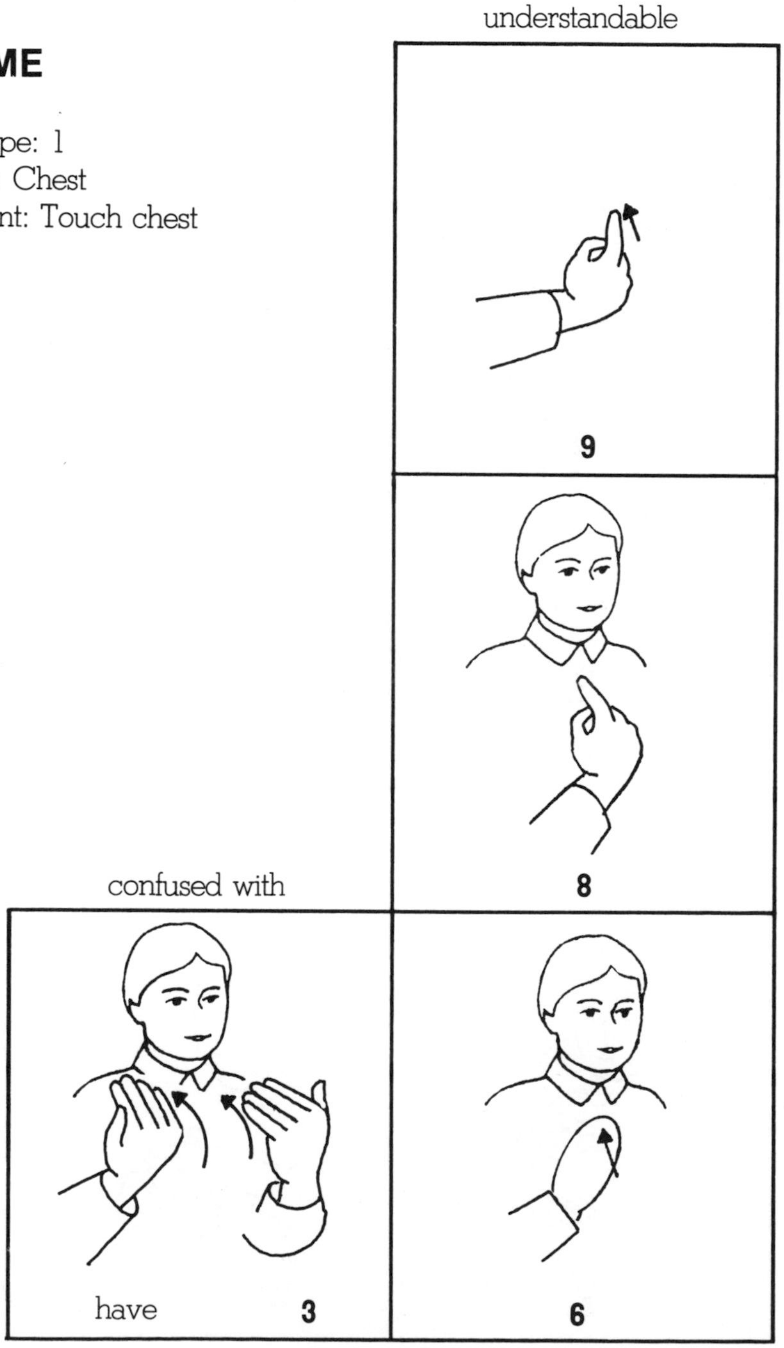

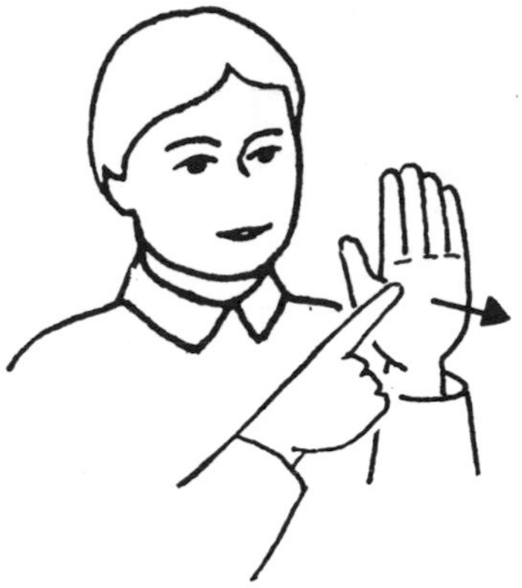

SHOW

Handshape: 1; open B
Location: In front of body
Movement: Move forward

understandable

9

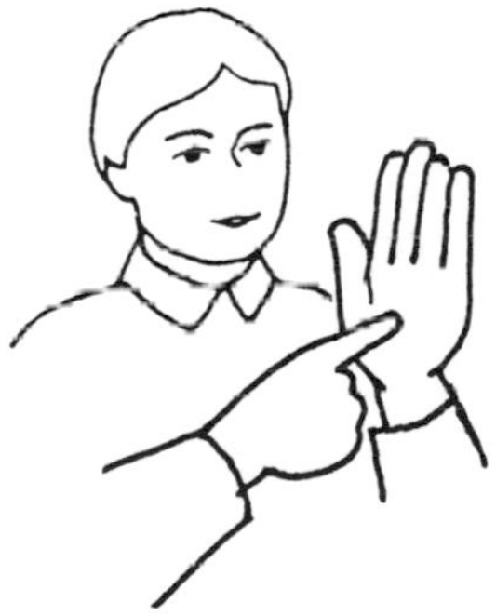

8

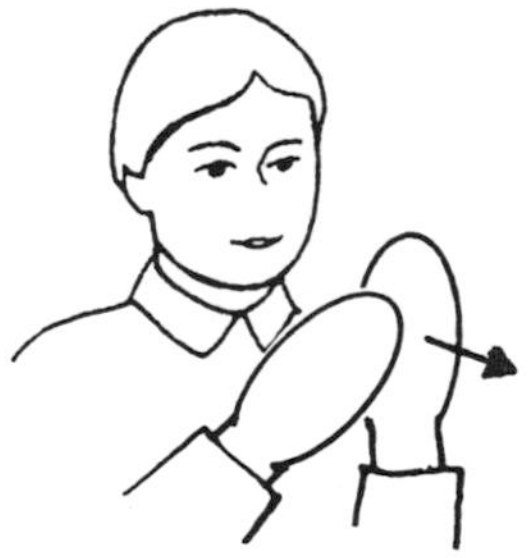

6

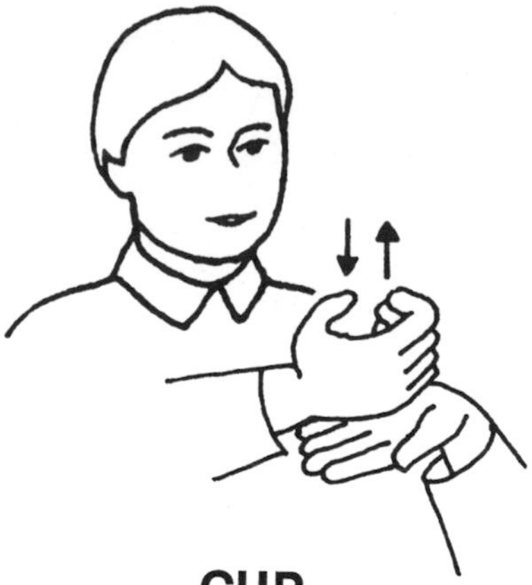

CUP

Handshape: C; open B
Location: In front of body
Movement: Tap palm

understandable

10

9

confused with

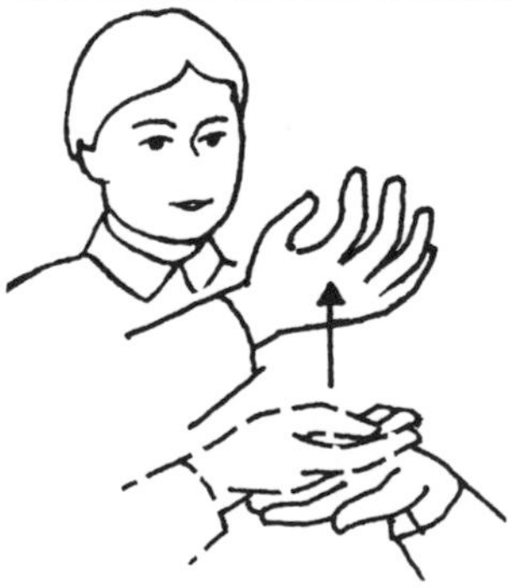

rich **1**

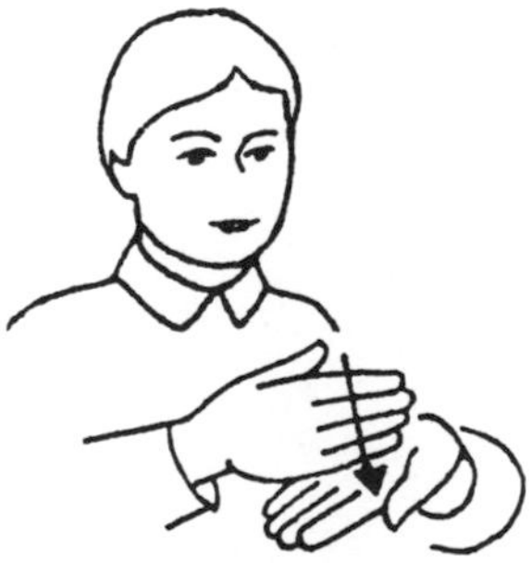

stop **2**

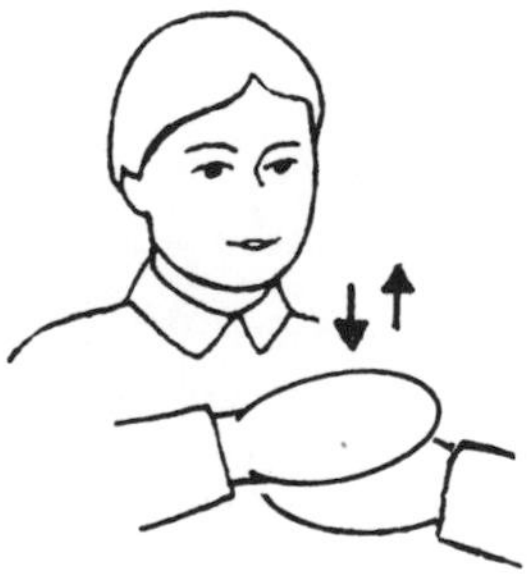

4

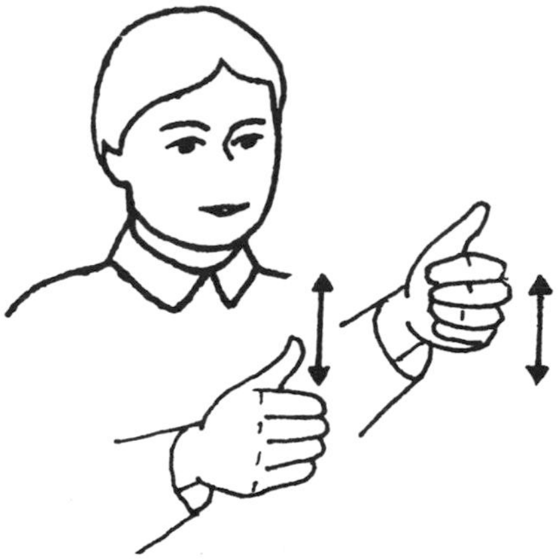

WHICH

Handshape: A, both
Location: In front of body
Movement: Alternately up and down

understandable

10

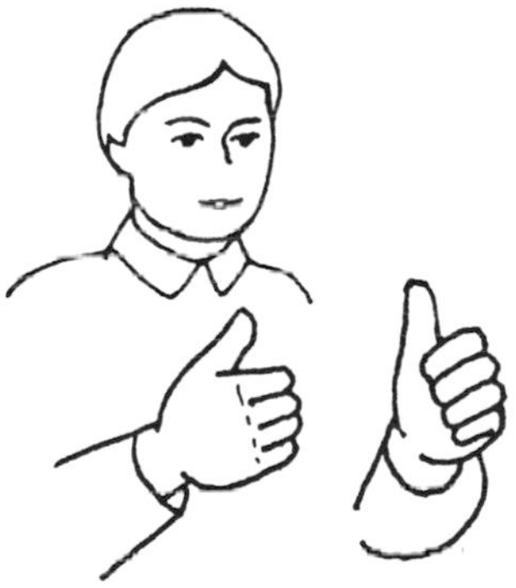

10

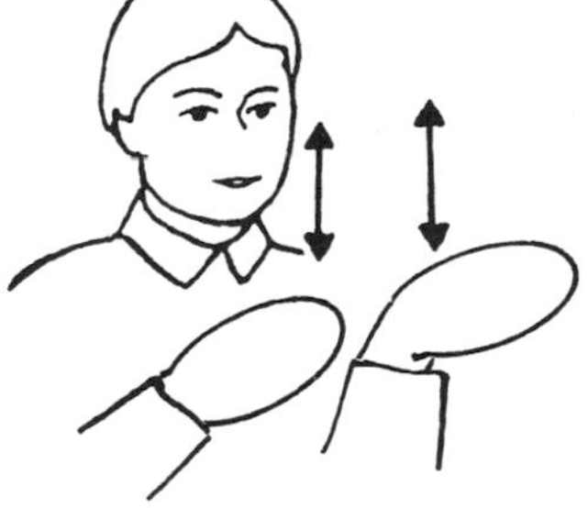

3

confused with

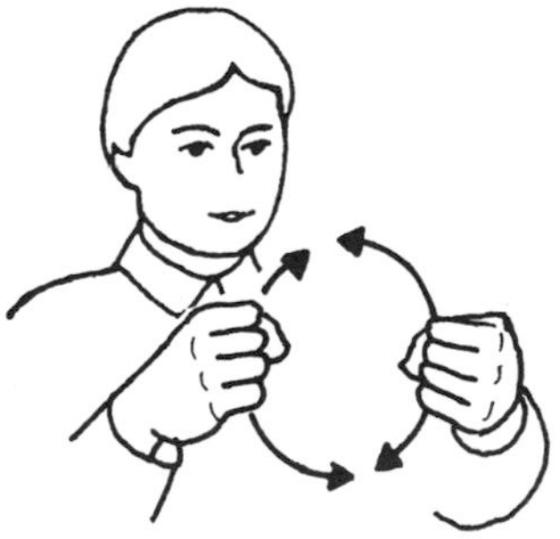

drive 2

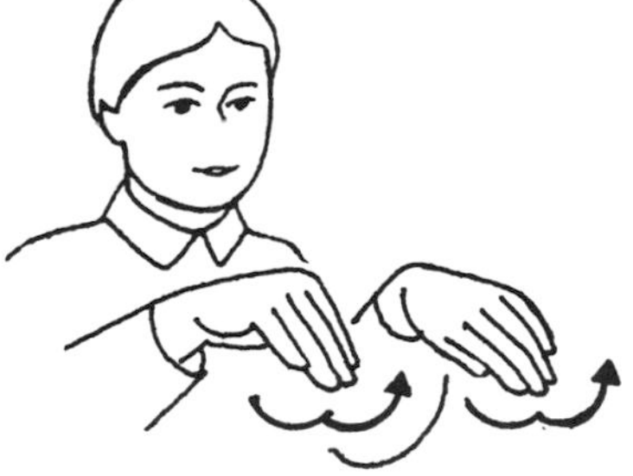

march 1

SEE

Handshape: V
Location: Eyes
Movement: Away from eyes

understandable

9

9

confused with

two 2

5

MAKE

Handshape: S, both
Location: In front of body
Movement: Tap, twist, tap again

understandable

10

9

confused with

4

hold **3**

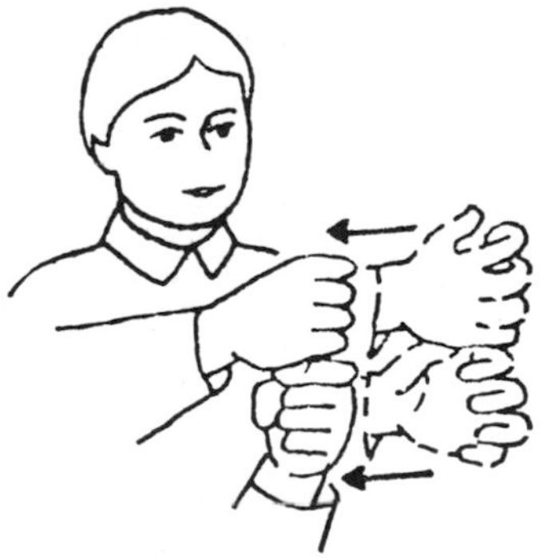

get **1**

BRING

Handshape: Open B, both
Location: In front of body
Movement: Toward body

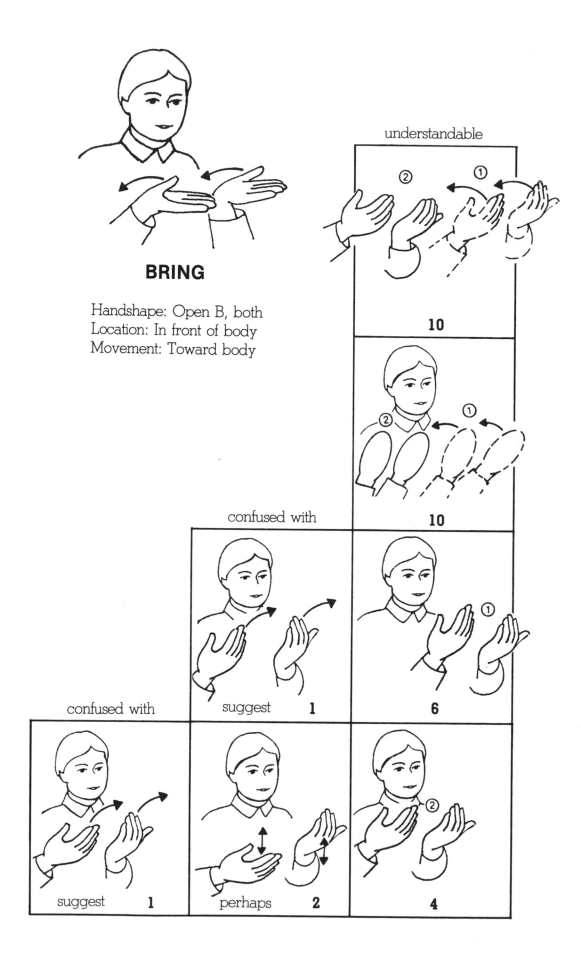

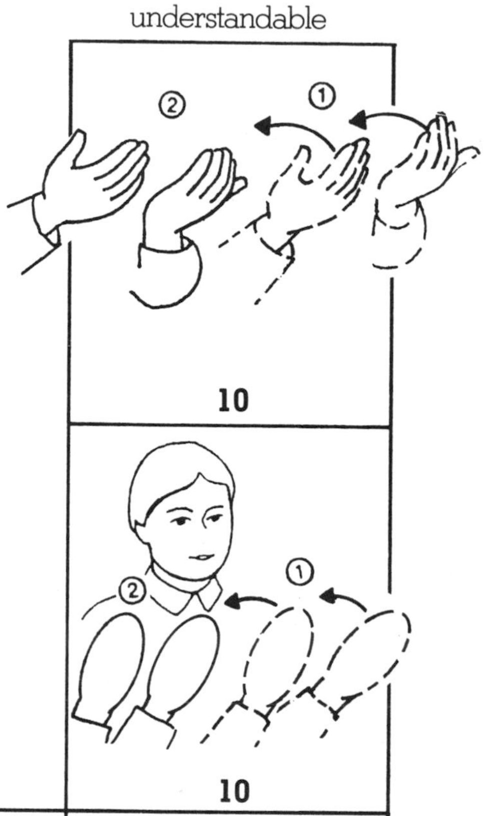

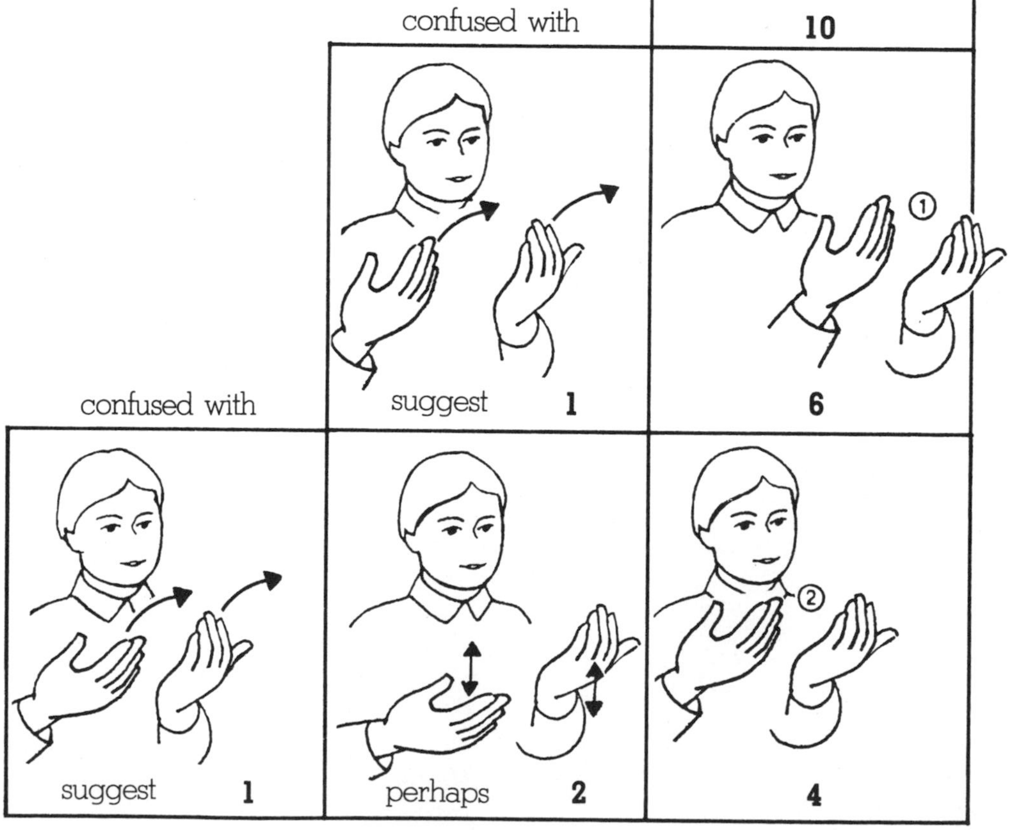

HOSPITAL

Handshape: H
Location: Upper arm
Movement: Trace a cross

understandable

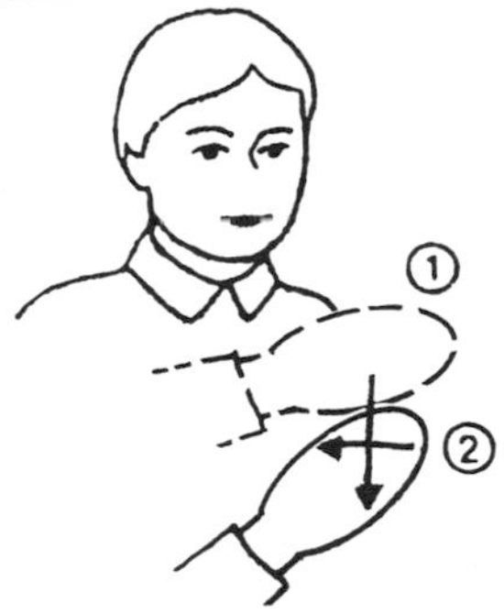

10

9

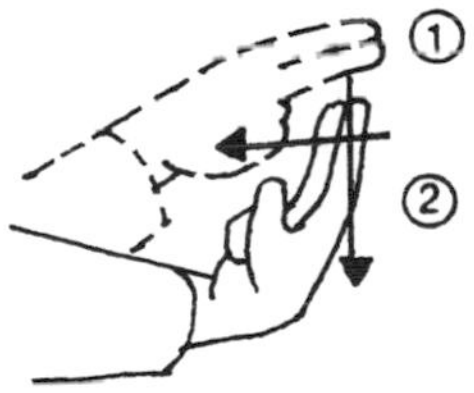

6

confused with

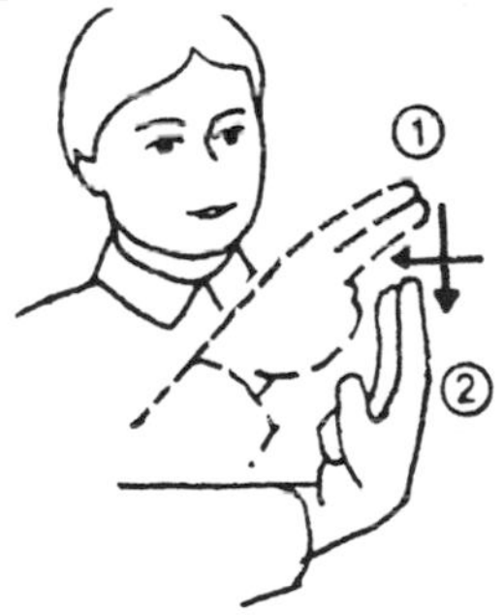

cross 1

5

us 3

confused with

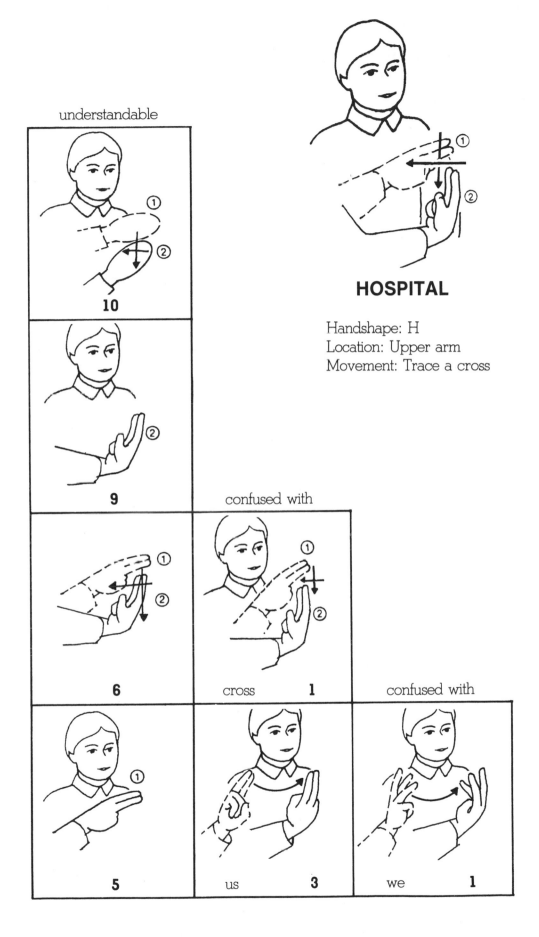

we 1

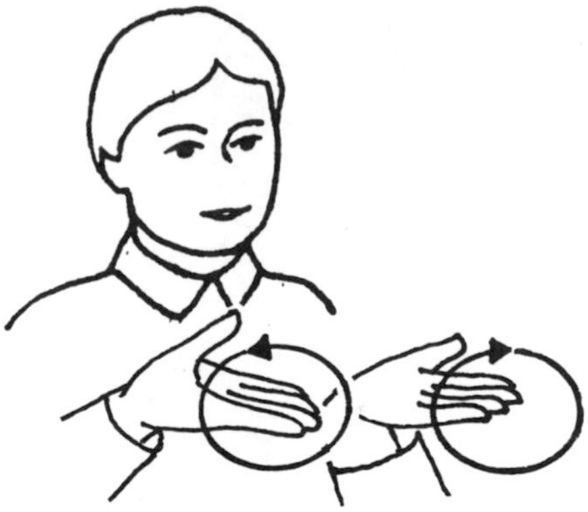

HERE

Handshape: Open B, both
Location: In front of body
Movement: Make tiny circles

understandable

9

9

confused with

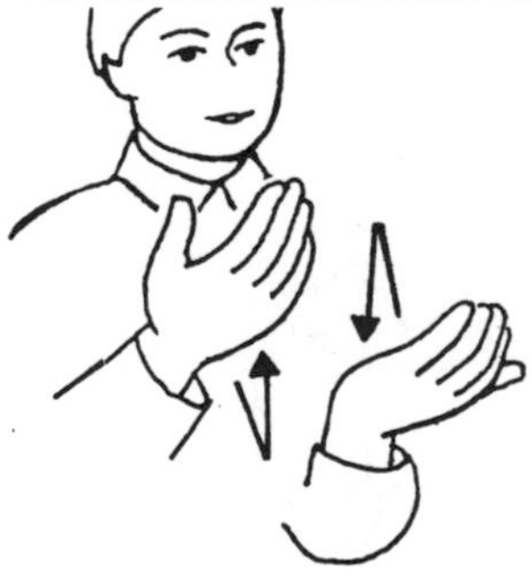

maybe 2

3

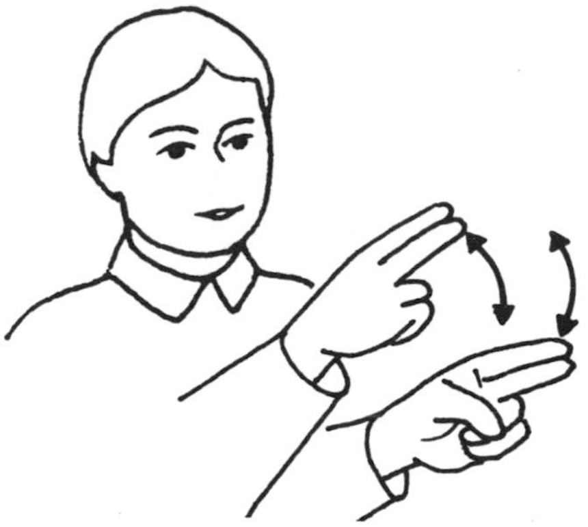

HURRY

Handshape: H, both
Location: In front of body
Movement: Shake up and down

understandable

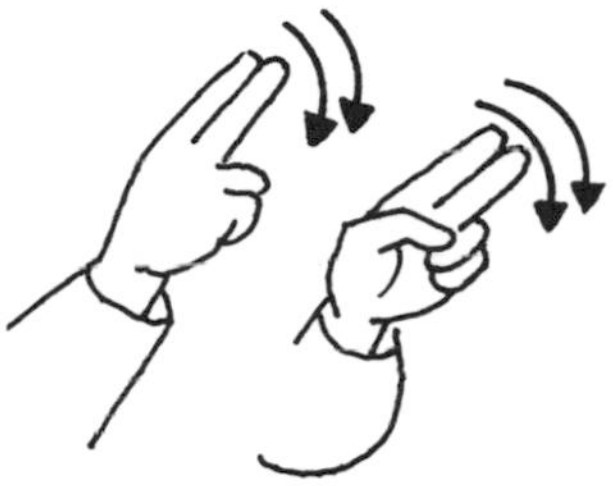

9

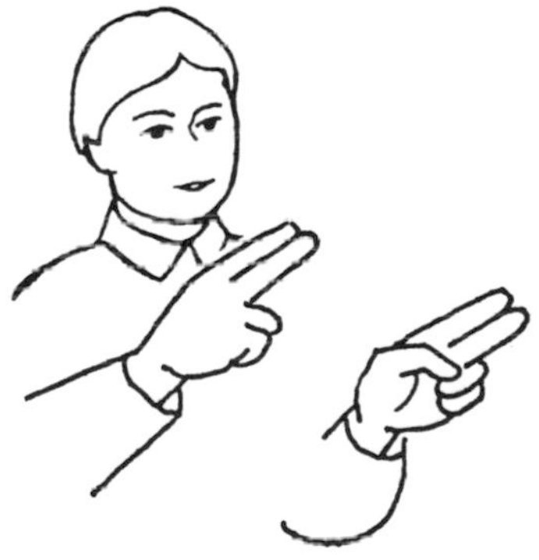

9

3

confused with

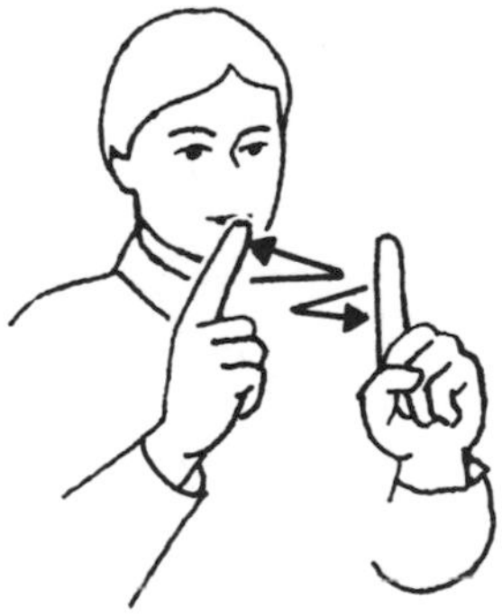

talk **3**

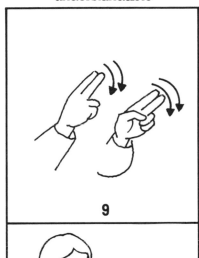

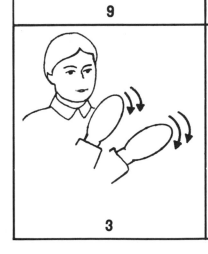

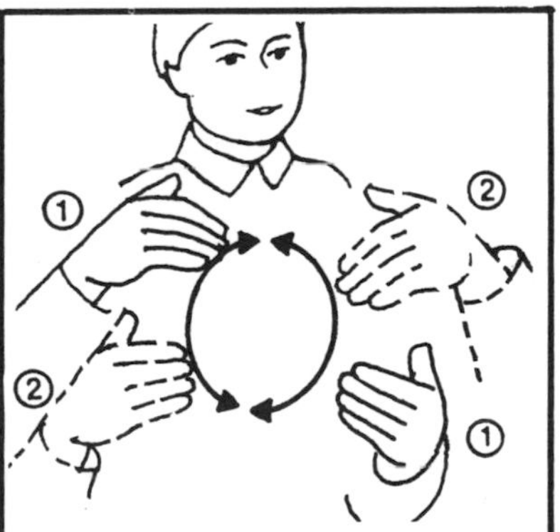

compare **2**

MONEY

Handshape: Open B, both
Location: In front of body
Movement: Tap palm twice

understandable

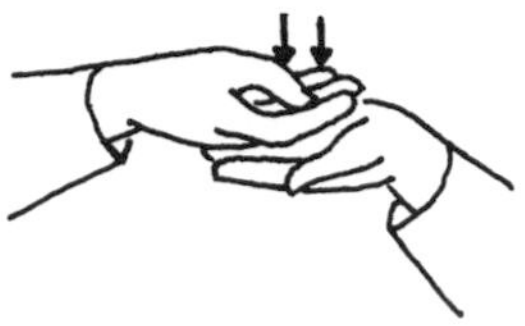

10

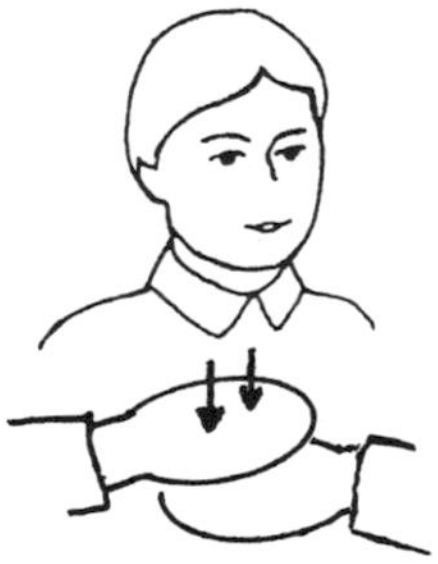

9

confused with

prove **1**

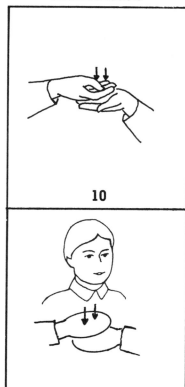

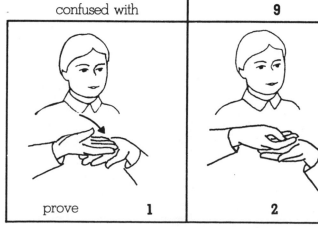

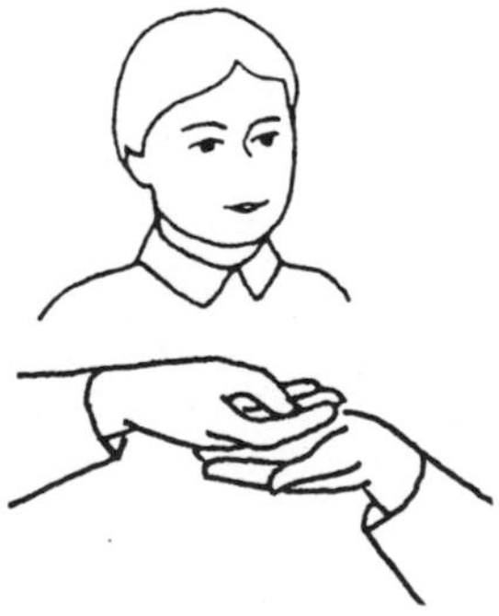

2

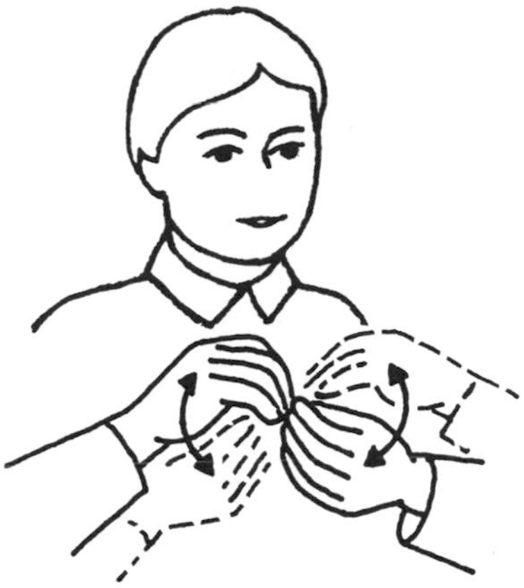

NUMBER

Handshape: Flat O, both
Location: In front of body
Movement: Touch, twist, touch again

understandable

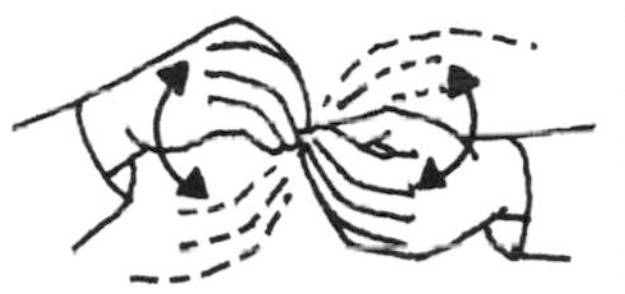

10

8

confused with

kiss 1

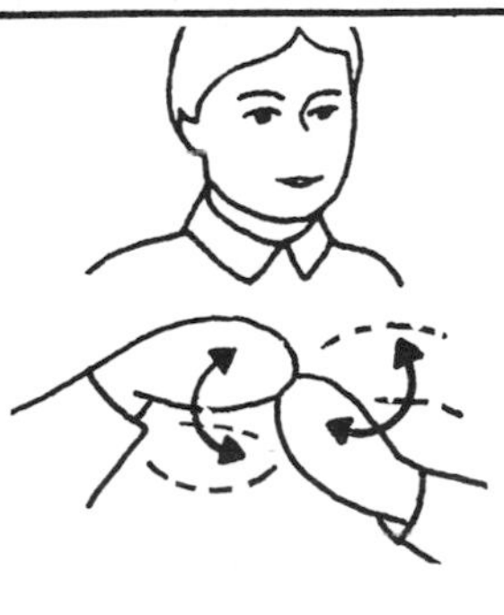

2

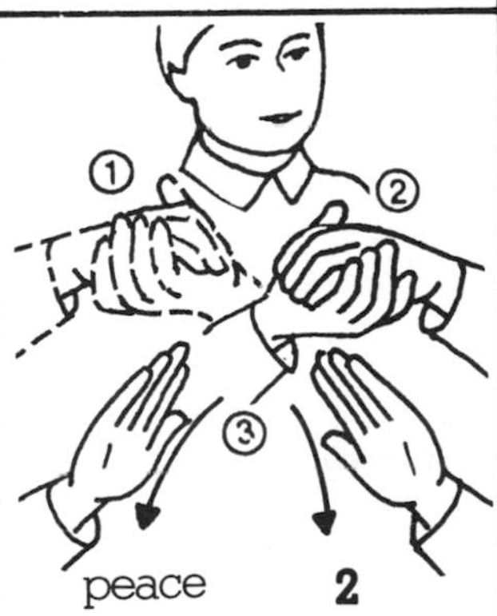

peace 2

confused with

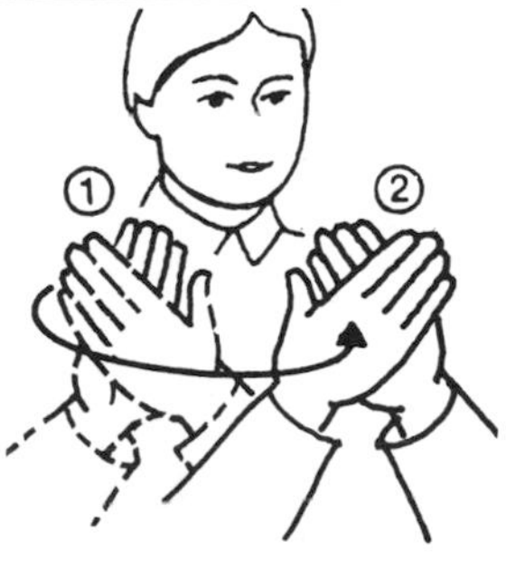

become 2

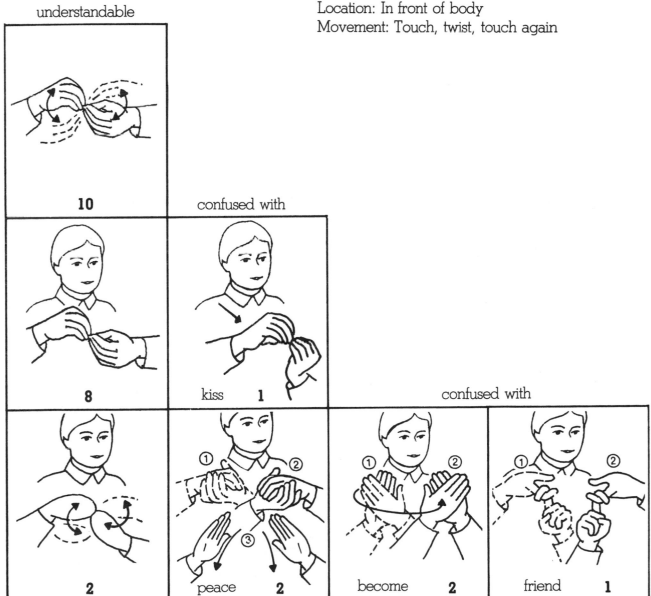

friend 1

FALL

Handshape: V
Location: In front of body
Movement: Invert and lower V

understandable

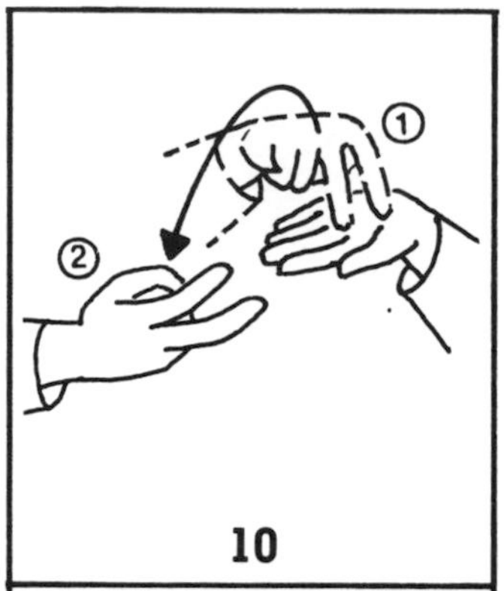

confused with

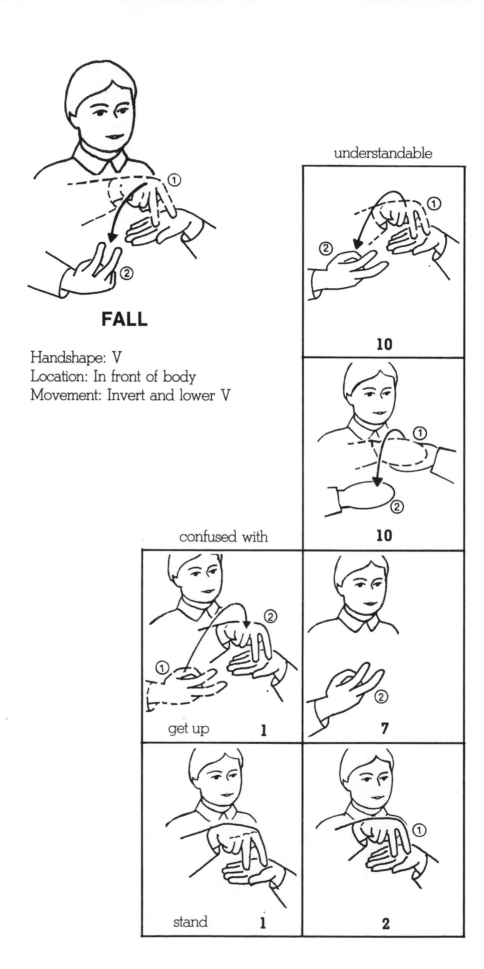

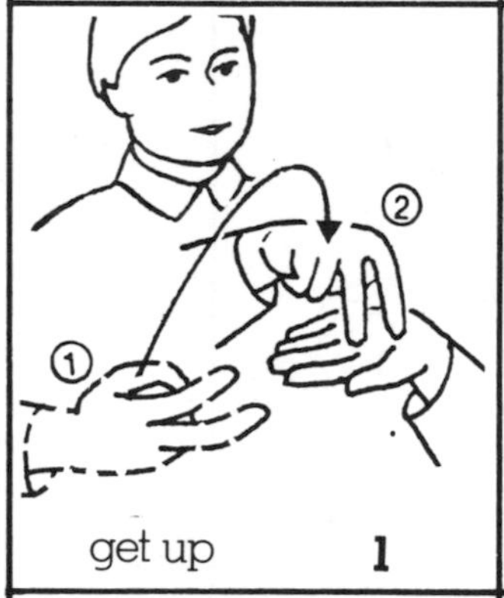

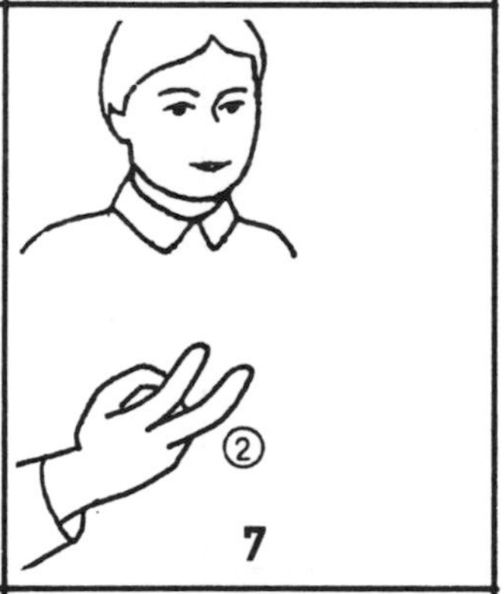

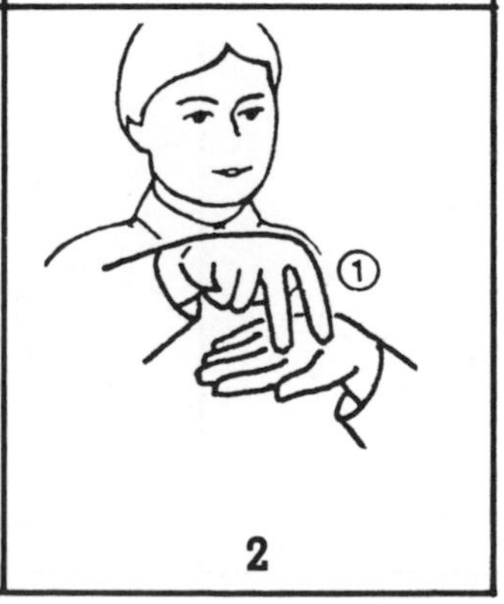

WHY

Handshape: Open B to Y
Location: Forehead
Movement: Outward, change to Y

understandable

confused with

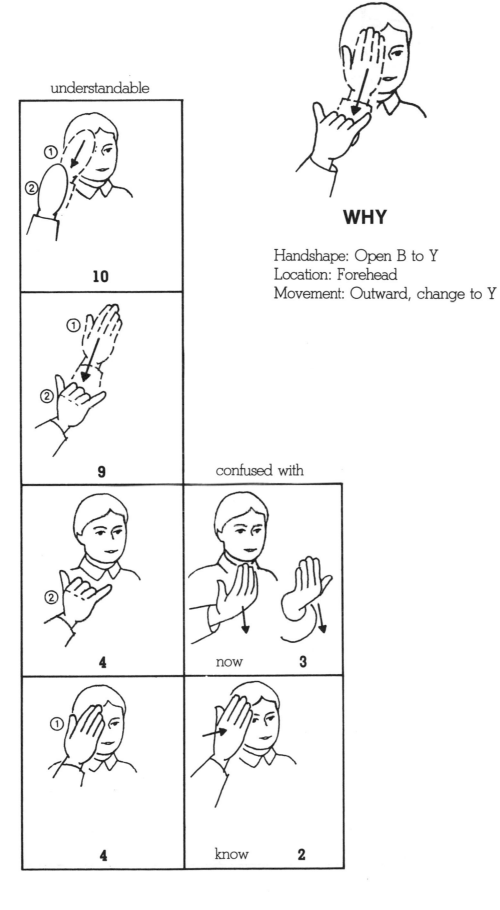

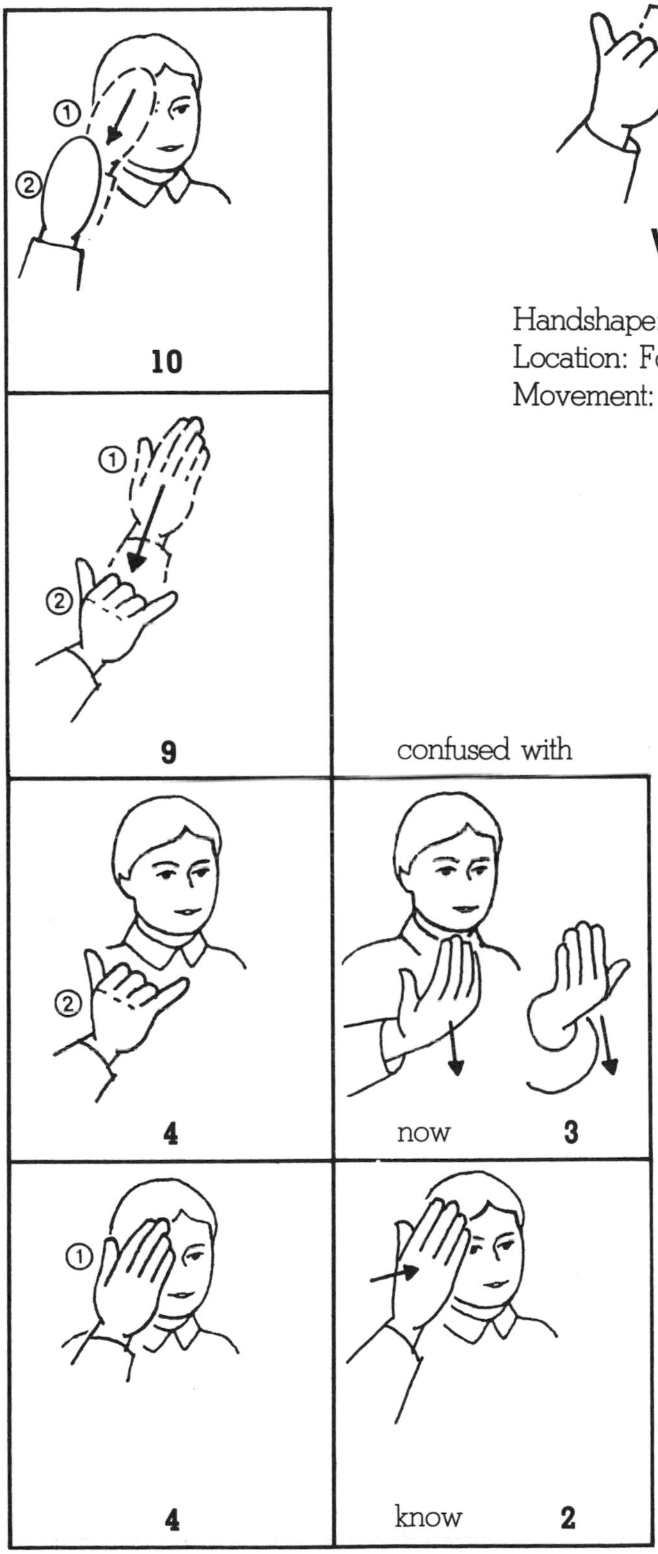

I

Handshape: I
Location: Chest
Movement: Tap thumb on chest

understandable

9

8

confused with

polite **6**

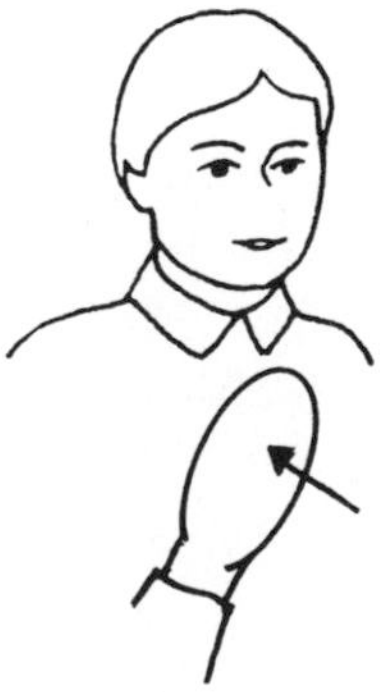

3

PLEASE

Handshape: Open B
Location: Chest
Movement: Make small circle

understandable

9

9

confused with

sorry 1

2

my 4

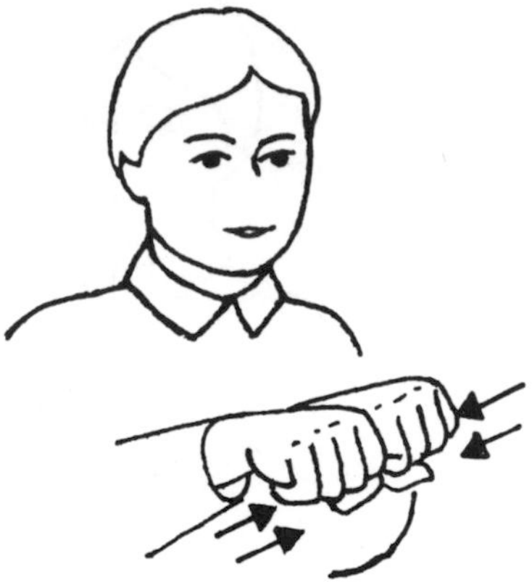

SHOE

Handshape: S, both
Location: In front of body
Movement: Tap thumbs twice

understandable

10

8

confused with

same **2**

confused with

close **2**

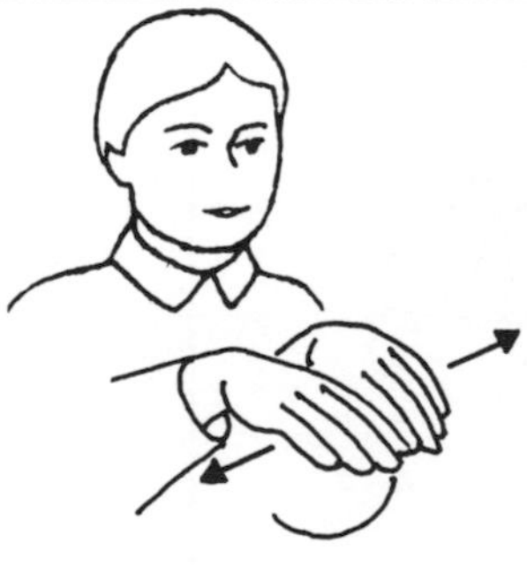

floor **2**

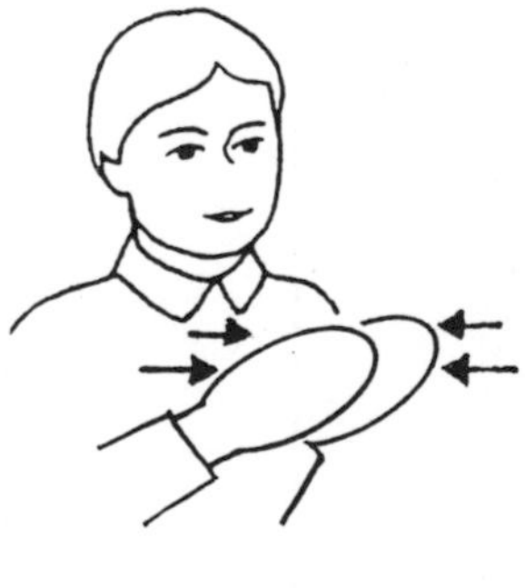

1

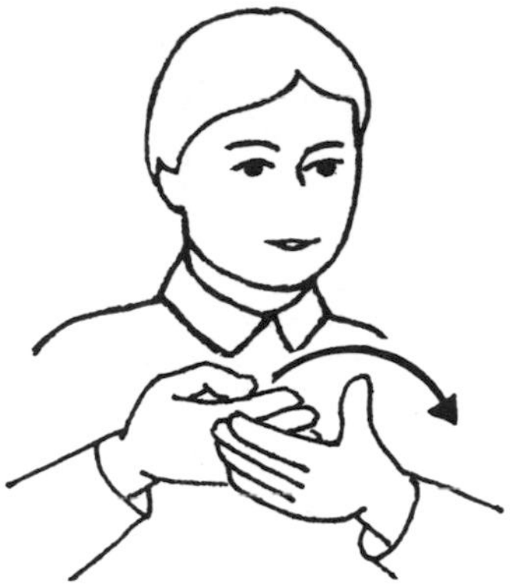

BUY

Handshape: Open B, both
Location: In front of body
Movement: Tap palm, move forward

understandable

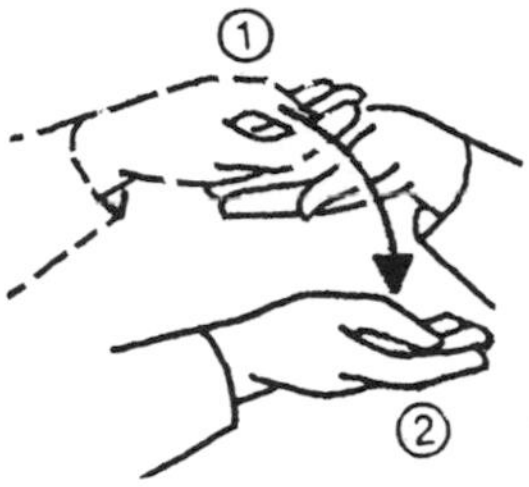

9

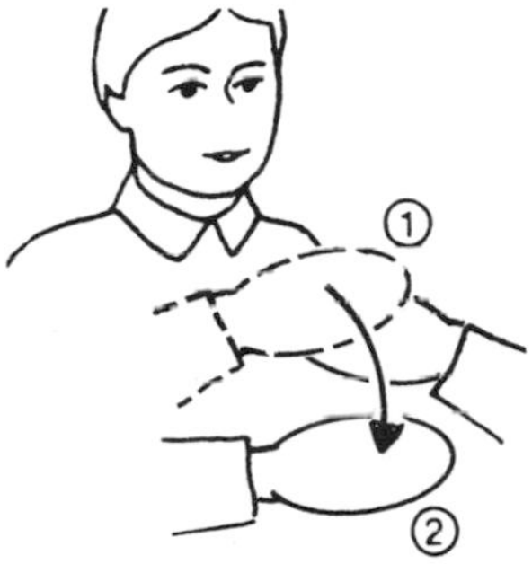

9

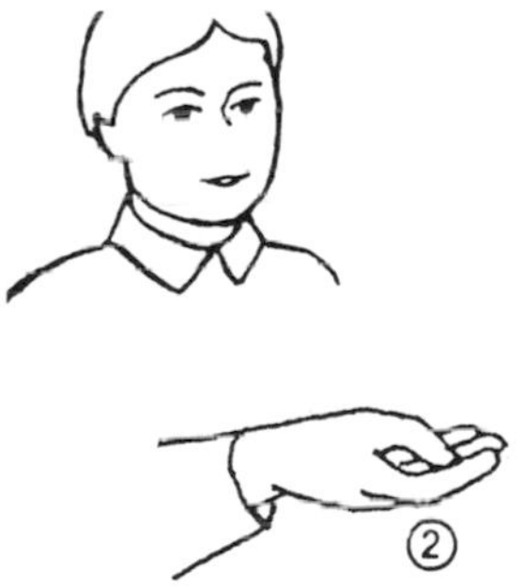

4

3

confused with

money 3

money 3

BOOK

Handshape: Open B, both
Location: In front of body
Movement: Open palms

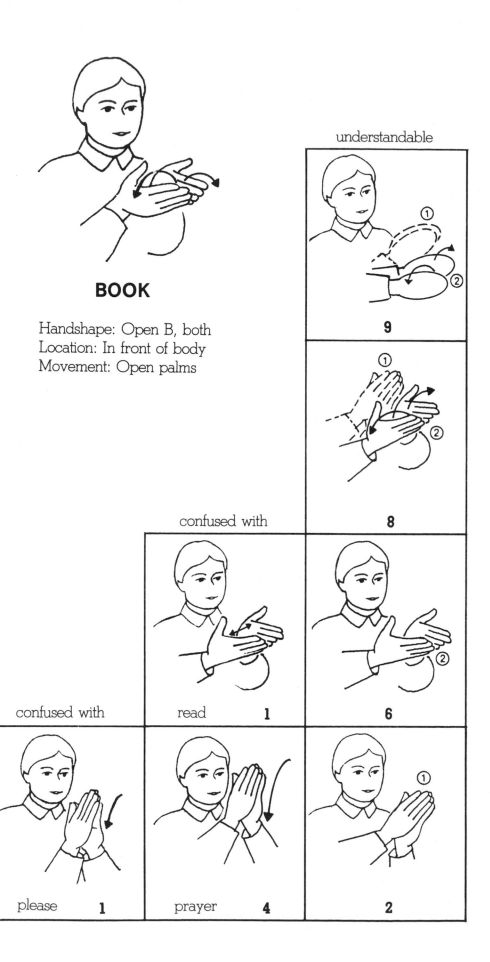

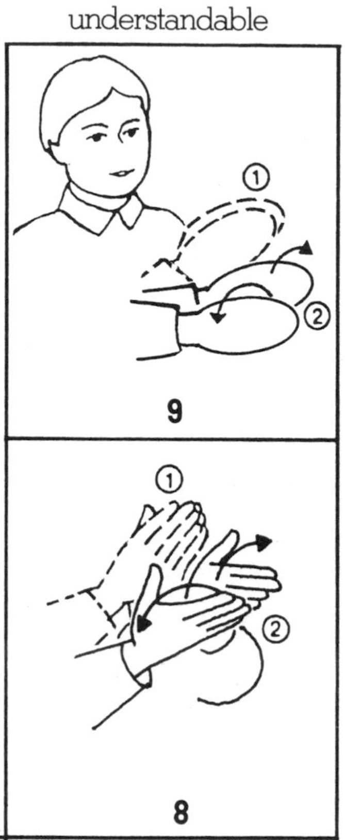

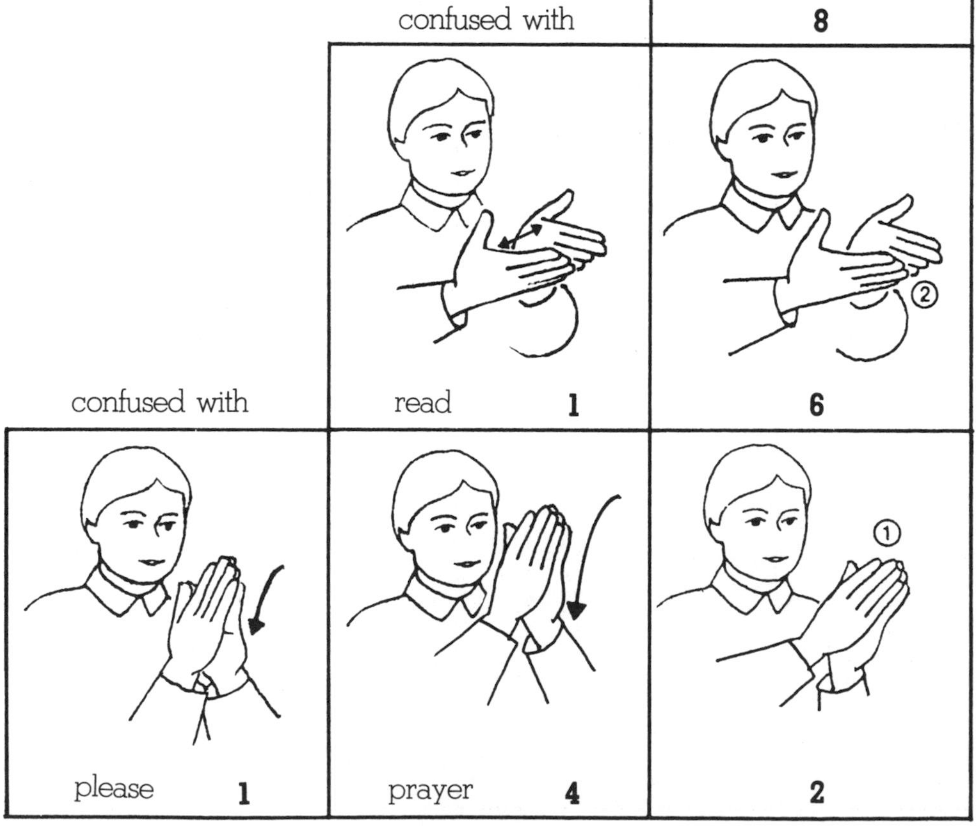

HOUSE

Handshape: Open B, both
Location: In front of body
Movement: Trace roof and walls

understandable

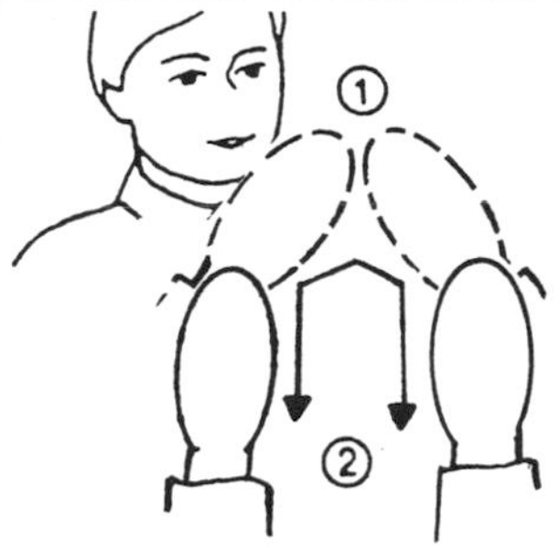

10

9

6

confused with

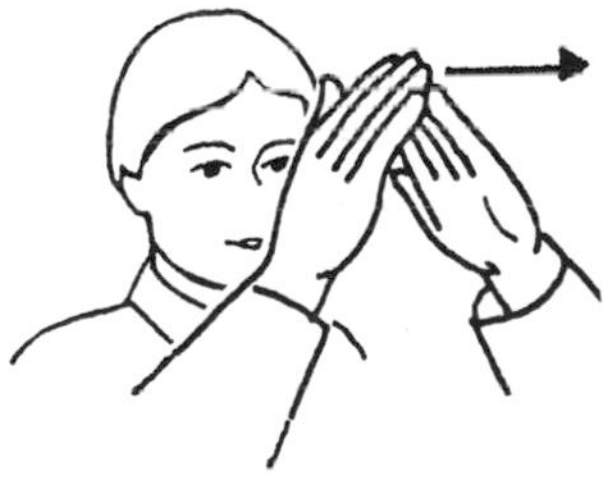

roof 1

0

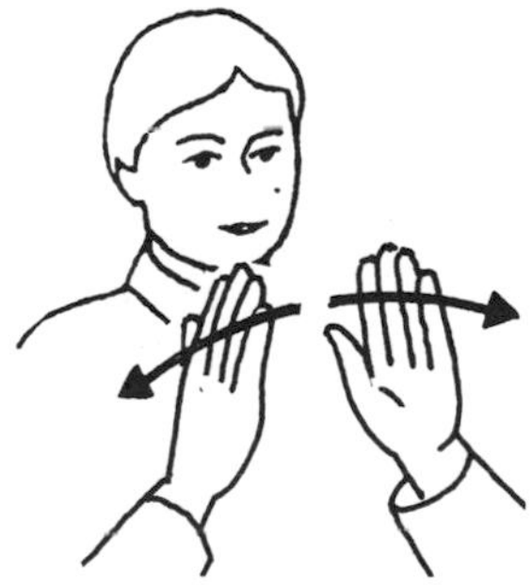

big 2

confused with

pay attention 2

BEFORE

Handshape: Open B, both
Location: In front of body
Movement: Move palm toward body

understandable

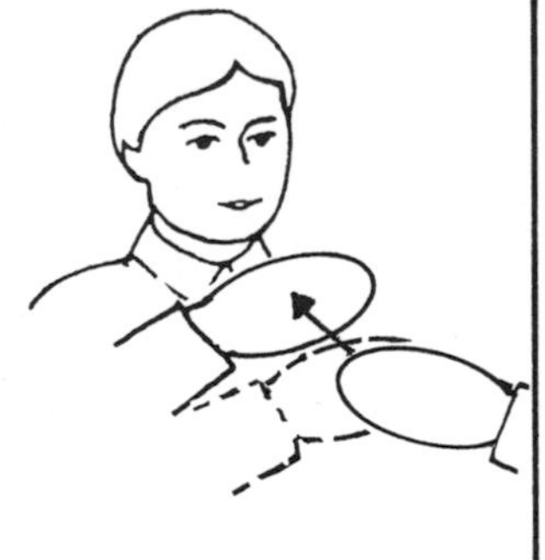

9

9

3

3

confused with

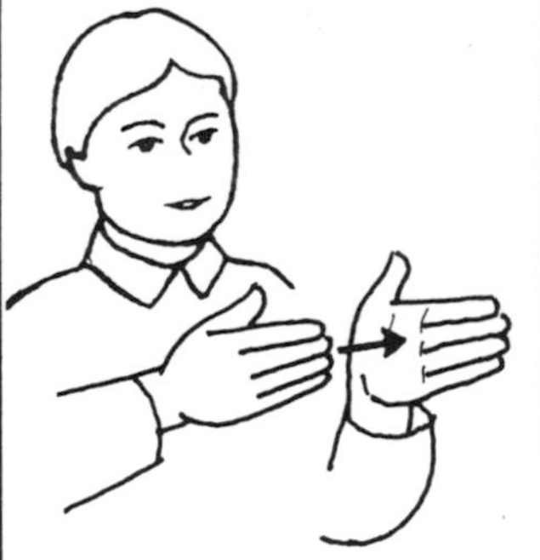

against 2

near 3

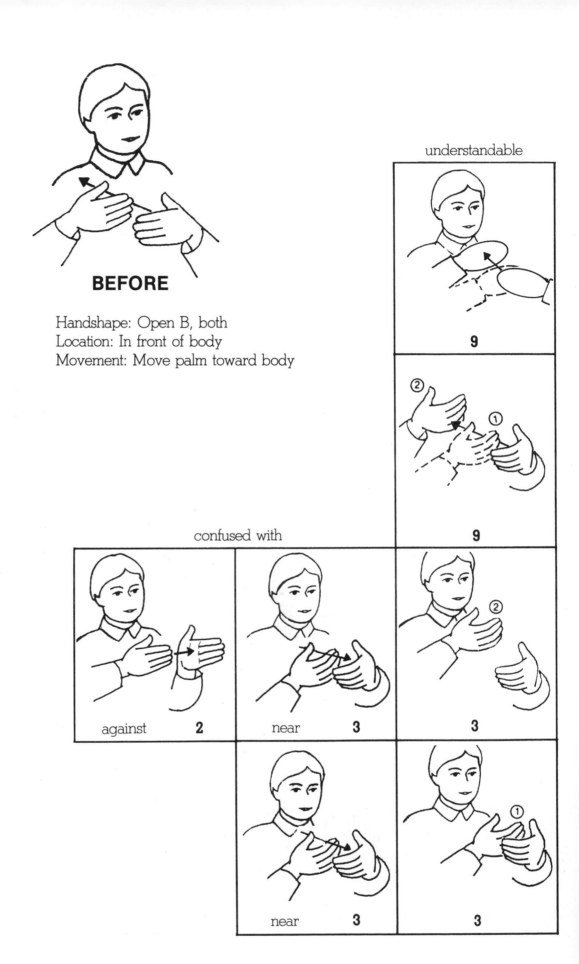

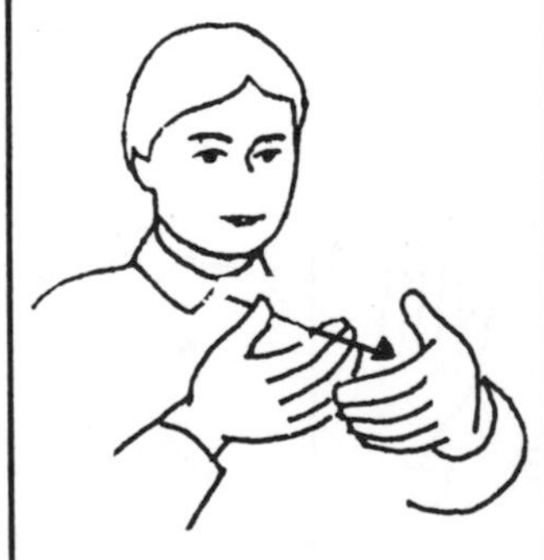

near 3

GOOD

Handshape: Open B, both
Location: Mouth
Movement: From mouth to palm

understandable

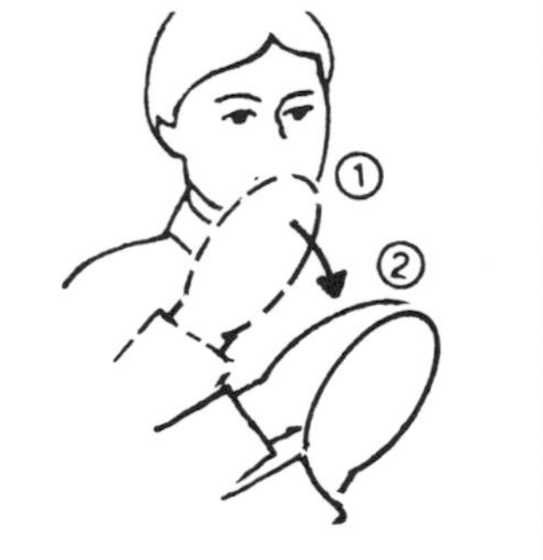

10

9

confused with

4

prove **5**

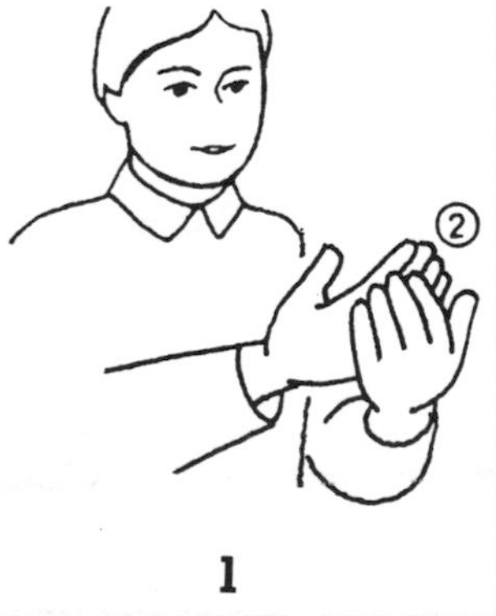

1

prove **3**

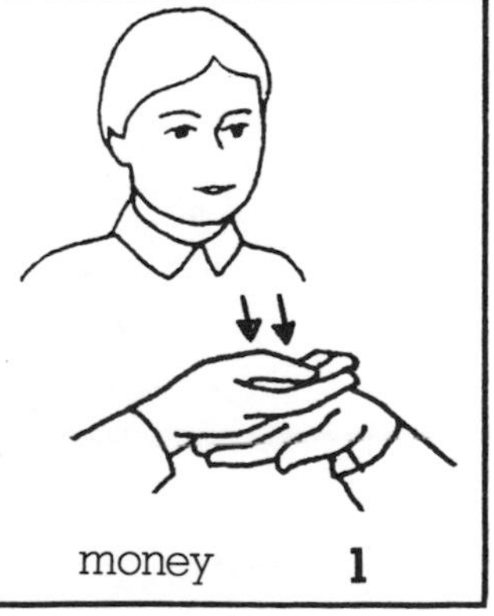

money **1**

JUMP

Handshape: V
Location: In front of body
Movement: Lift and bend V

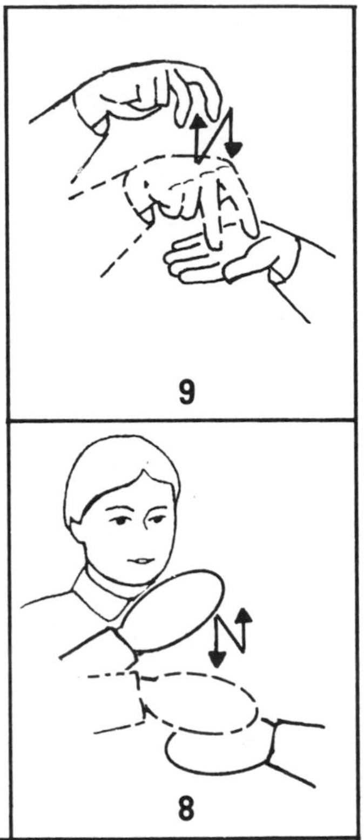

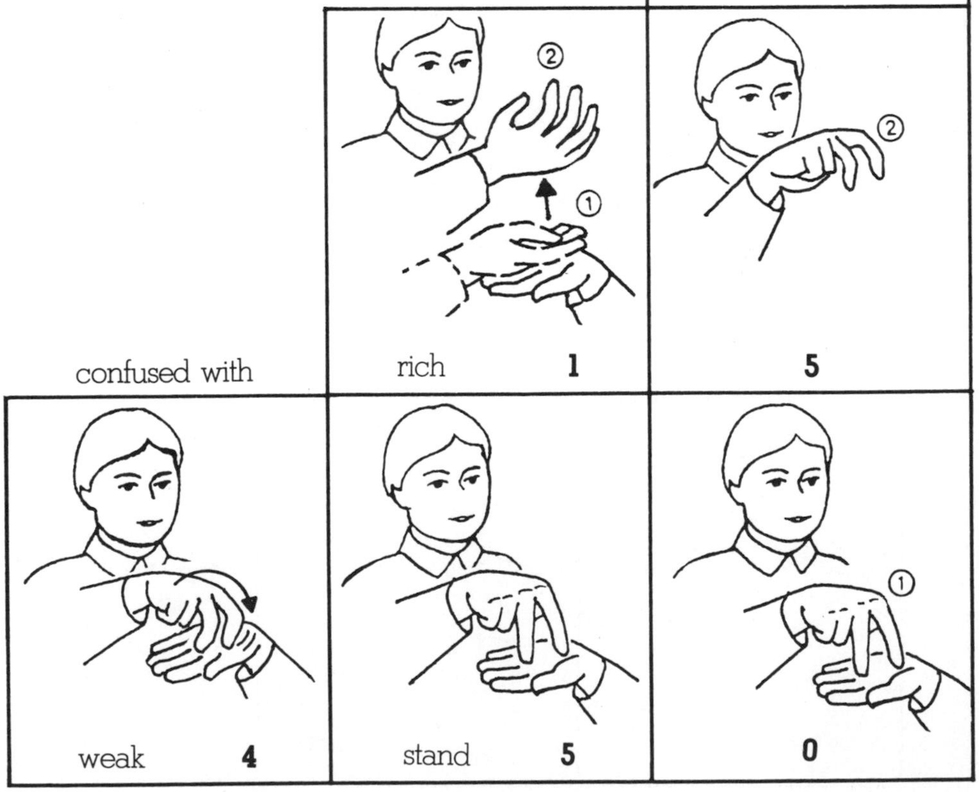

SODA POP

Handshape: F; S
Location: In front of body
Movement: Remove finger and tap

understandable

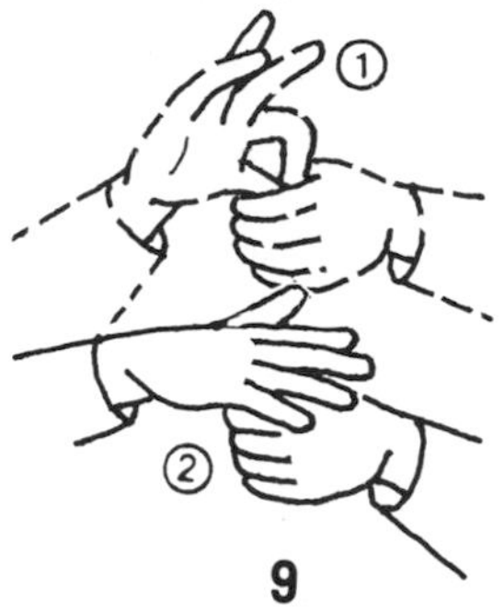

9

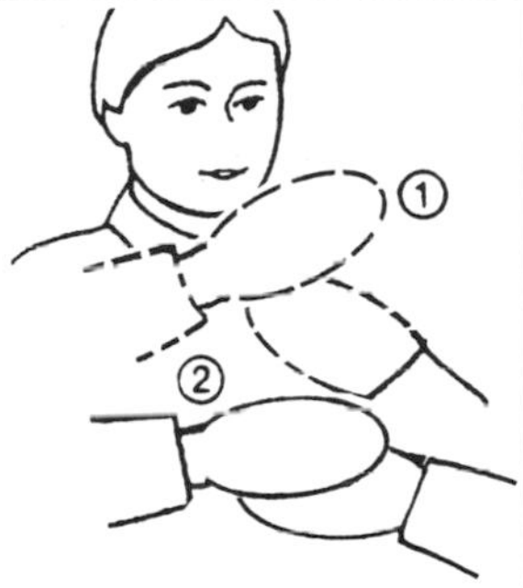

9

confused with

3

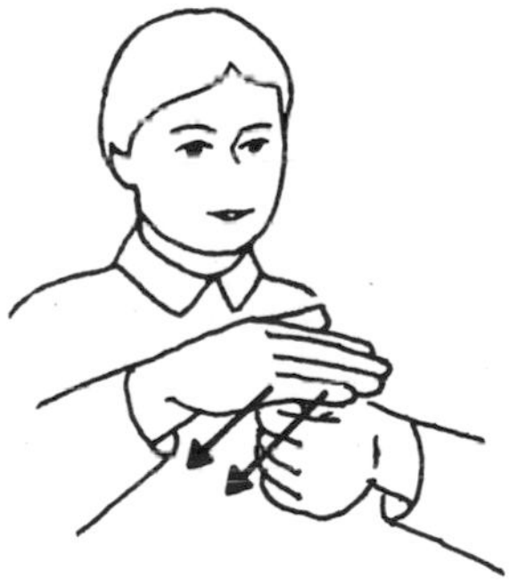

enough 3

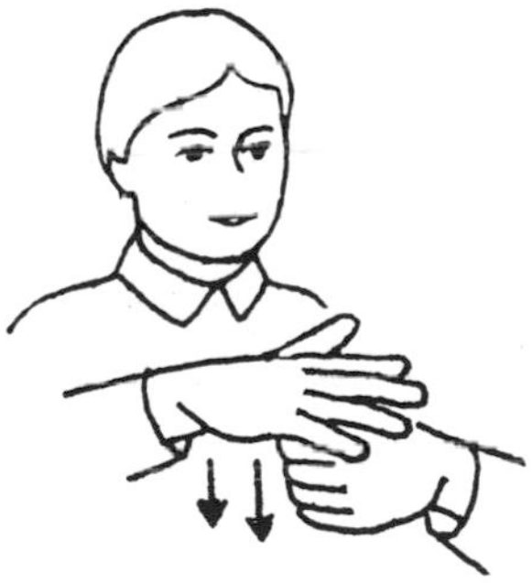

pressure 2

1

tea 5

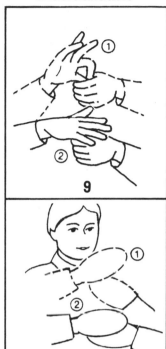

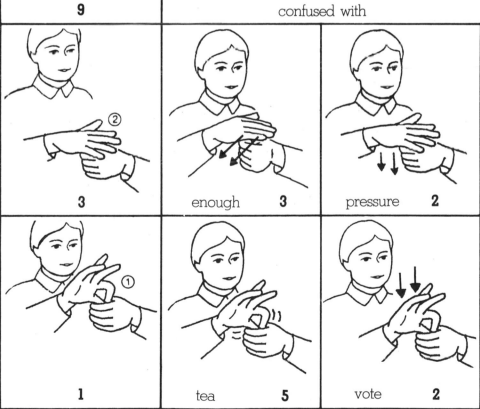

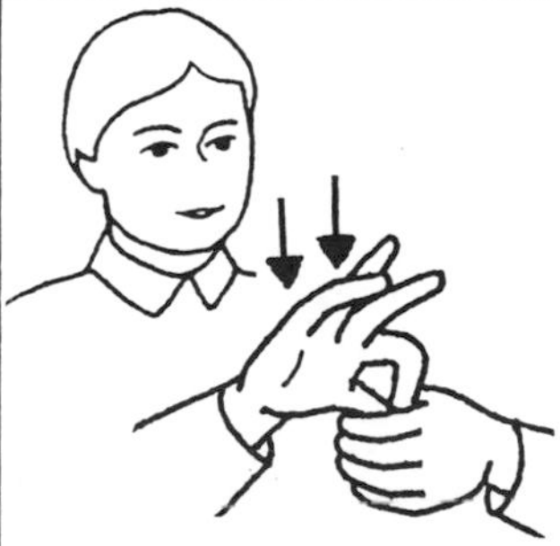

vote 2

BOWL

Handshape: Bent open B, both
Location: In front of body
Movement: Trace sides of bowl

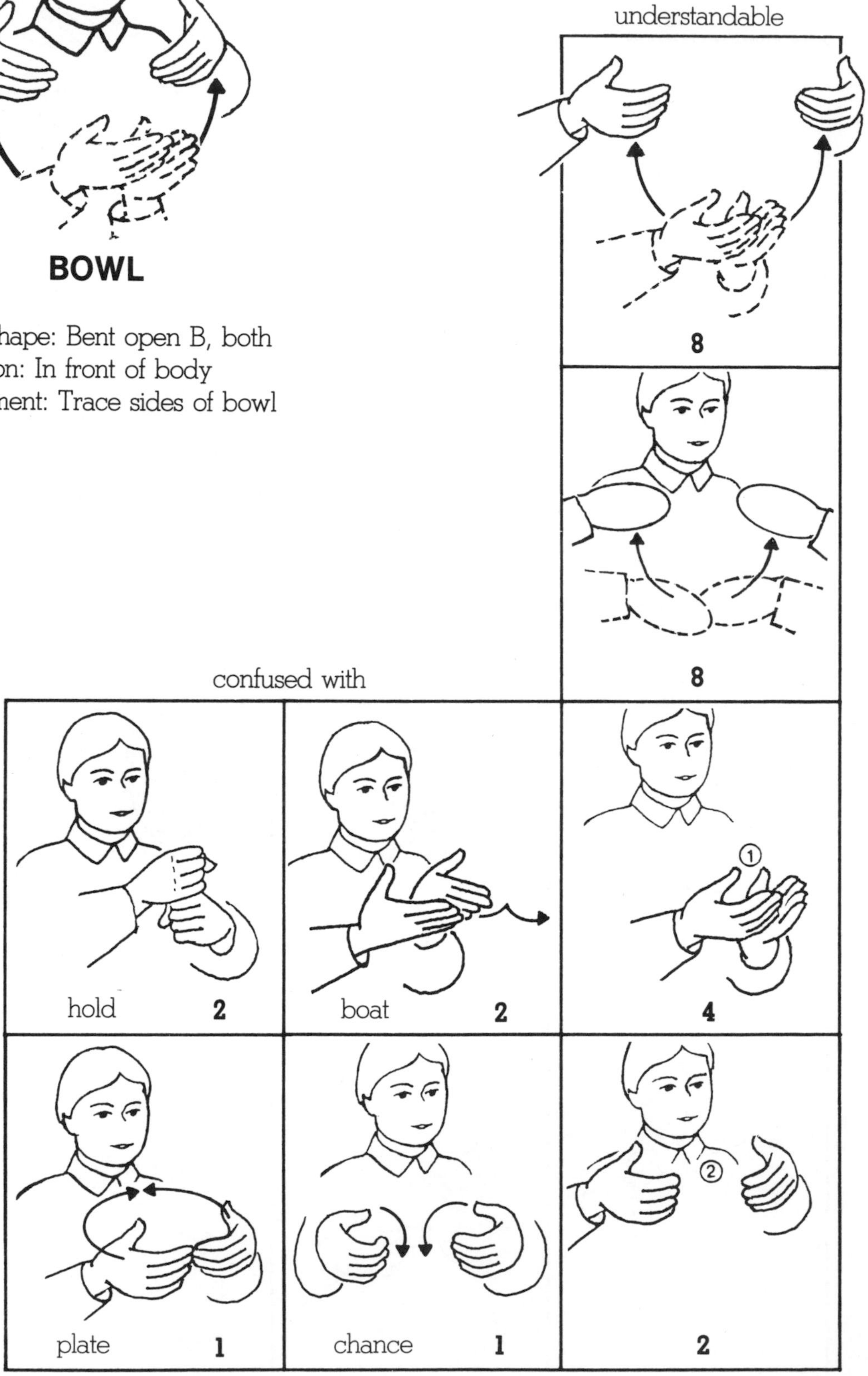

DOUBLE ROBUST

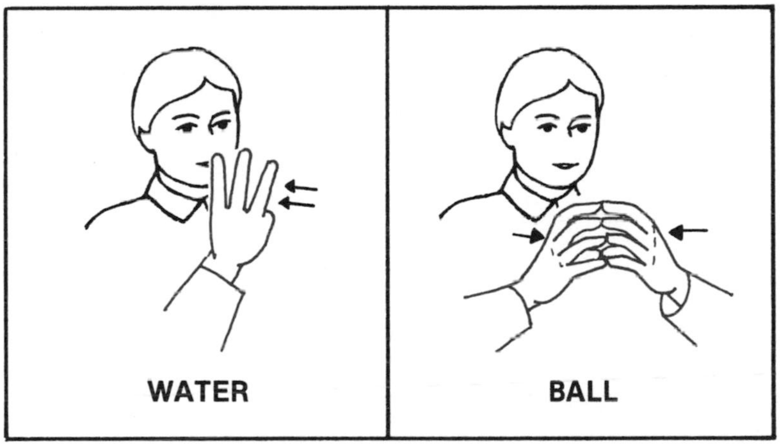

SINGLE FRAGILE

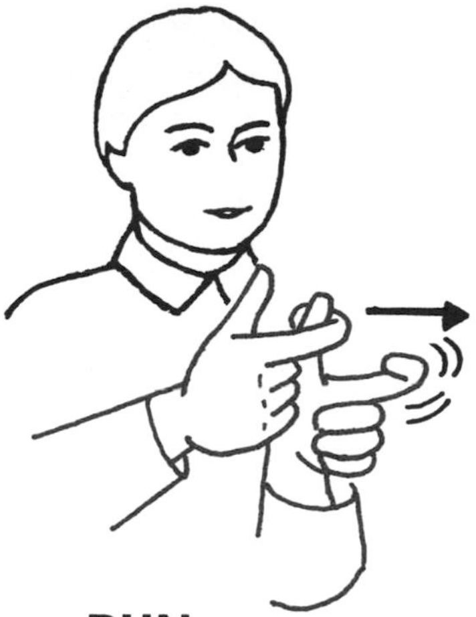

RUN

Handshape: L, both
Location: In front of body
Movement: Forward, wiggling

understandable

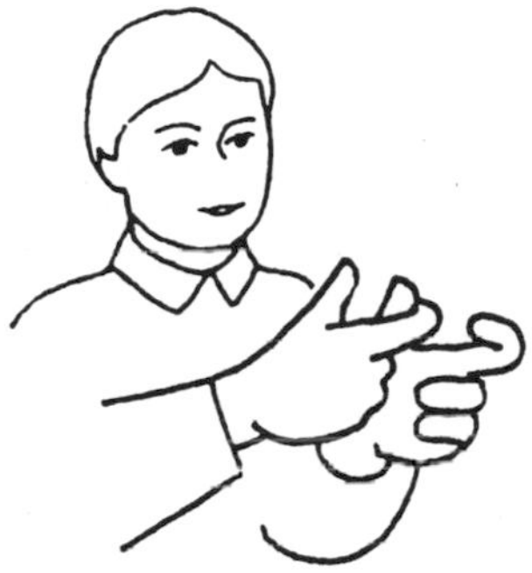

10

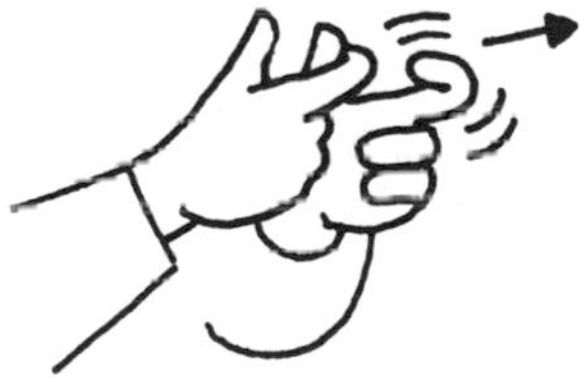

10

confused with

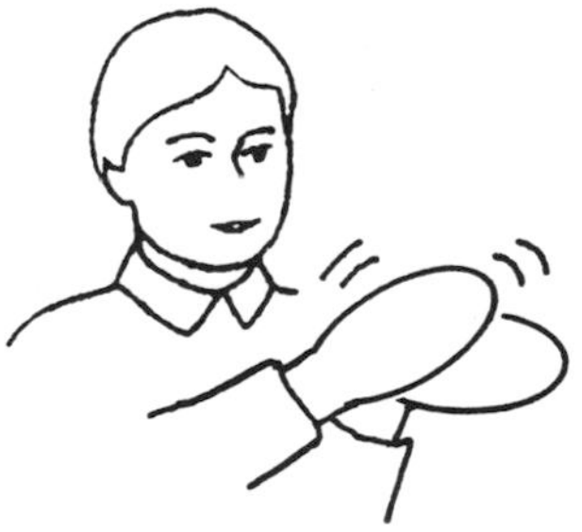

1

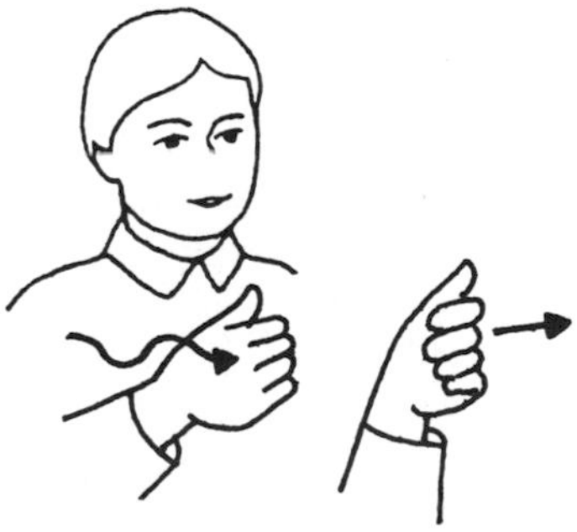

follow 5

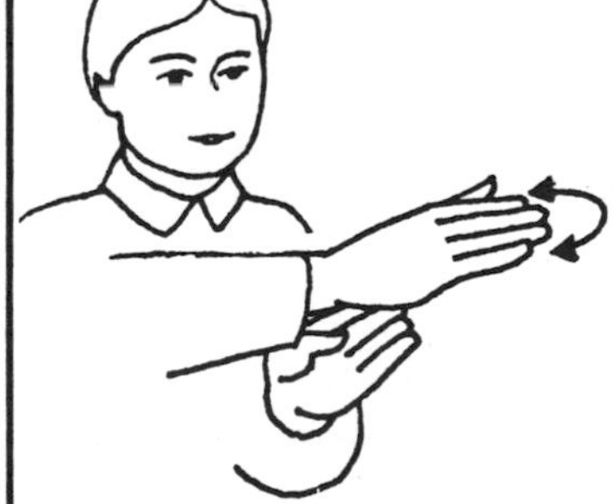

fish 2

DO

Handshape: C, both
Location: In front of body
Movement: Back and forth

understandable

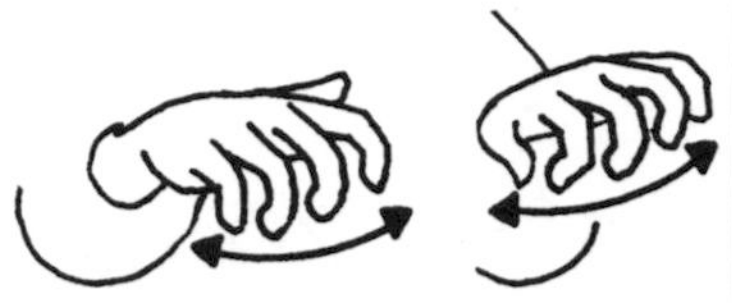

10

9

2

confused with

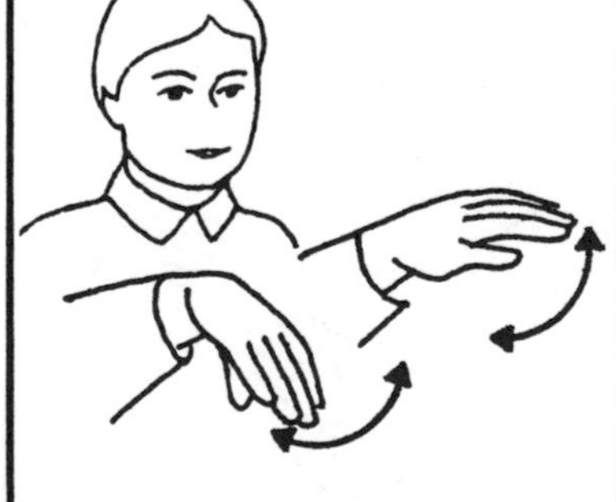

walk 2

behavior 5

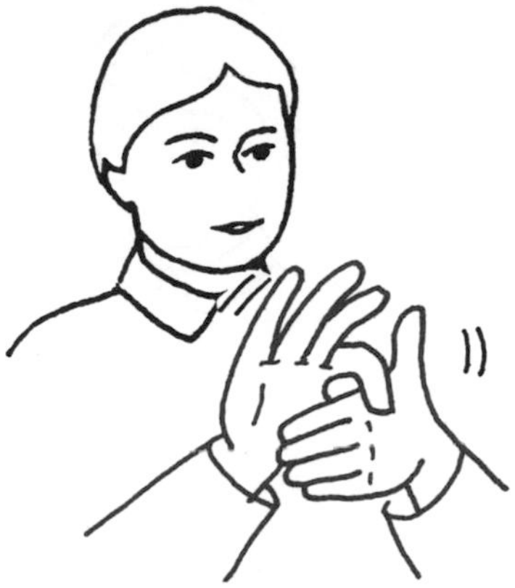

MEAT

Handshape: Open B
Location: In front of body
Movement: Pinch

understandable

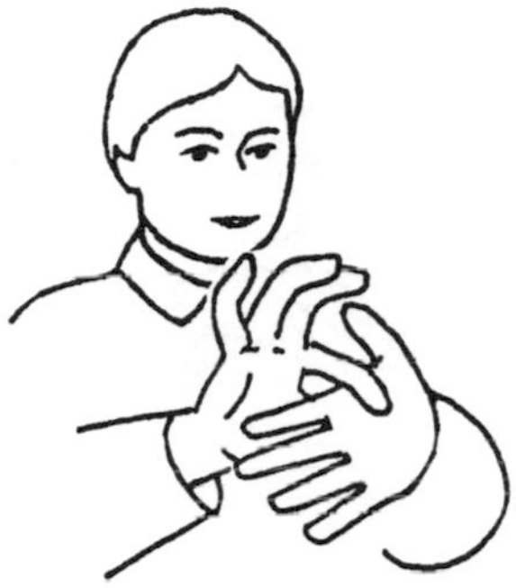

10

9

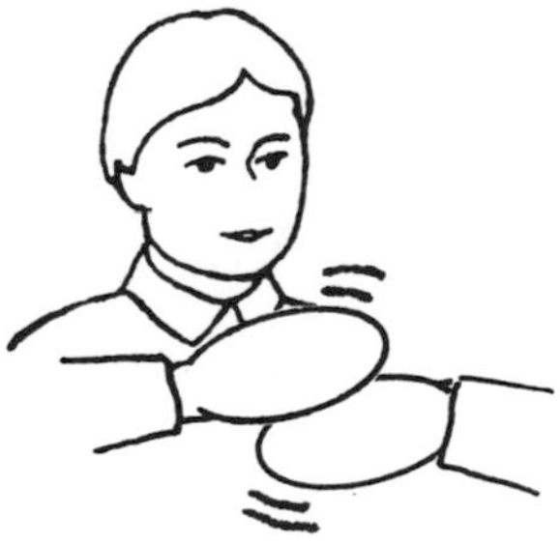

1

confused with

enough 1

CHURCH

Handshape: C
Location: In front of body
Movement: Tap on wrist twice

understandable

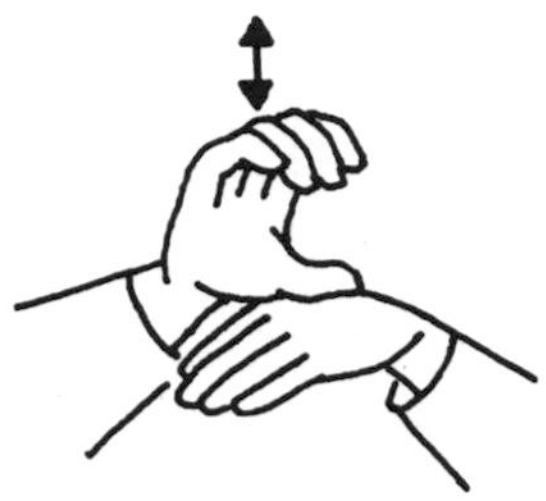

10

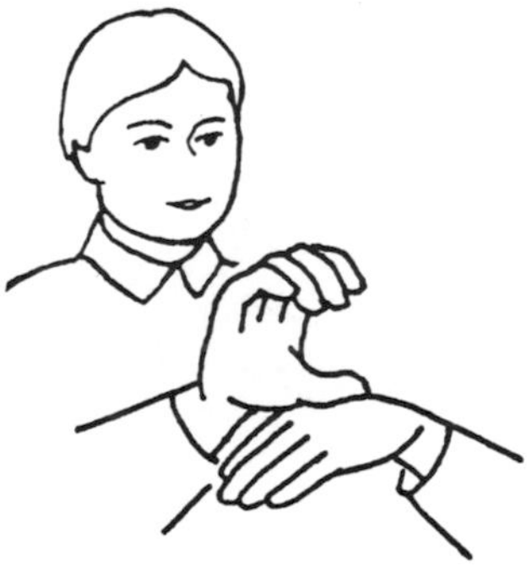

9

confused with

business 6

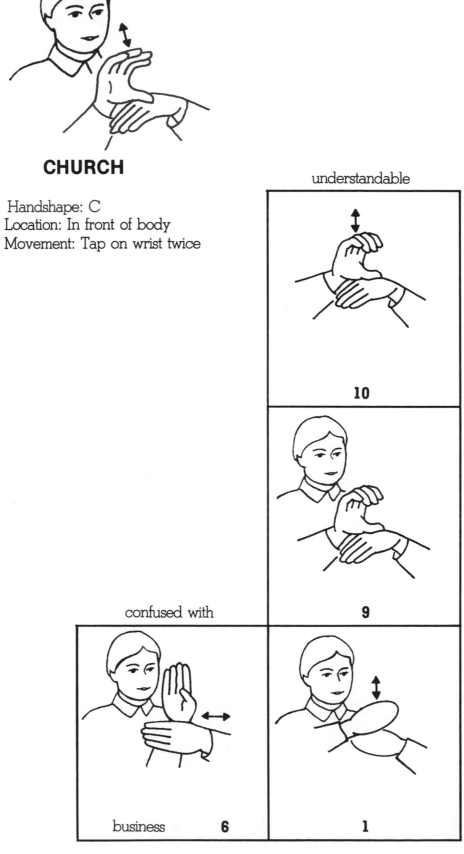

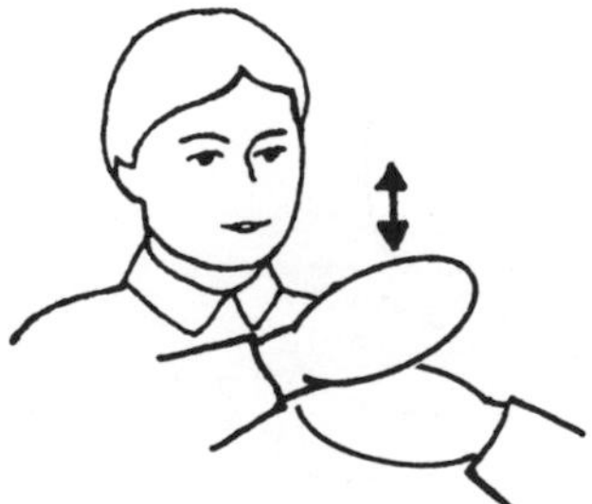

1

ALL

Handshape: Open B, both
Location: In front of body
Movement: Circle hand and tap palm

understandable

10

10

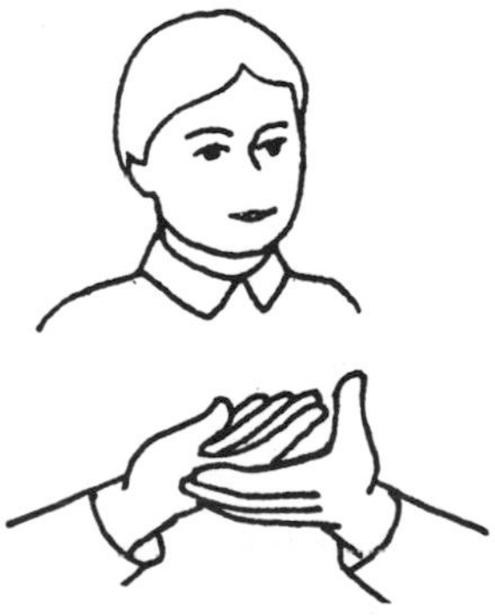

0

confused with

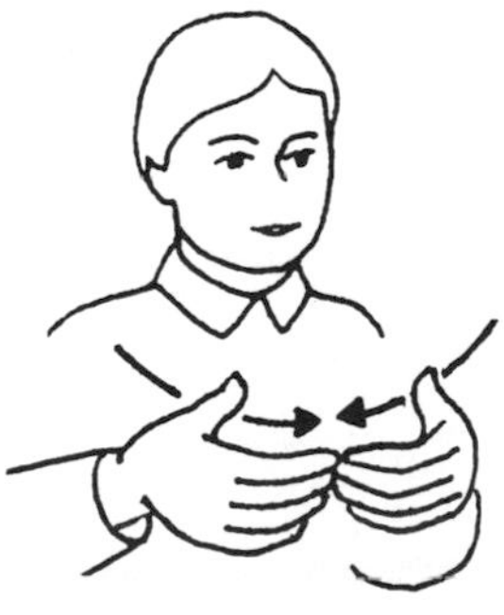

introduce **2**

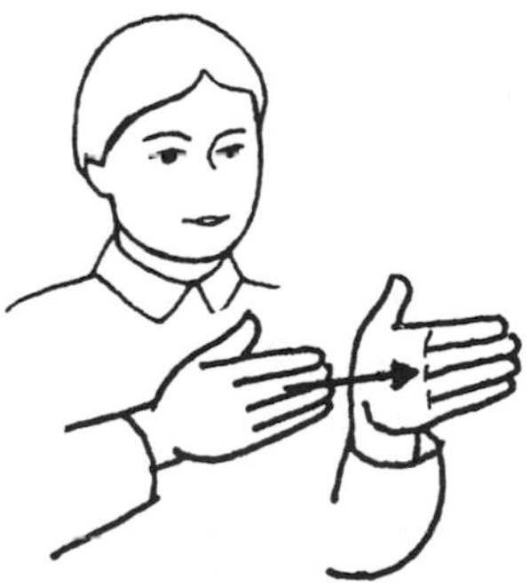

against **1**

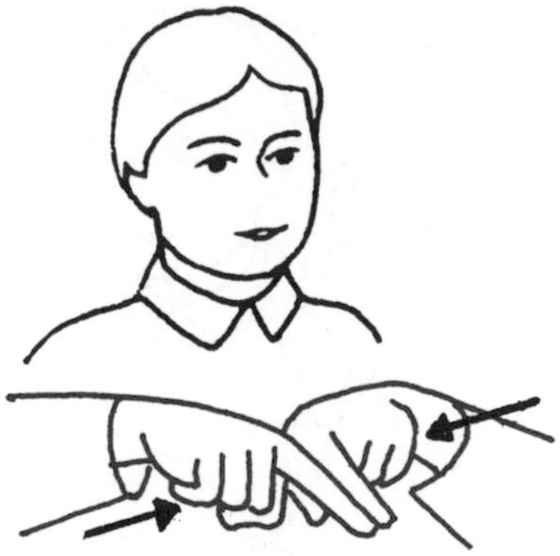

SAME

Handshape: 1, both
Location: In front of body
Movement: Tap twice

understandable

confused with

confused with

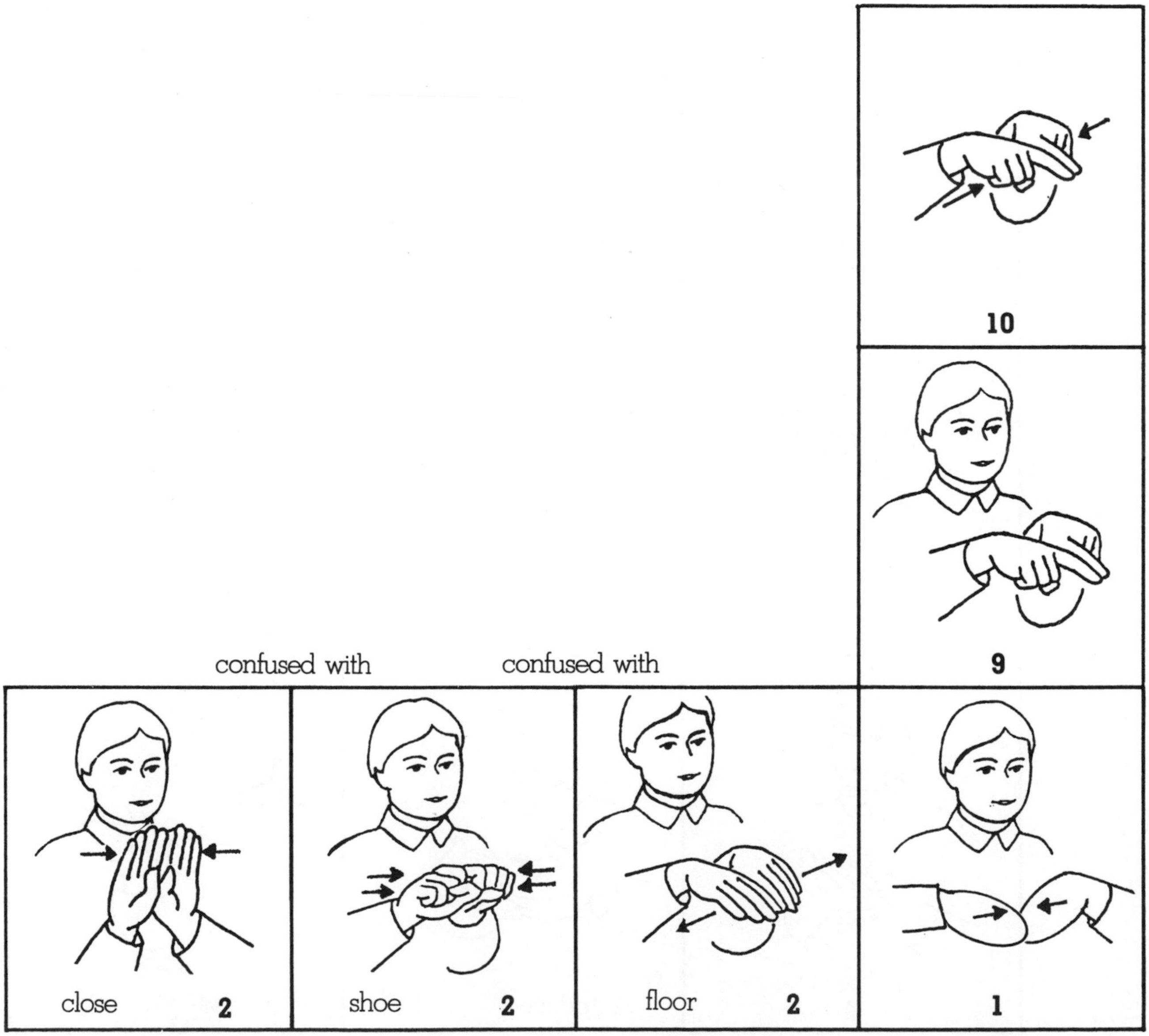

WORK

Handshape: S, both
Location: In front of body
Movement: Tap wrists twice

understandable

10

9

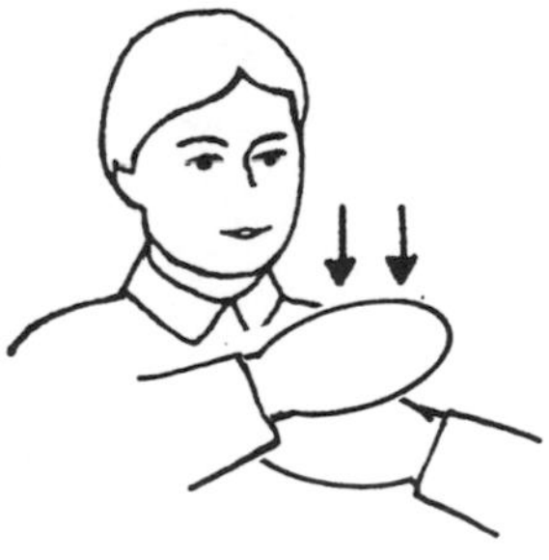

1

confused with

night 3

confused with

afternoon 2

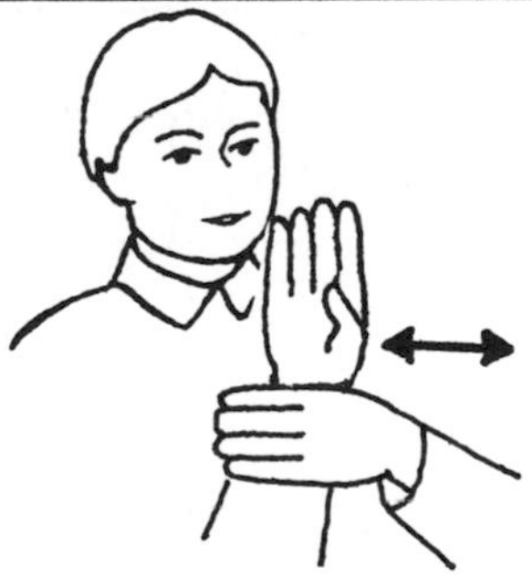

business 1

OPEN

Handshape: B, both
Location: In front of body
Movement: Open palms

understandable

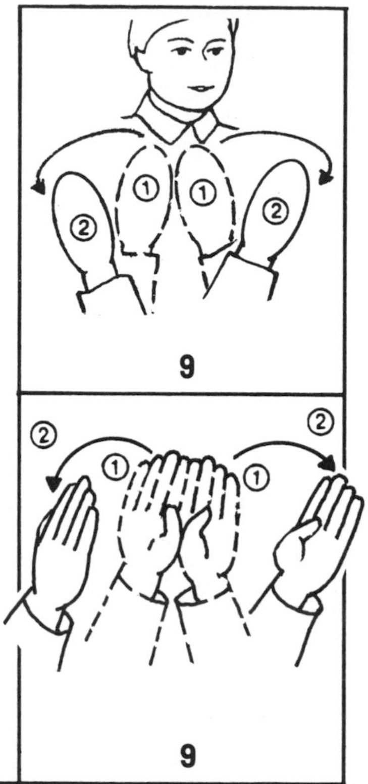

confused with

confused with

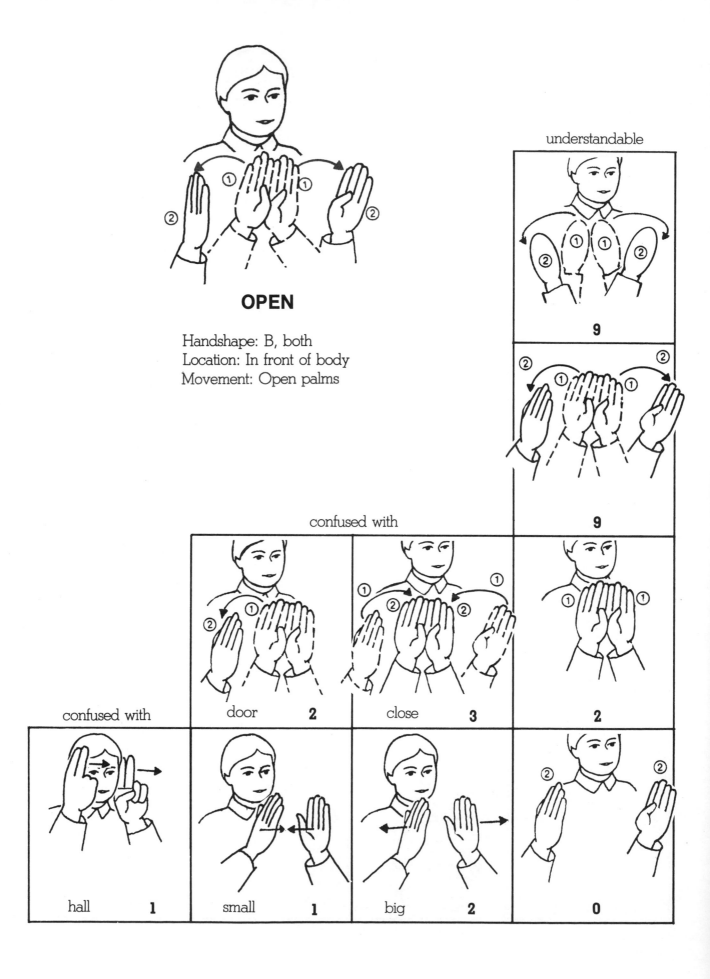

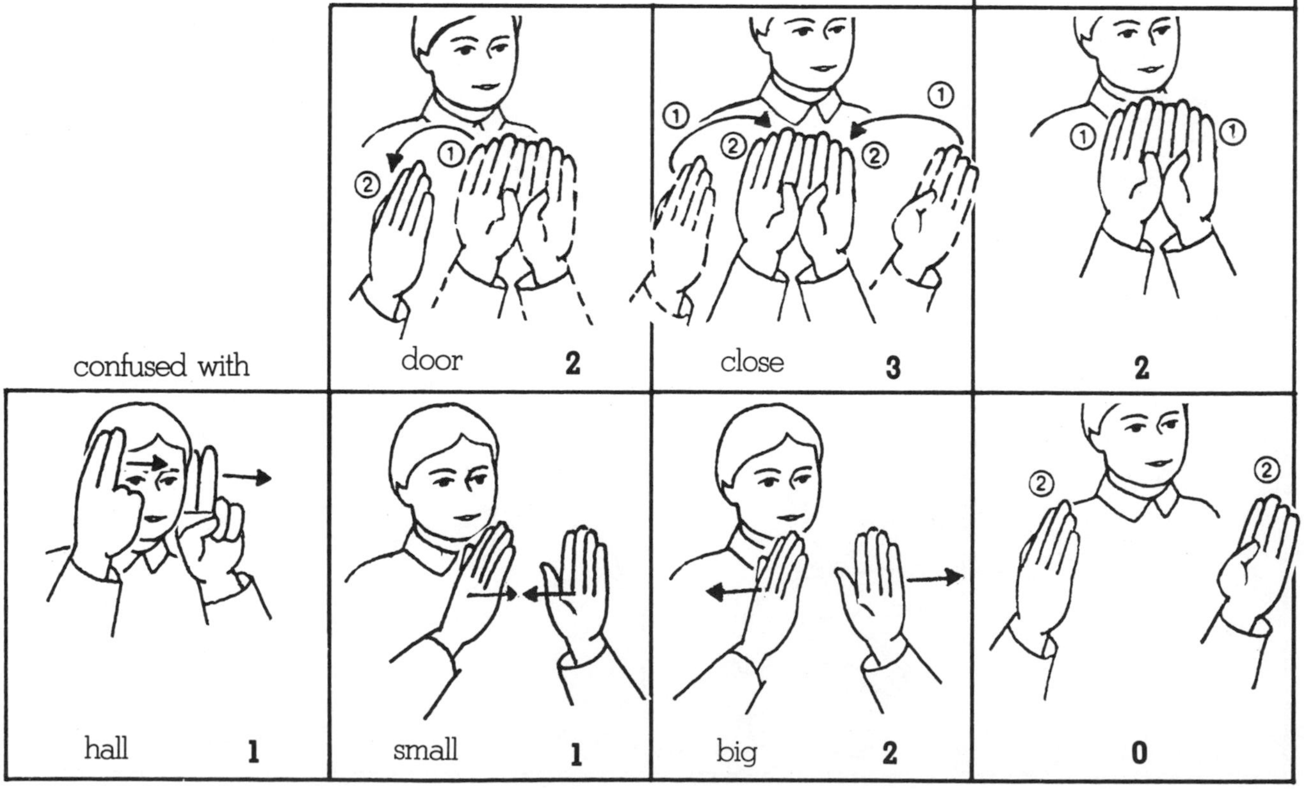

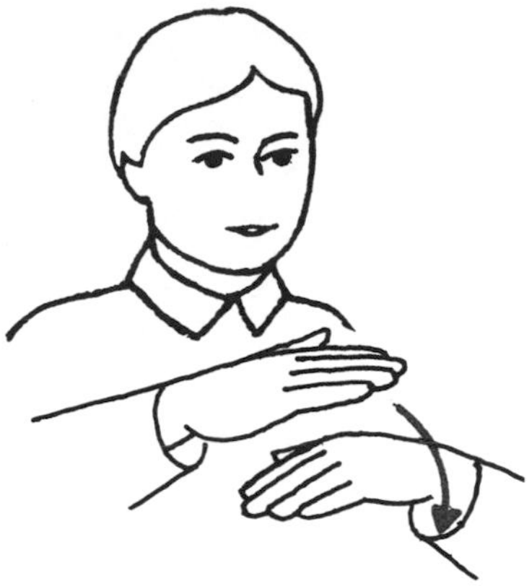

OVER

Handshape: Open B, both
Location: In front of body
Movement: Circle over hand

understandable

9

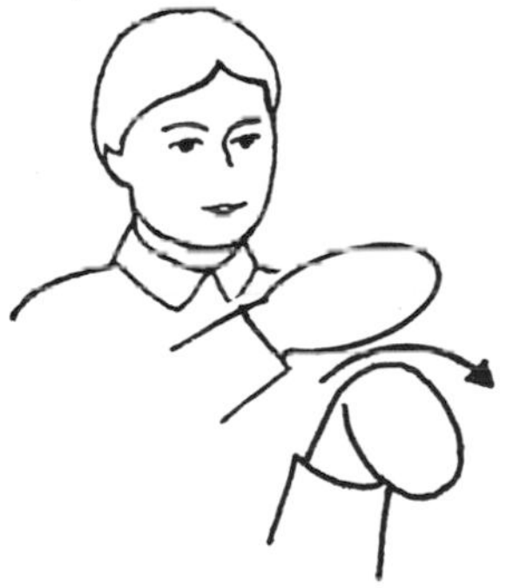

8

confused with

after **1**

2

trouble **1**

confused with

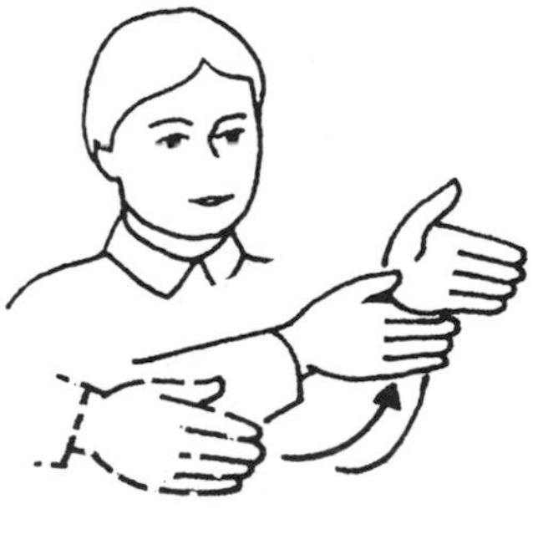

kick **1**

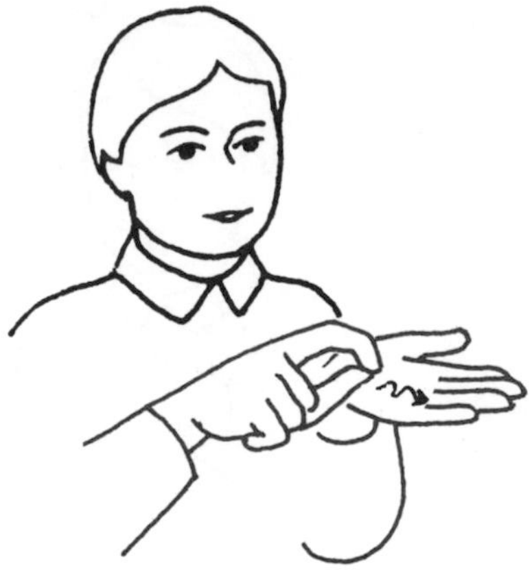

WRITE

Handshape: Closed G
Location: In front of body
Movement: Write on palm

understandable

10

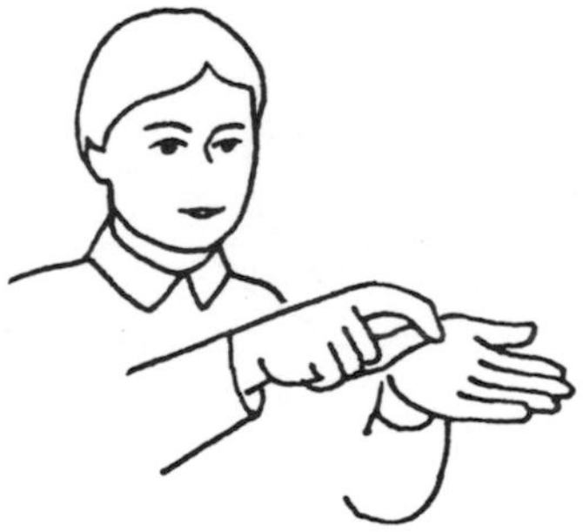

10

confused with

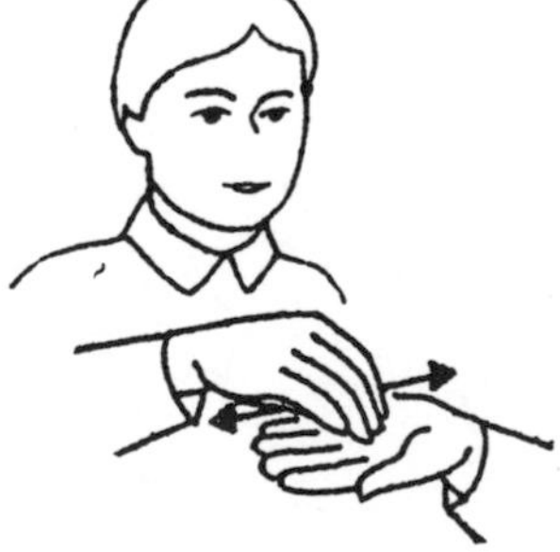

excuse **1**

nice **9**

0

BALL

Handshape: Open B, both
Location: In front of body
Movement: Shape a ball

understandable

10

10

confused with

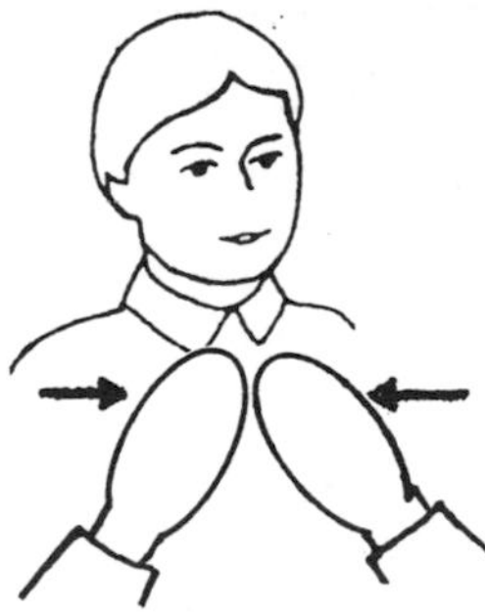

0

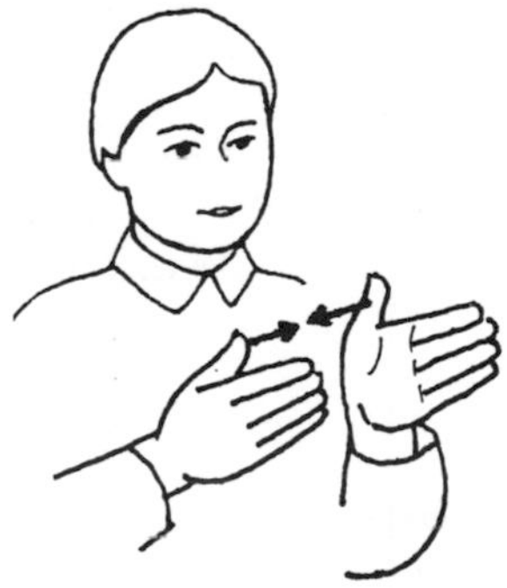

small **9**

THREE

Handshape: 3
Location: In front of body
Movement: None

understandable

10

10

confused with

your 8

0

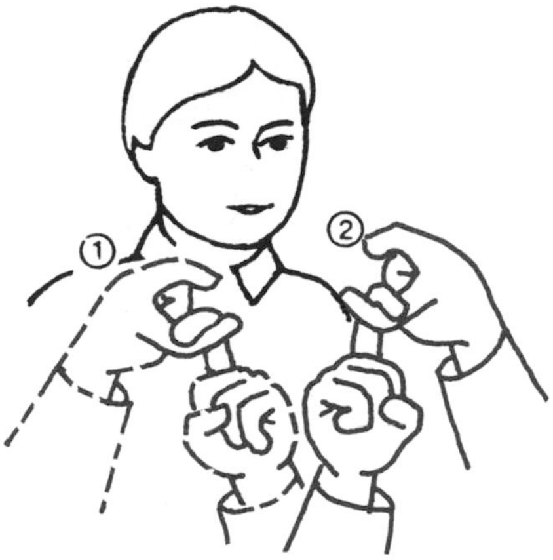

FRIEND

Handshape: 1, both
Location: In front of body
Movement: Reverse fingers

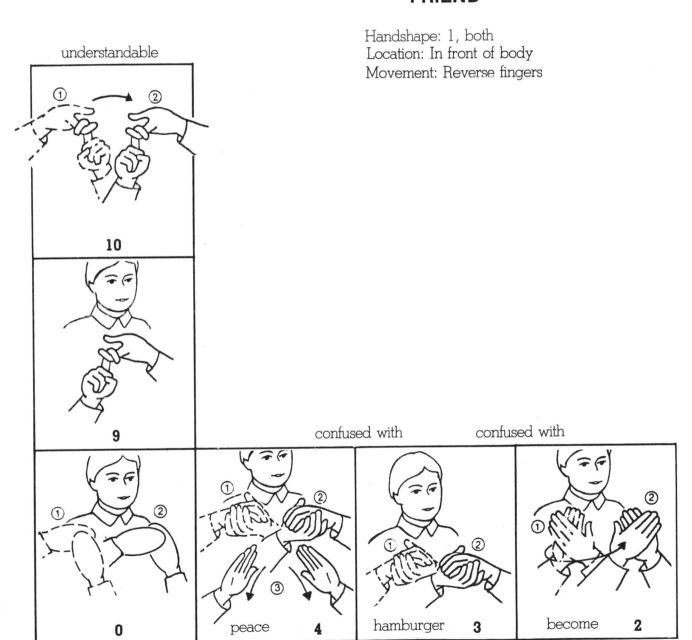

understandable

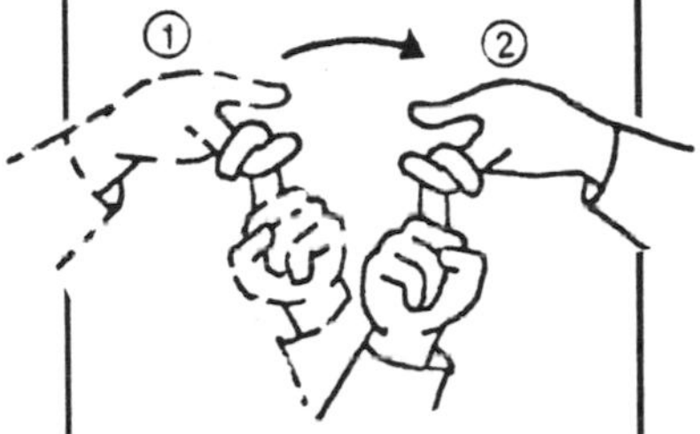

10

9

0

confused with

peace **4**

hamburger **3**

confused with

become **2**

FOUR

Handshape: 4
Location: In front of body
Movement: None

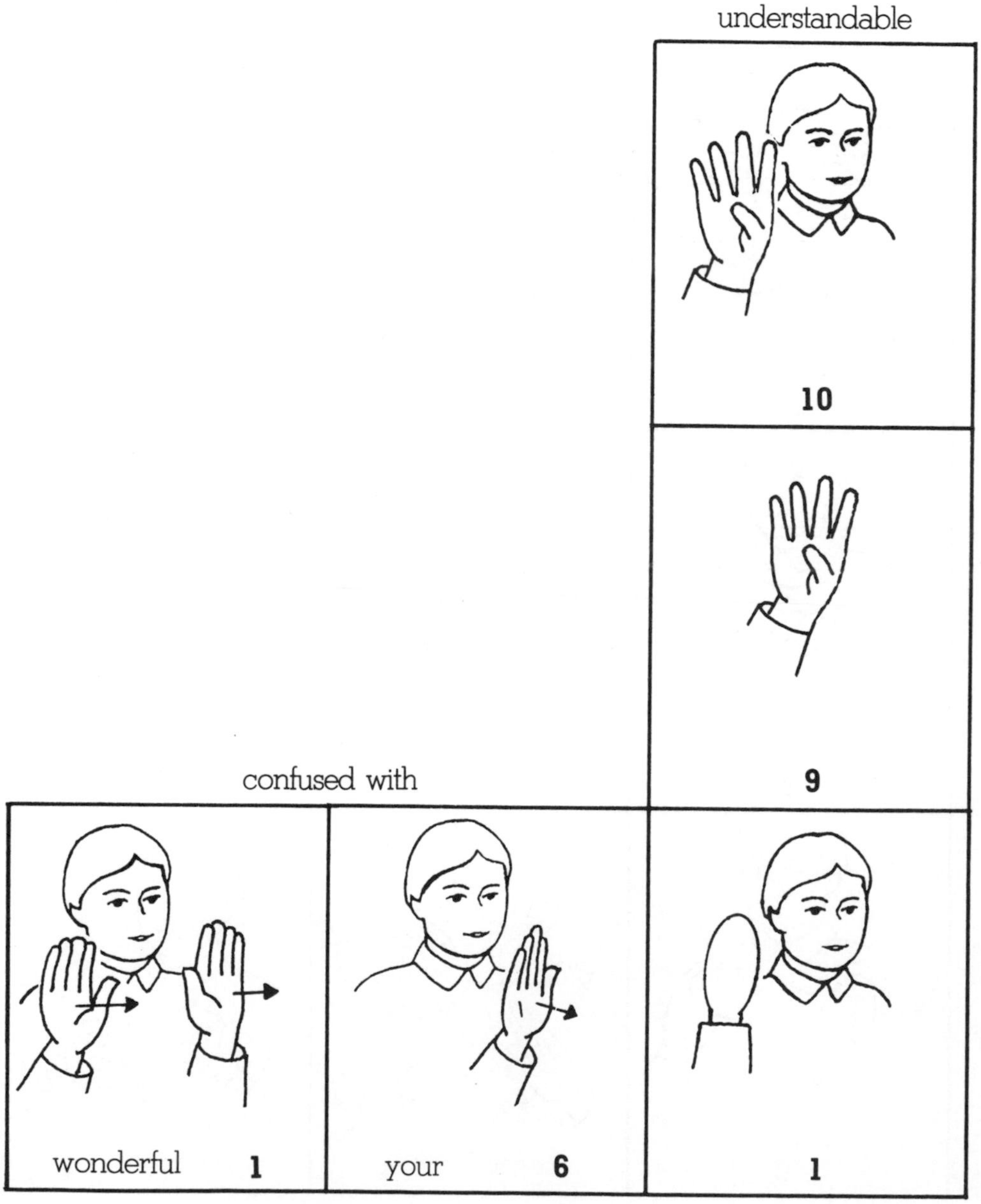

RIGHT

Handshape: 1, both
Location: In front of body
Movement: Tap once

understandable

9

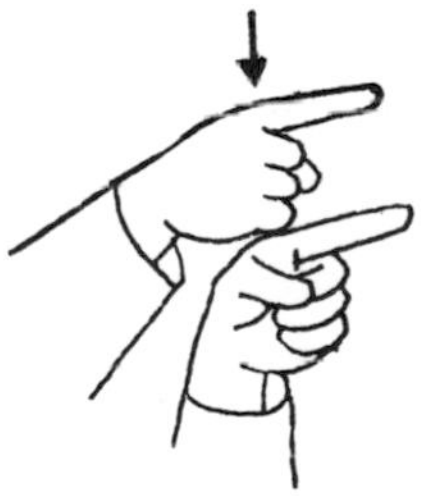

9

0

confused with

prevent 4

confused with

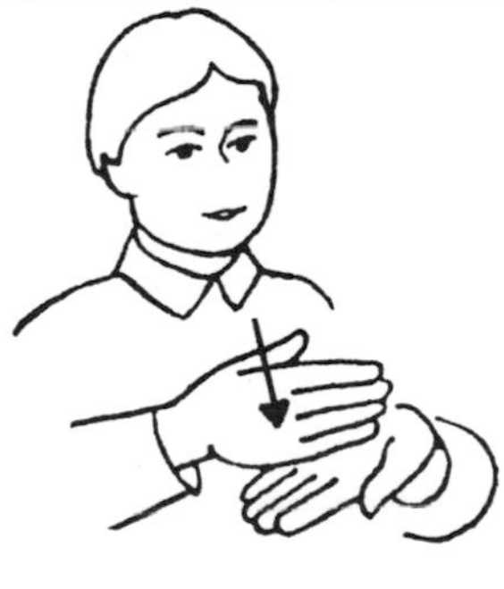

stop 3

pledge 2

PAPER

Handshape: Open B, both
Location: In front of body
Movement: Tap heel of palm twice

understandable

9

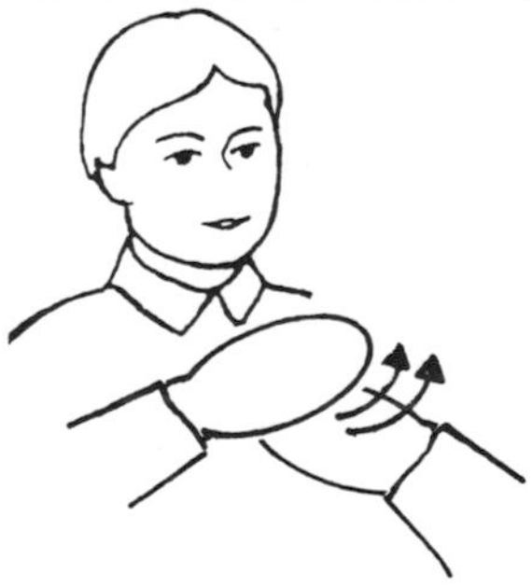

9

confused with

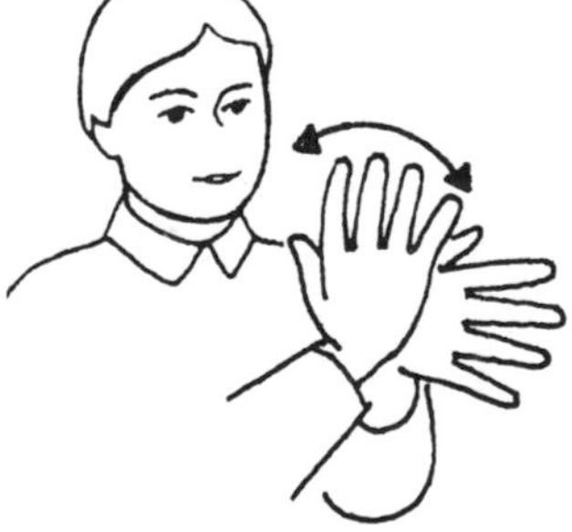

movie **1**

nice / clean **4**

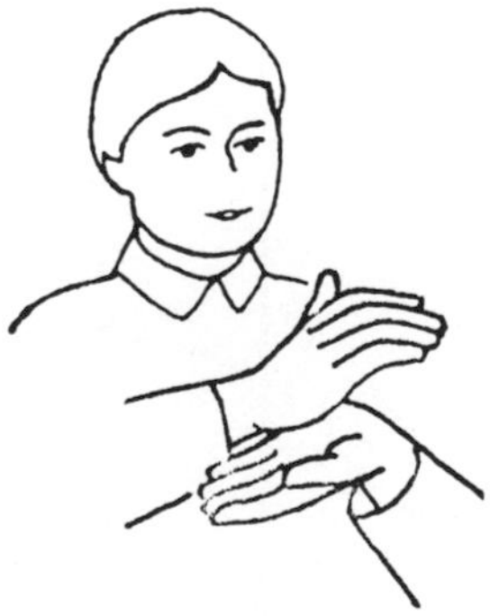

0

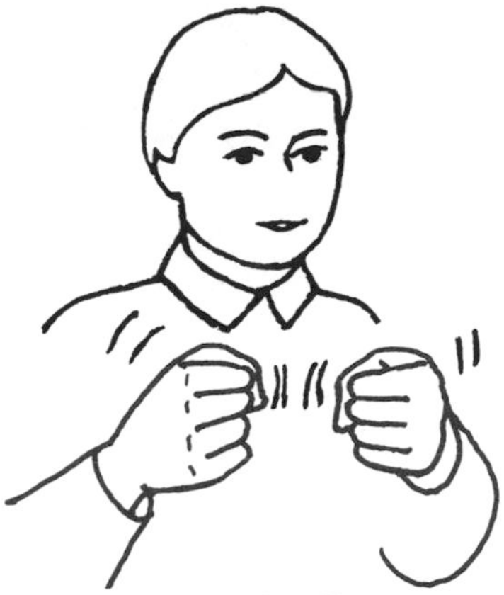

COLD

Handshape: S, both
Location: At chest
Movement: Shiver

understandable

10

8

confused with

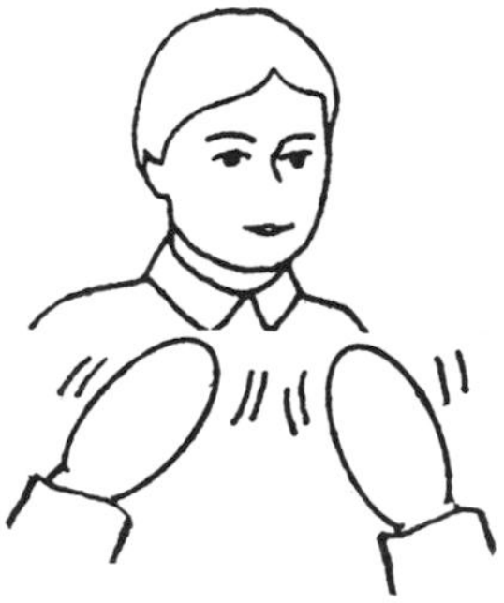

1

small **8**

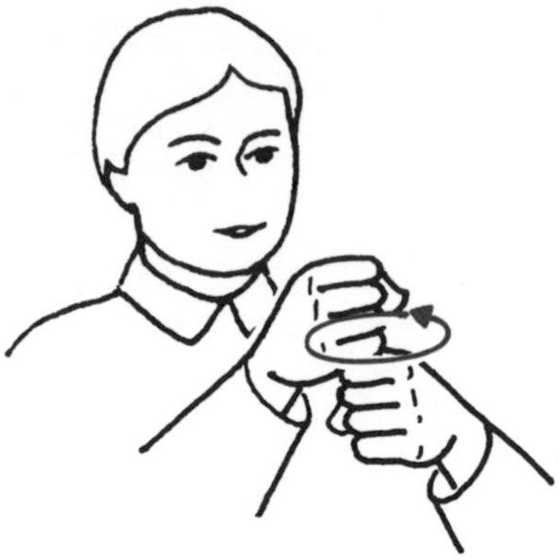

COFFEE

Handshape: S, both
Location: In front of body
Movement: Circle top hand

understandable

10

8

confused with

make **3**

hold **4**

0

GREEN

Handshape: G
Location: In front of body
Movement: Shake slightly

understandable

10

9

0

confused with

yellow **1**

confused with

blue **1**

insane **1**

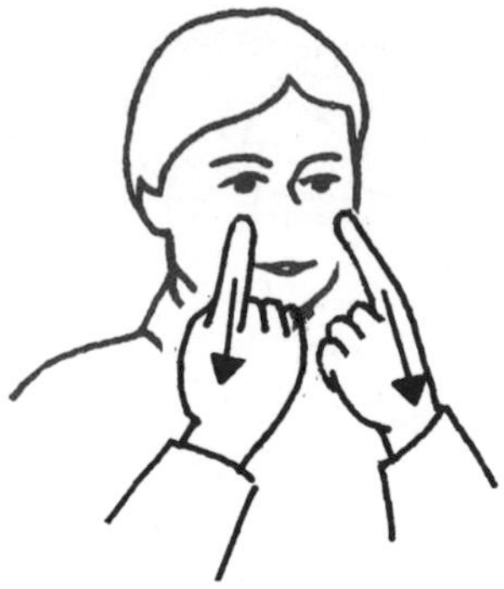

CRY

Handshape: 1, both
Location: Cheeks
Movement: Alternately trace tears

understandable

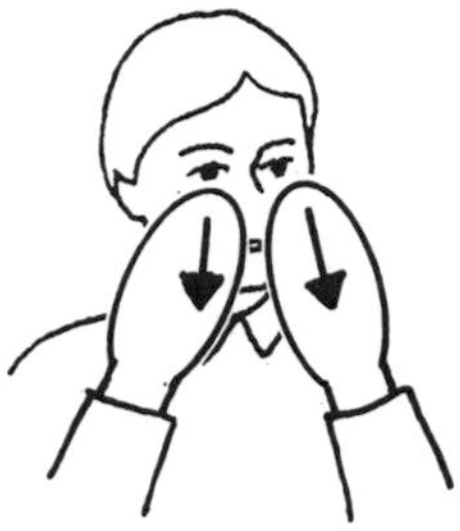

9

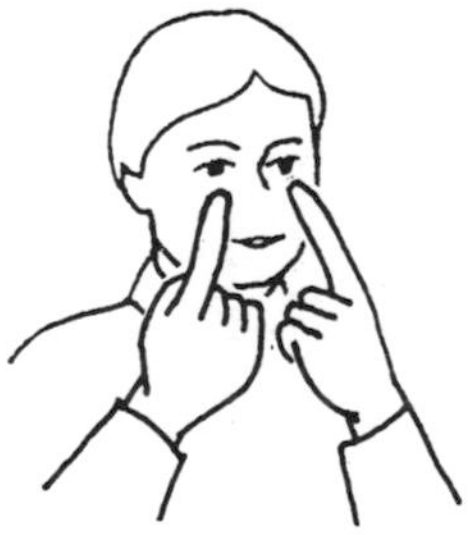

9

confused with

bitter 2

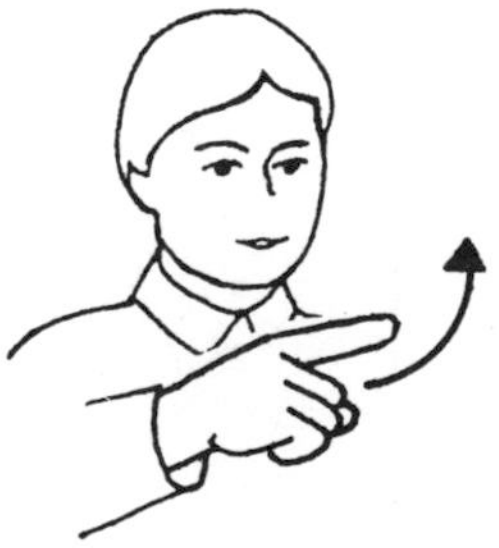

insult 3

1

SALT

Handshape: V, both
Location: In front of body
Movement: Alternately tap fingers

understandable

9

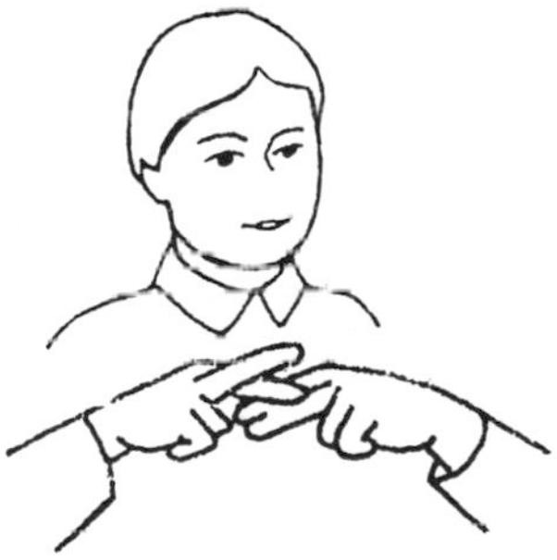

8

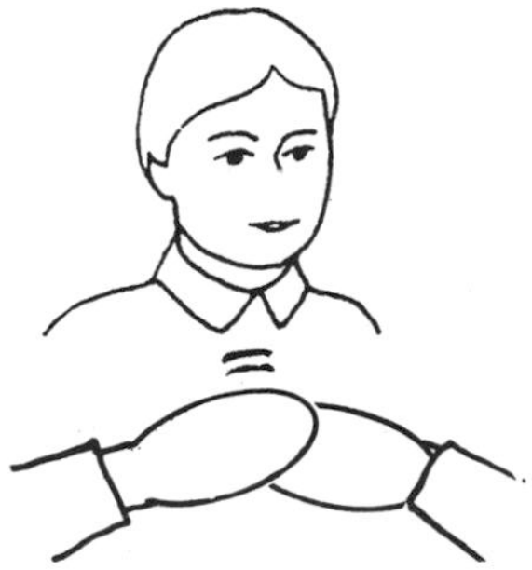

0

confused with

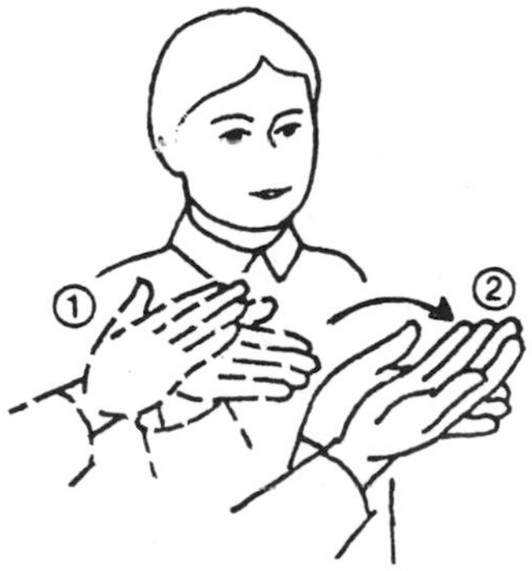

cook **4**

WATER

Handshape: W
Location: Mouth
Movement: Tap twice

understandable

10

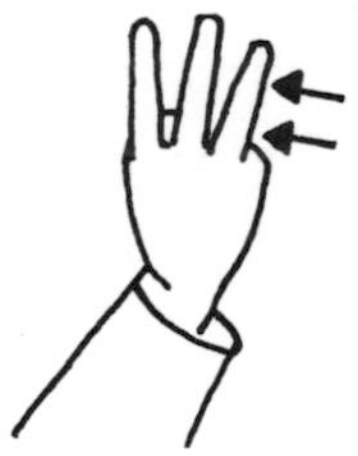

9

confused with

confused with

breakfast **2**

mother **2**

talk **4**

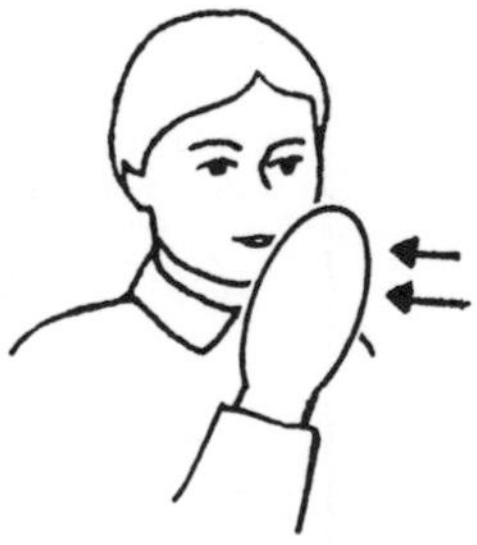

0

TWO

Handshape: V
Location: In front of body
Movement: None

understandable

9

9

0

confused with

your **7**

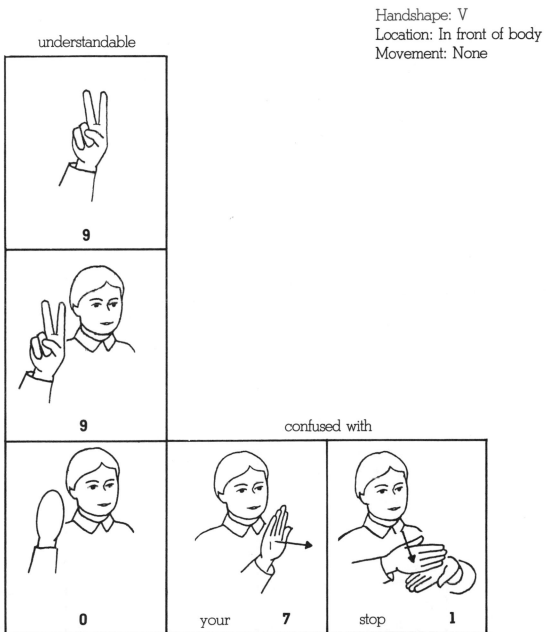

stop **1**

MEDICINE

Handshape: 5, bent middle finger
Location: In front of body
Movement: Stir in center of palm

confused with	understandable
 poison **1**	 **8**
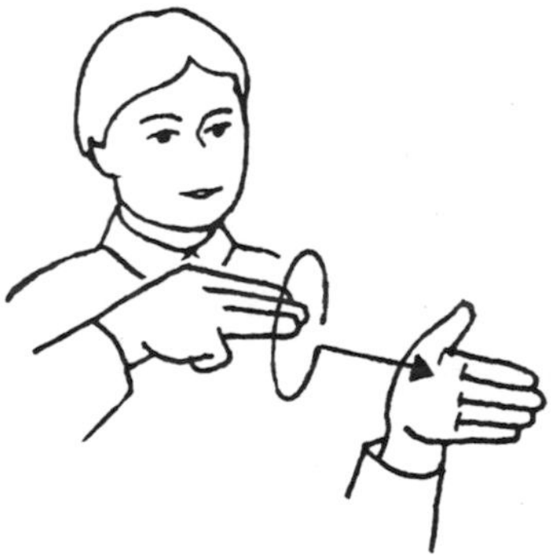 middle **1**	 **8**
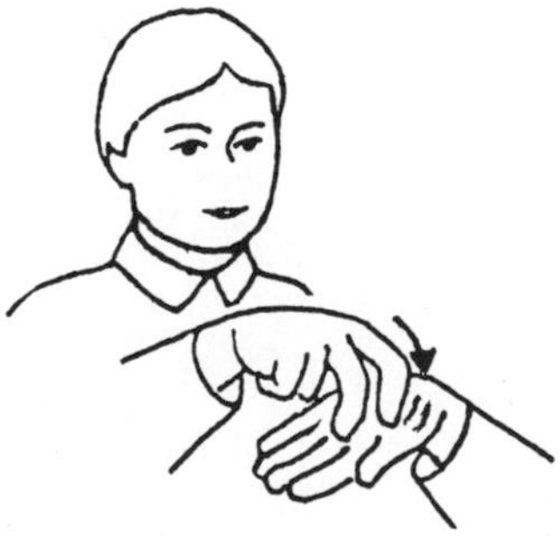 weak **8**	 **1**

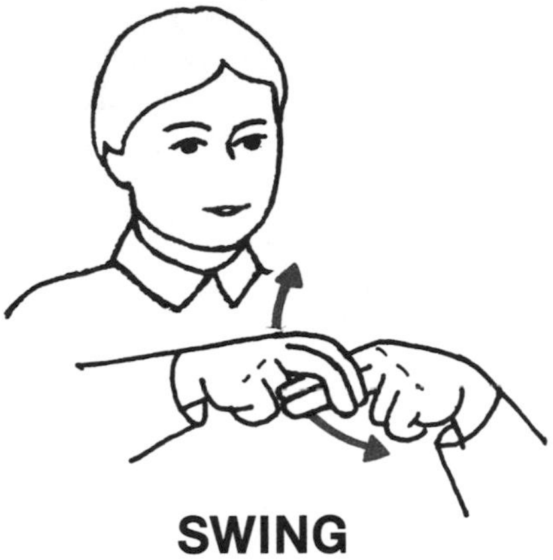

SWING

Handshape: H, both
Location: In front of body
Movement: Swing in and out

understandable

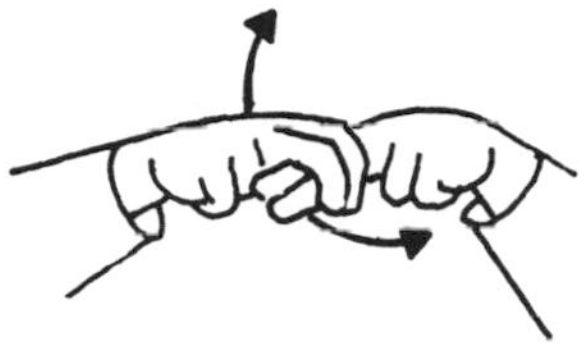

8

8

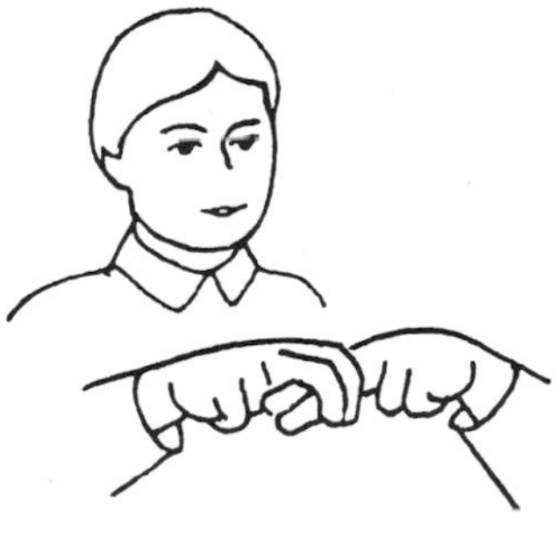

1

confused with

sit **7**

train **1**

COOK

Handshape: Open B, both
Location: In front of body
Movement: Tap, flip hand, tap again

understandable

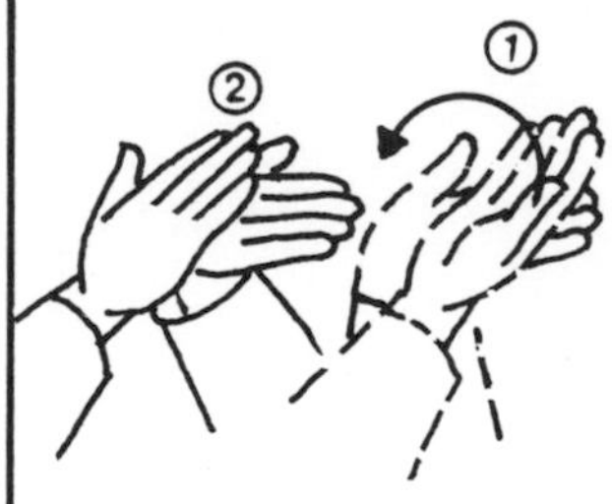

9

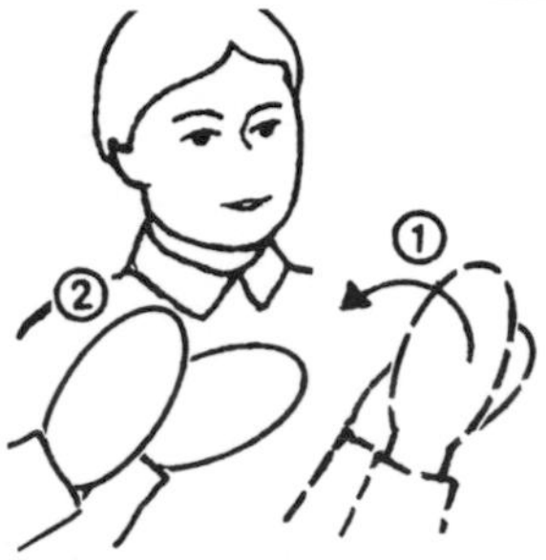

9

1

0

confused with

clean 2

nice 3

buy 1

money 4

confused with

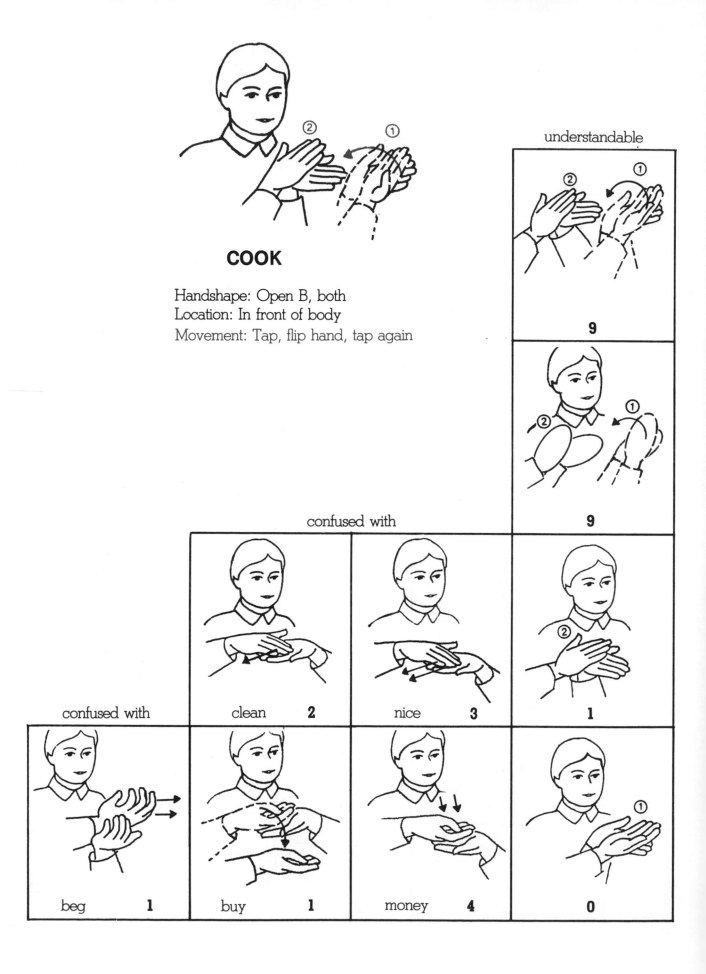

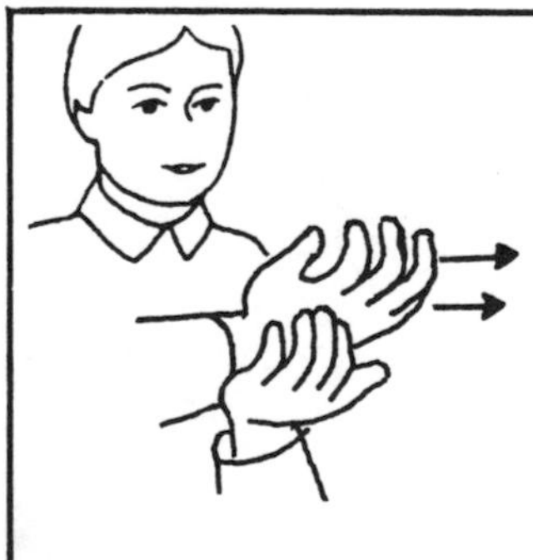

beg 1

HOT

Handshape: C
Location: Mouth
Movement: Twist out

understandable

9

8

confused with

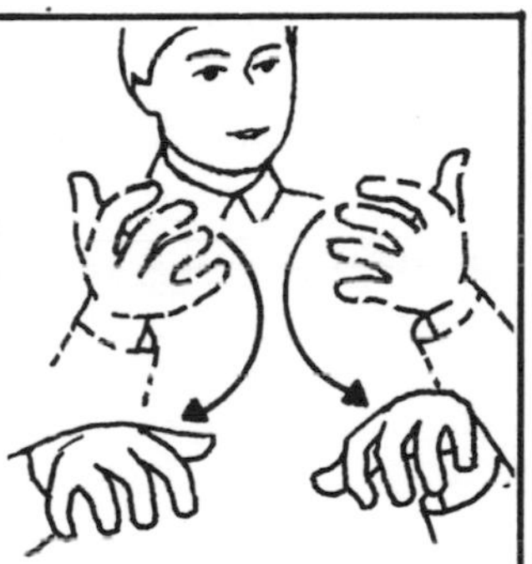

don't want **1**

2

1

confused with

bad **8**

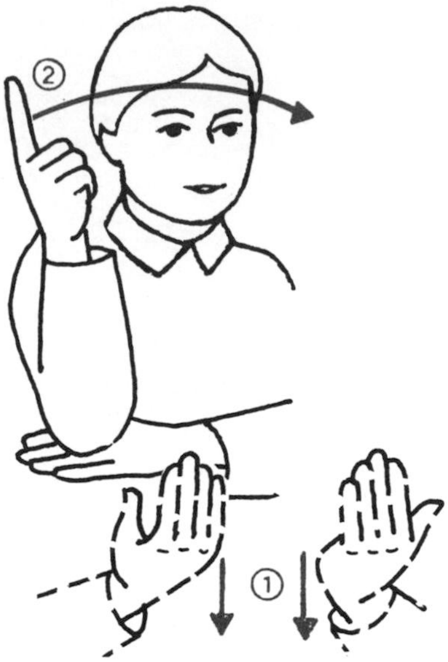

TODAY

Handshape: Bent open B, both; 1, both
Location: In front of body
Movement: Combine now and day

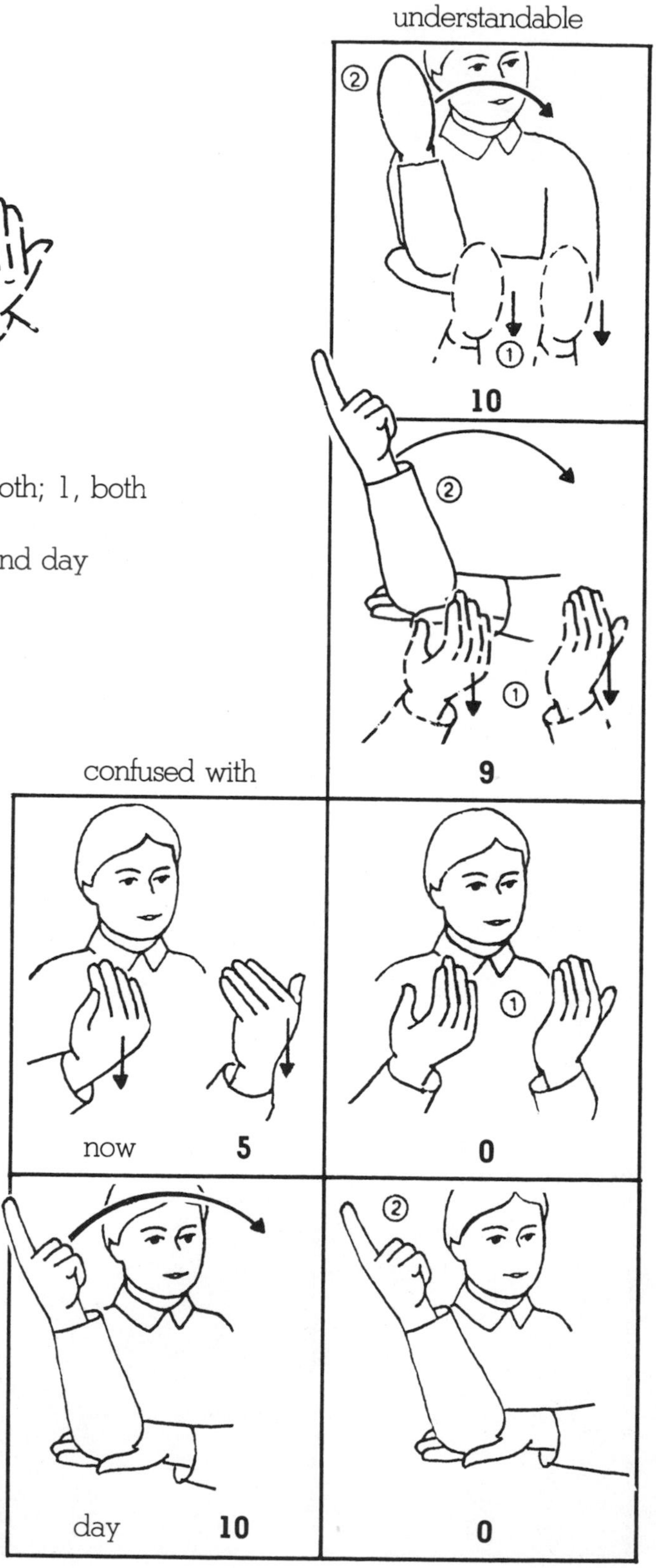

GOOD MORNING

Handshape: Open B, both
Location: Mouth and in front of body
Movement: Combine good and morning

understandable

10

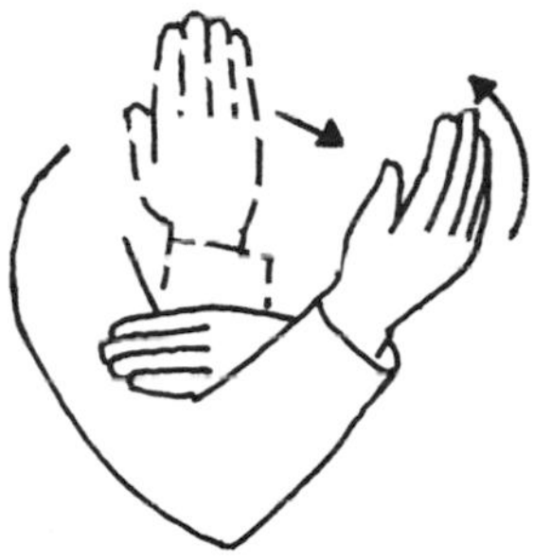

9

confused with

0

prove **3**

money **1**

0

morning **10**

BAD

Handshape: Open B, both
Location: Mouth
Movement: Flip hand and lower to palm

understandable

9

8

confused with

good **9**

0

nice / clean **3**

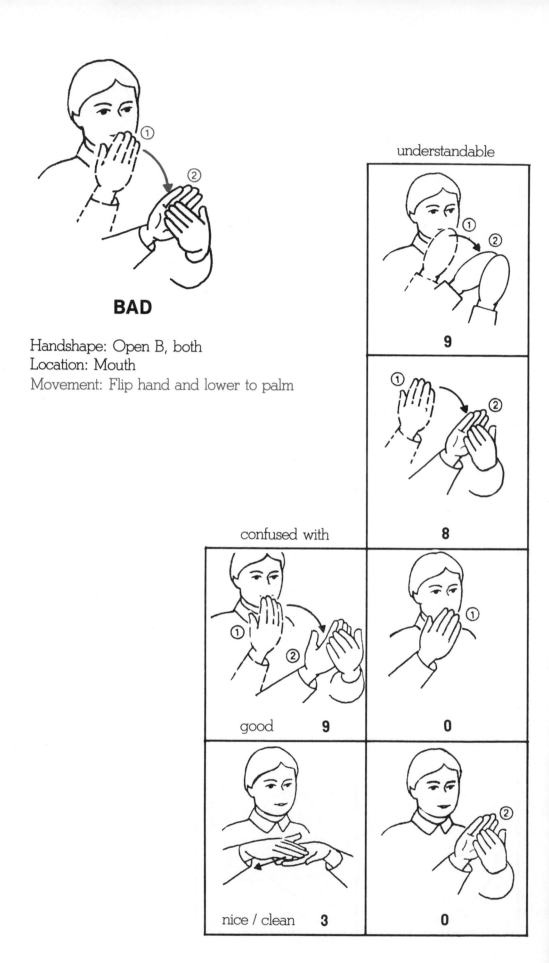

0

SINGLE ROBUST

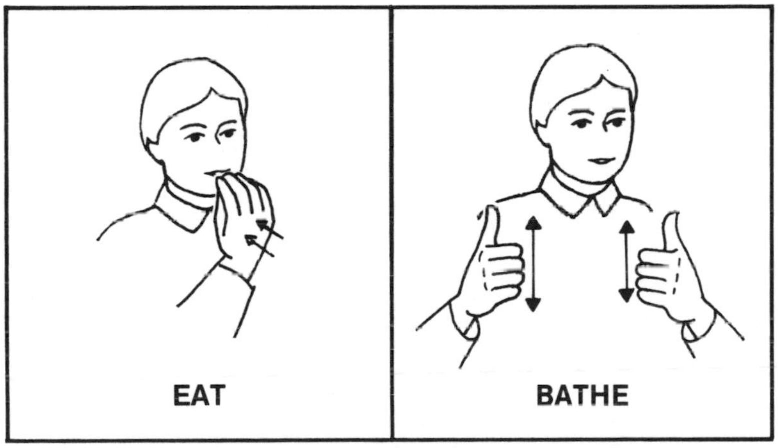

ZERO FRAGILE

PAIN

Handshape: 1, both
Location: In front of body
Movement: Toward each other and back, twice

understandable confused with

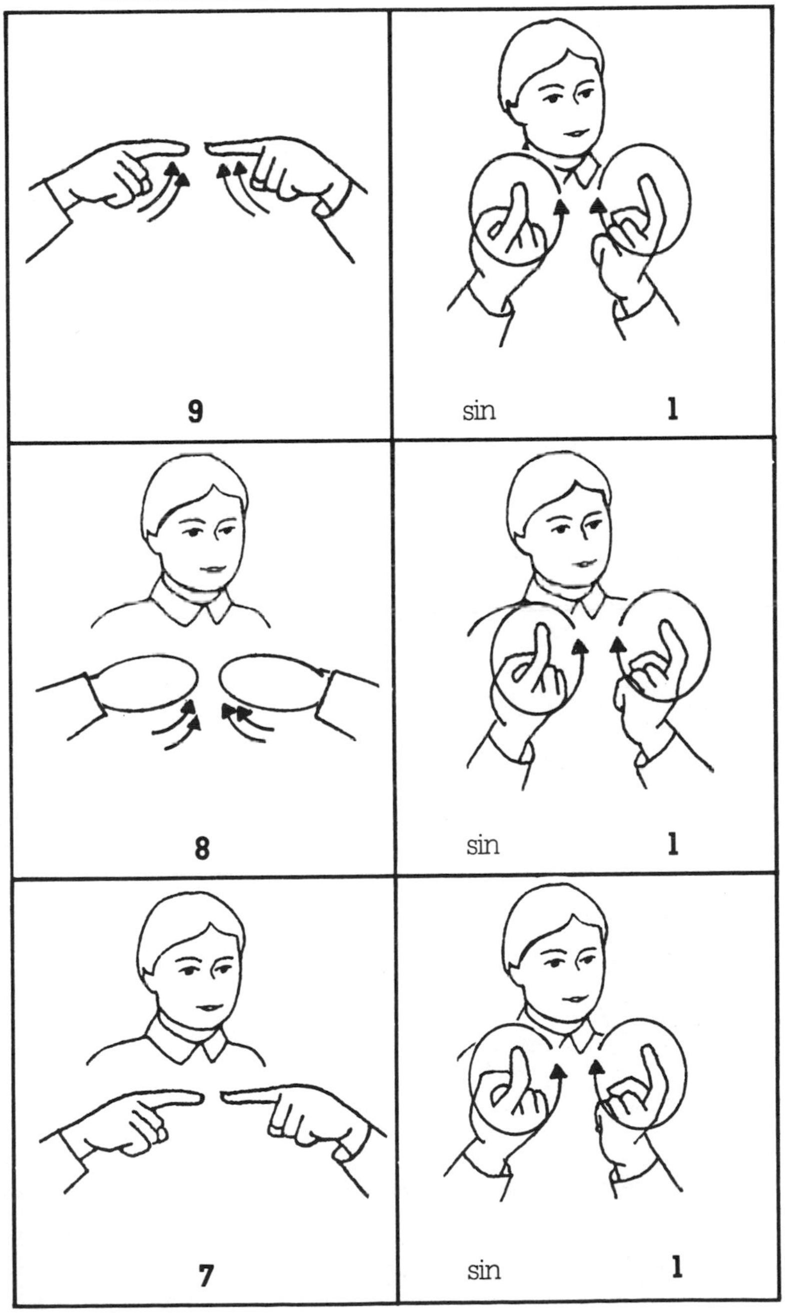

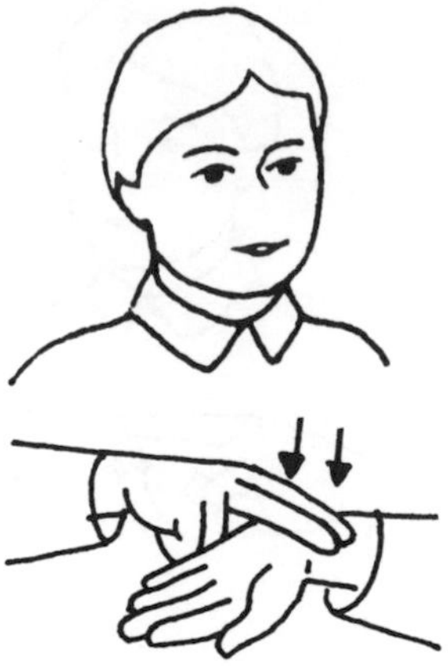

NURSE

Handshape: N
Location: In front of body
Movement: Tap wrist twice

understandable

10

confused with

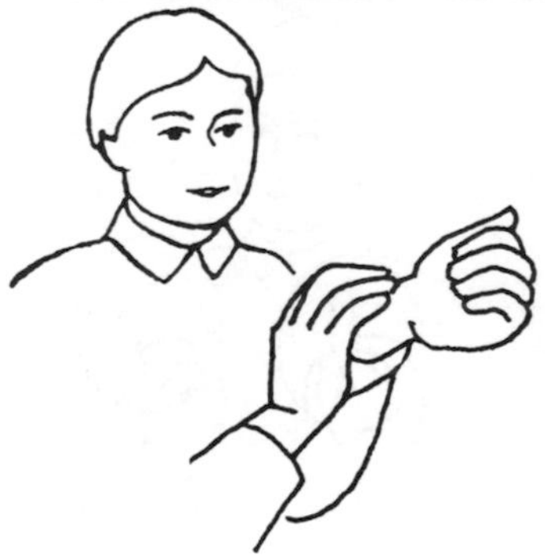

pulse **1**

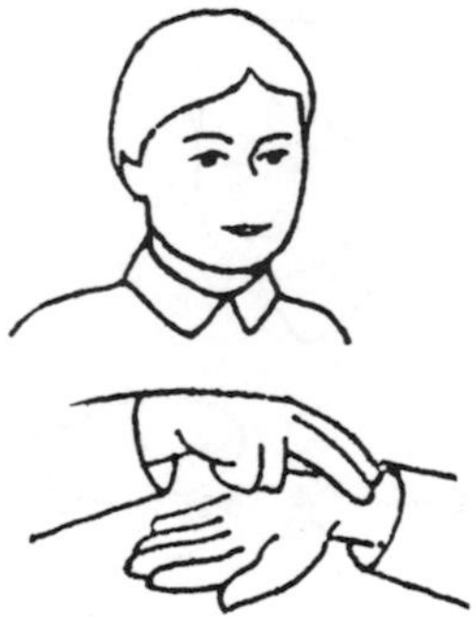

7

school **2**

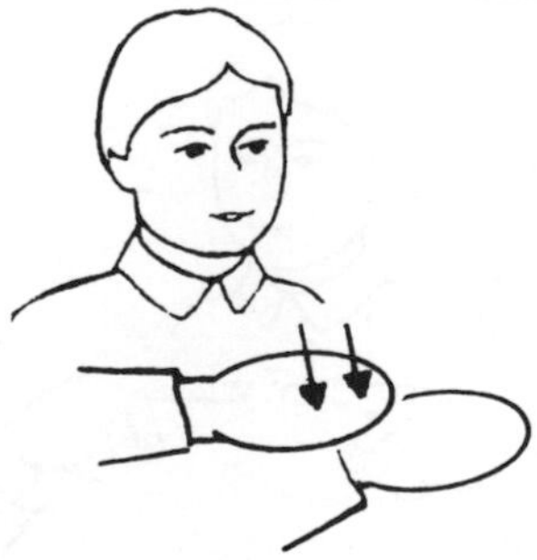

6

MY

Handshape: Open B
Location: Chest
Movement: Tap Chest

understandable

9

confused with

7

please **1**

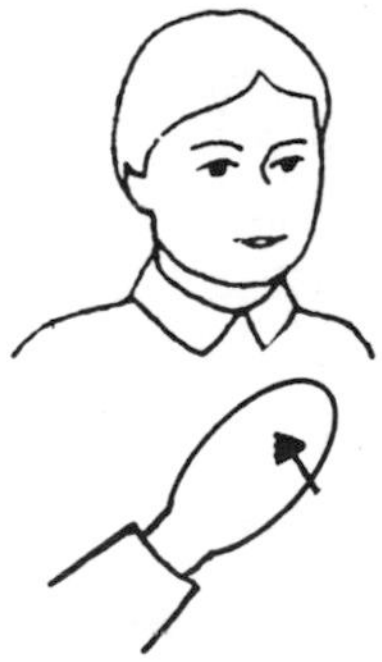

7

complain **1**

NOW

Handshape: Bent open B, both
Location: In front of body
Movement: Lower slightly

understandable

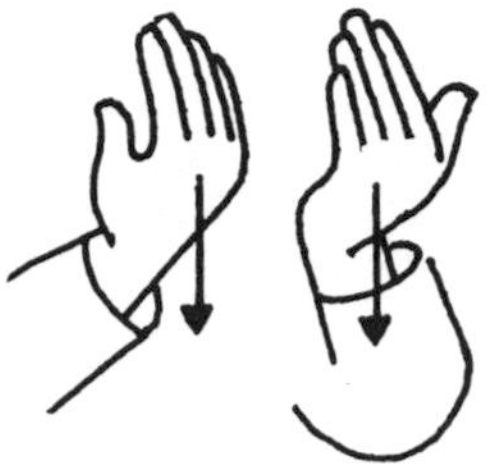

10

confused with

play **1**

7

sad **2**

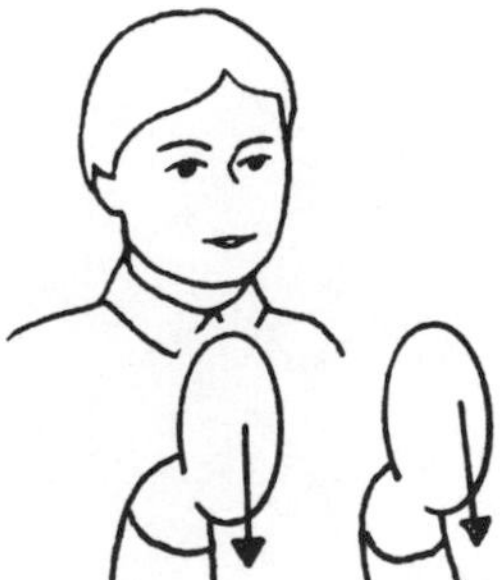

7

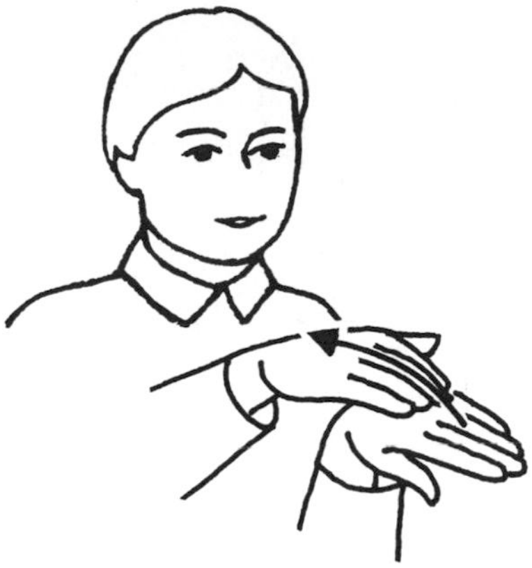

SLOW

Handshape: Open B, both
Location: In front of body
Movement: Across back of hand

understandable

10

8

6

NOON

Handshape: B, both
Location: In front of body
Movement: Rest elbow on back of hand

confused with	understandable

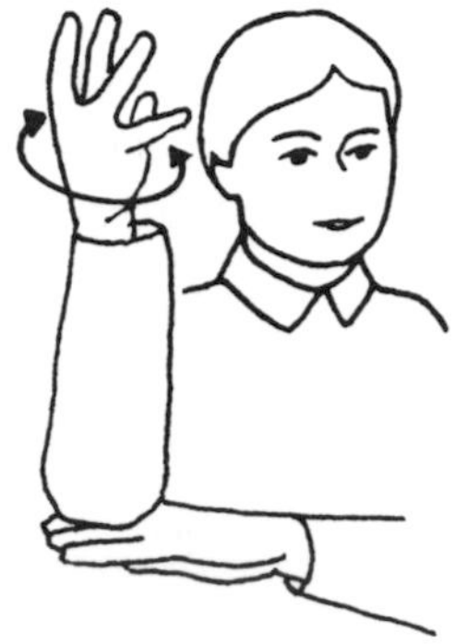

tree **1**

8

tree **1**

8

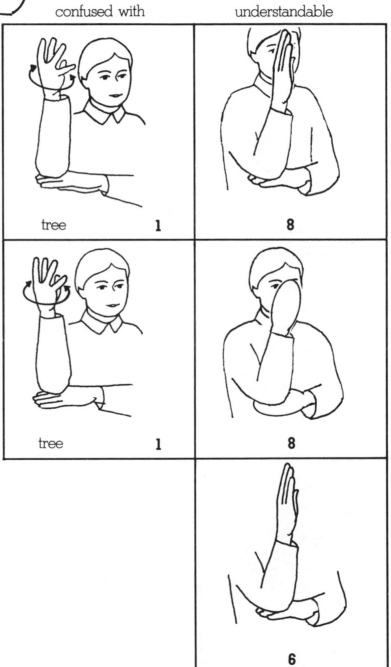

6

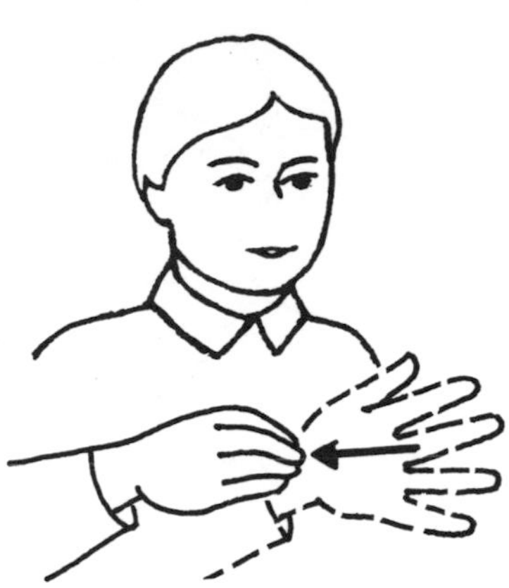

AND

Handshape: 5 to flat O
Location: In front of body
Movement: Across and close hand

understandable

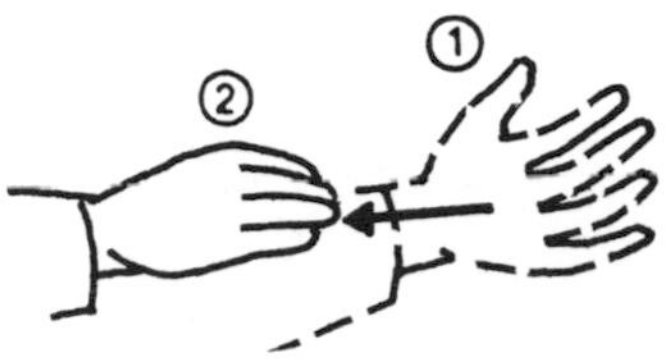

10

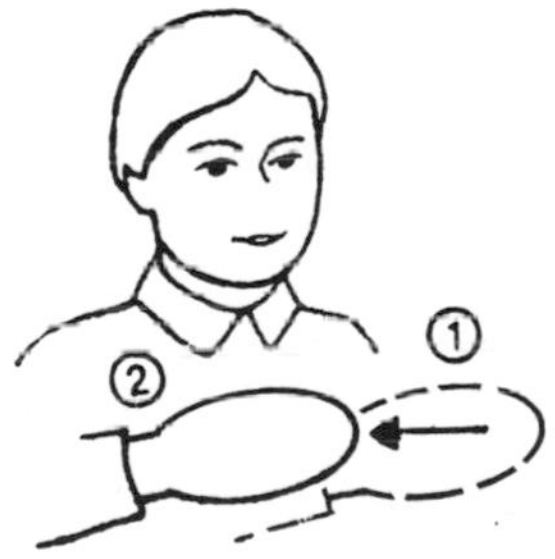

7

confused with

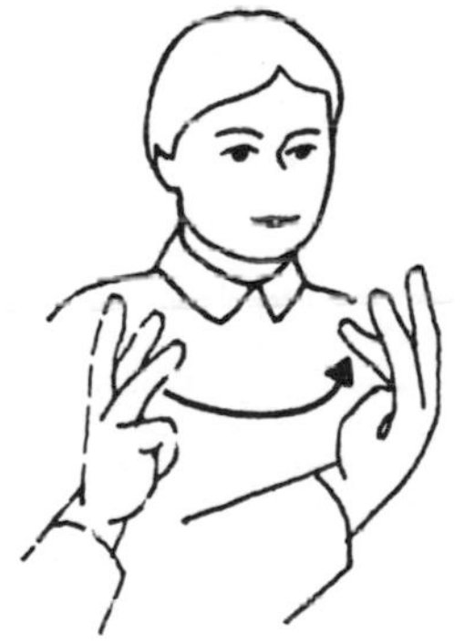

we **1**

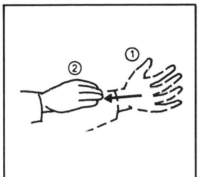

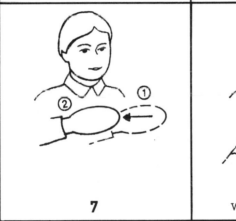

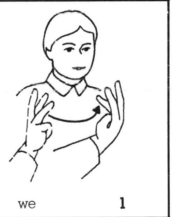

5

HOW

Handshape: Open B, both
Location: In front of body
Movement: Turn in and up

understandable

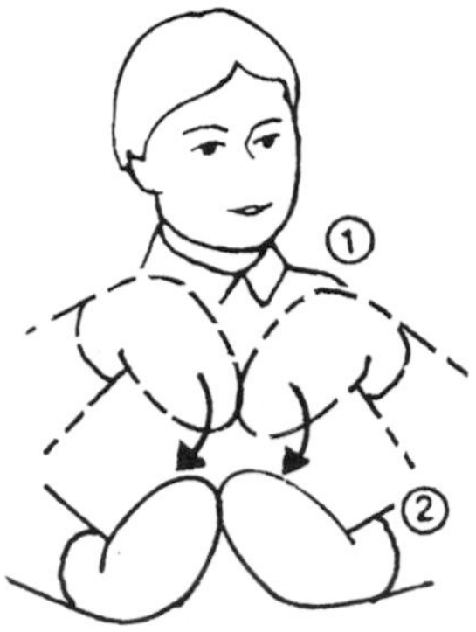

10

confused with

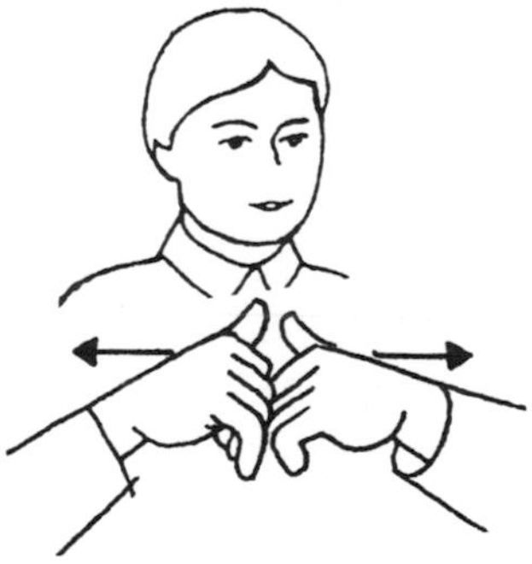

separate **2**

6

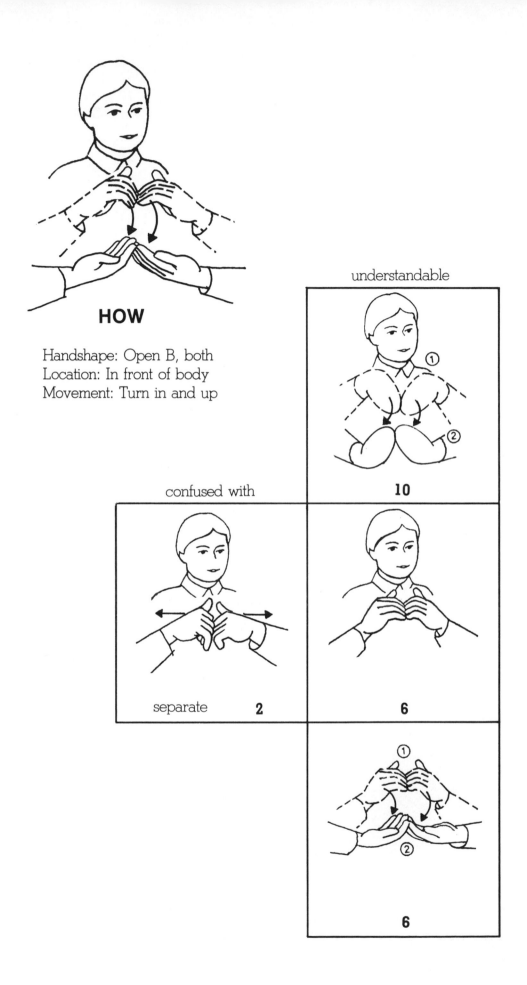

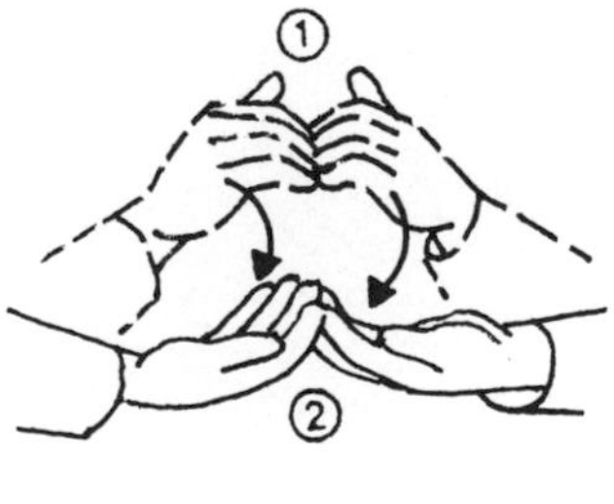

6

ANGRY

Handshape: Bent 5, both
Location: Chest
Movement: Forcefully toward shoulders

understandable

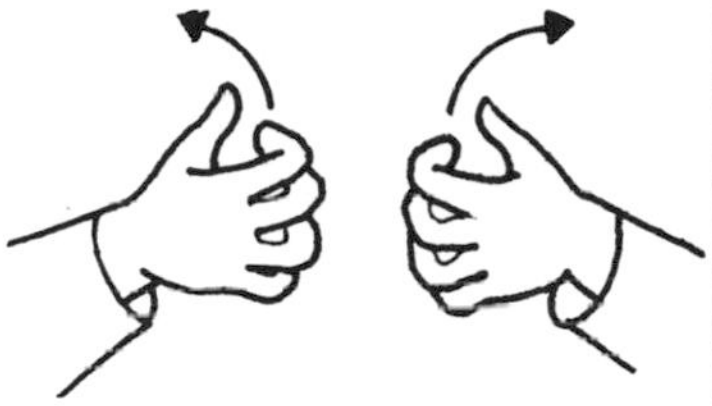

10

7

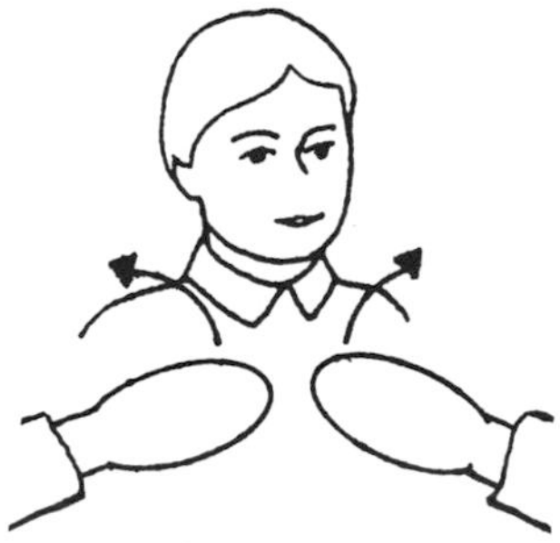

5

confused with

what's up **1**

happy **1**

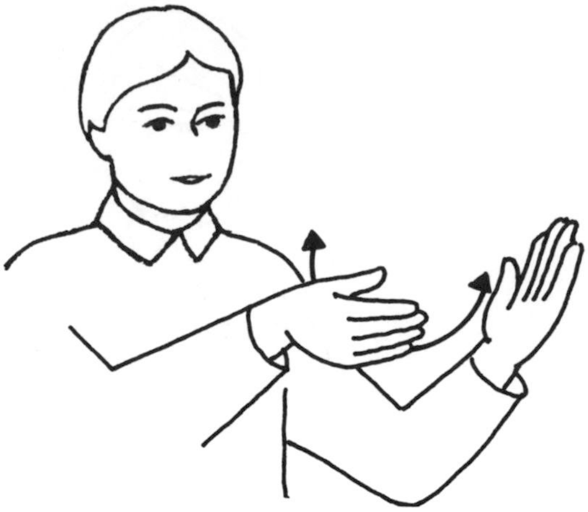

MUSIC

Handshape: Open B, both
Location: In front of body
Movement: Swing back and forth

understandable

8

confused with

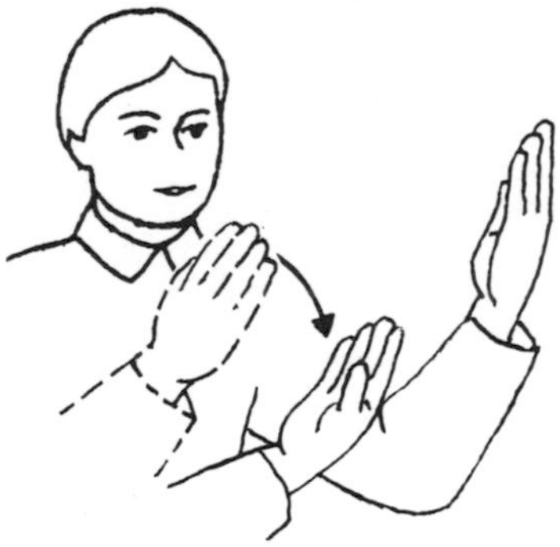

birthday **1**

dance **2**

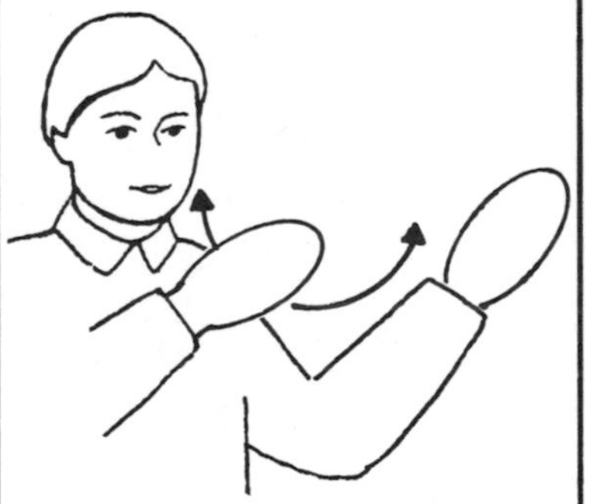

7

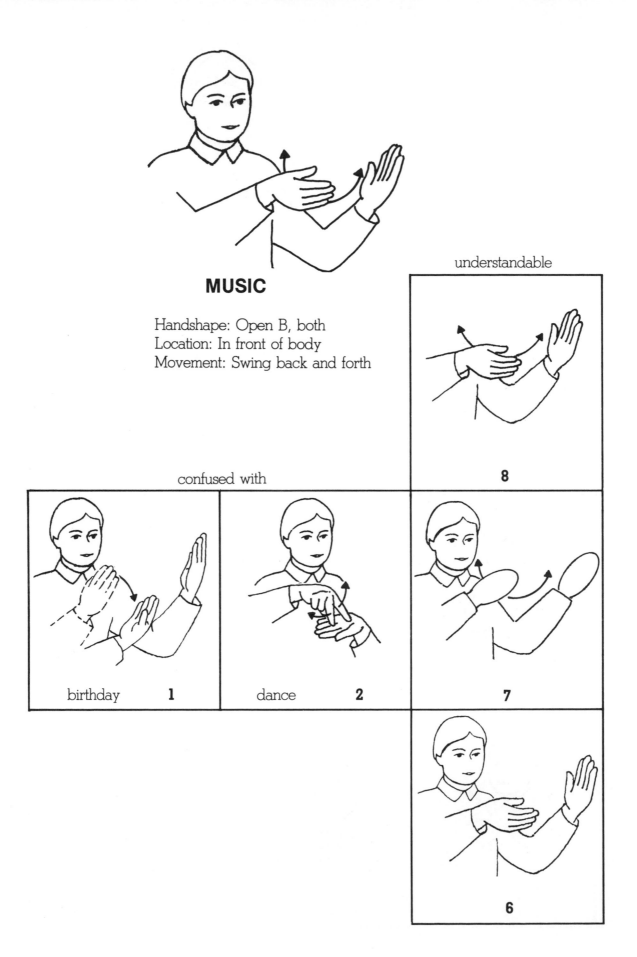

6

UNDER

Handshape: A, thumb extended
Location: In front of body
Movement: Pass under palm

understandable

9

6

confused with

basement **2**

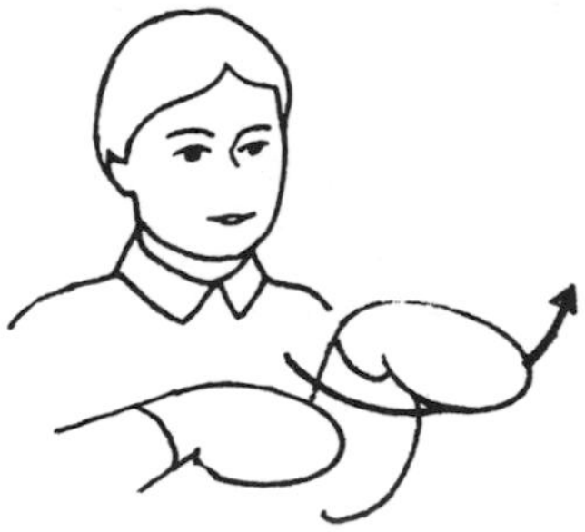

6

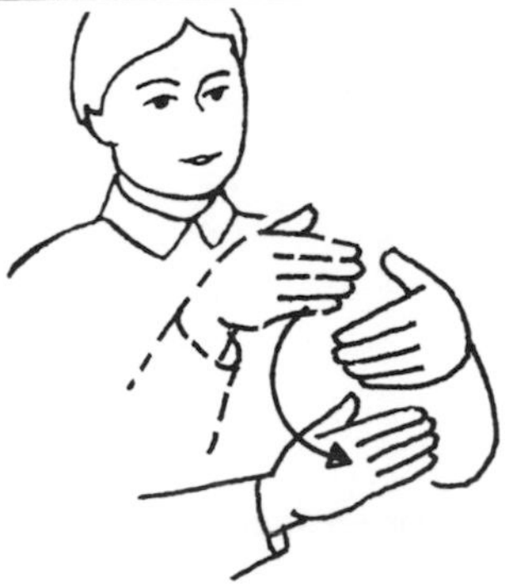

born **2**

YOU

Handshape: 1
Location: In front of body
Movement: Point

understandable

10

7

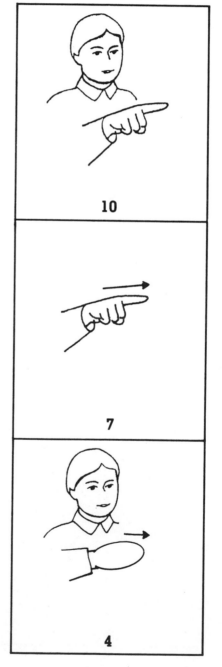

4

RABBIT

Handshape: H, both
Location: In front of body
Movement: Cross wrists, flick H's

understandable

10

7

4

confused with

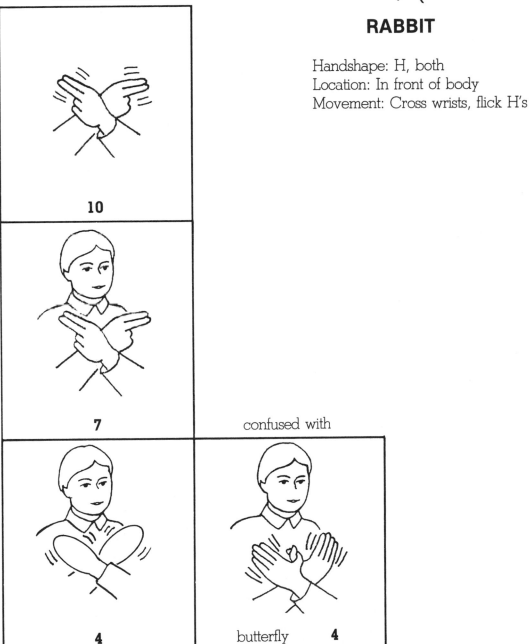

butterfly **4**

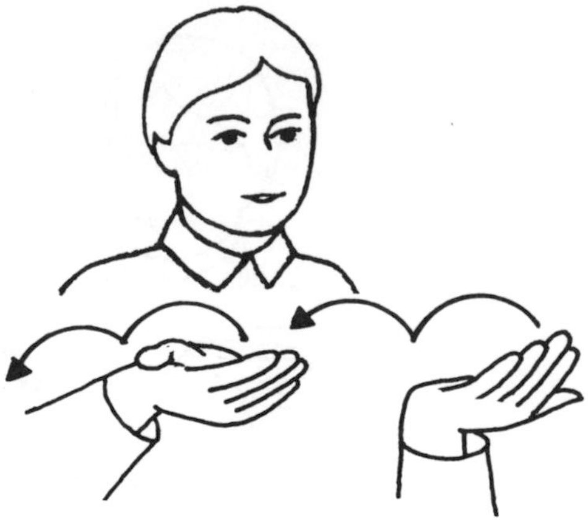

CARRY

Handshape: Open B, both
Location: In front of body
Movement: Across in front of body

understandable

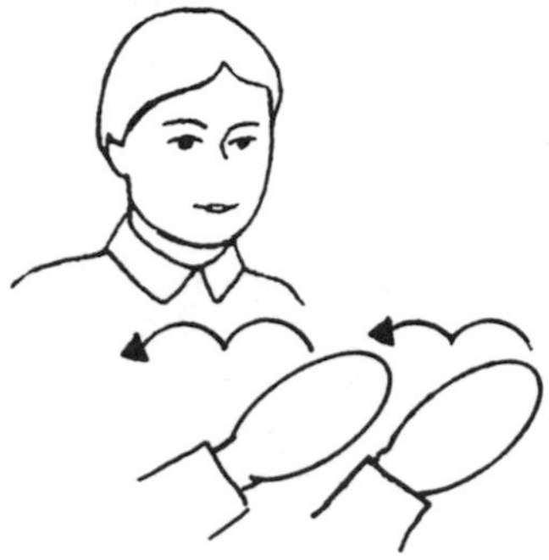

9

6

6

PARTY

Handshape: P, both
Location: In front of body
Movement: Swing from wrist

understandable

10

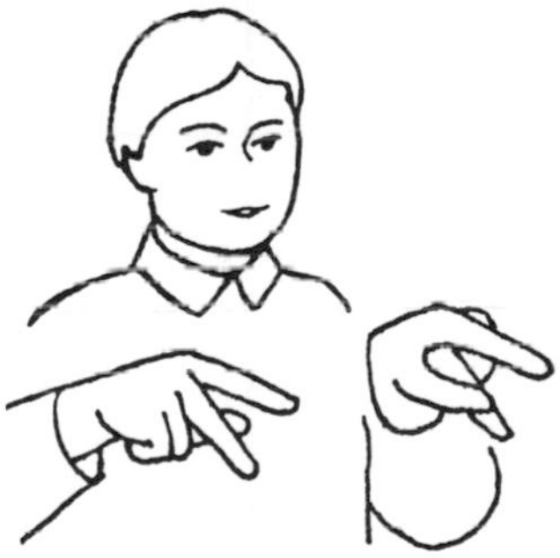

8

confused with

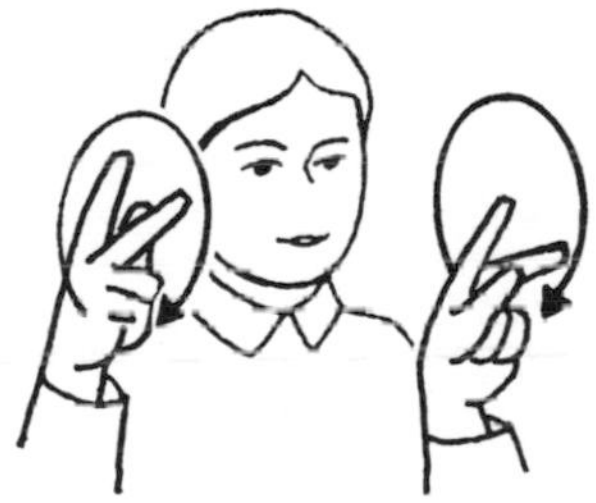

people **1**

3

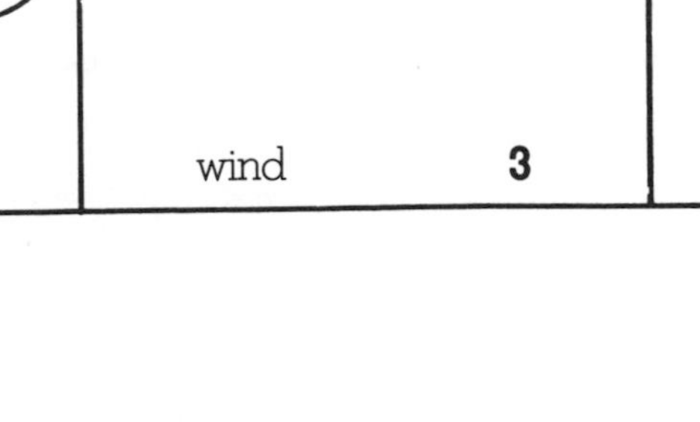

wind **3**

confused with

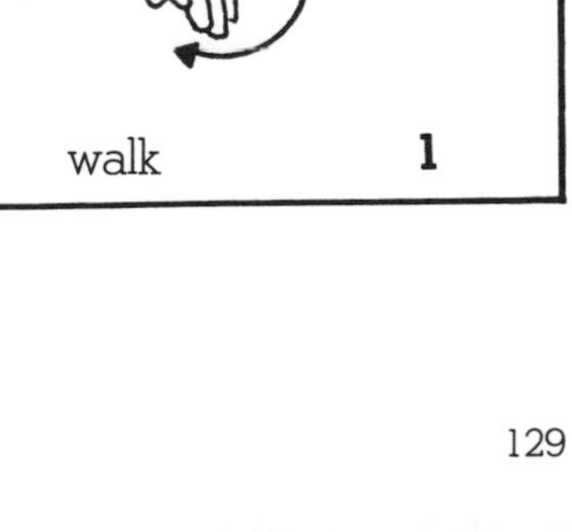

walk **1**

HOME

Handshape: Flat O
Location: Mouth
Movement: From mouth to cheek

confused with	understandable
 kiss 1	 8
 eat 1	 7
	 7
	 6

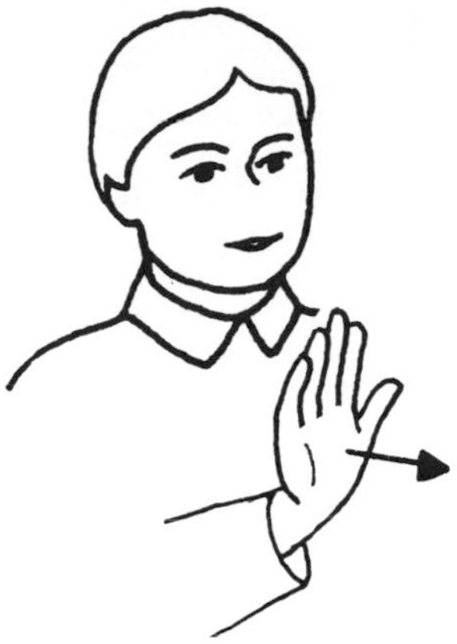

YOUR

Handshape: Open B
Location: In front of body
Movement: Out slightly

understandable

9

7

confused with

stop **2**

confused with

5

stop **3**

wait **1**

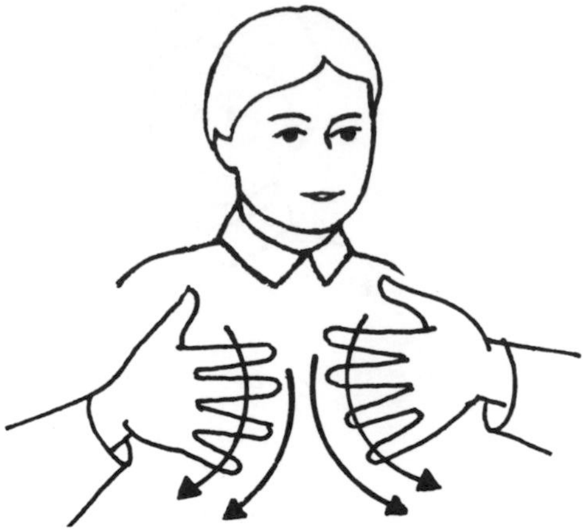

CLOTHING

Handshape: 5, both
Location: Chest
Movement: Brush down twice

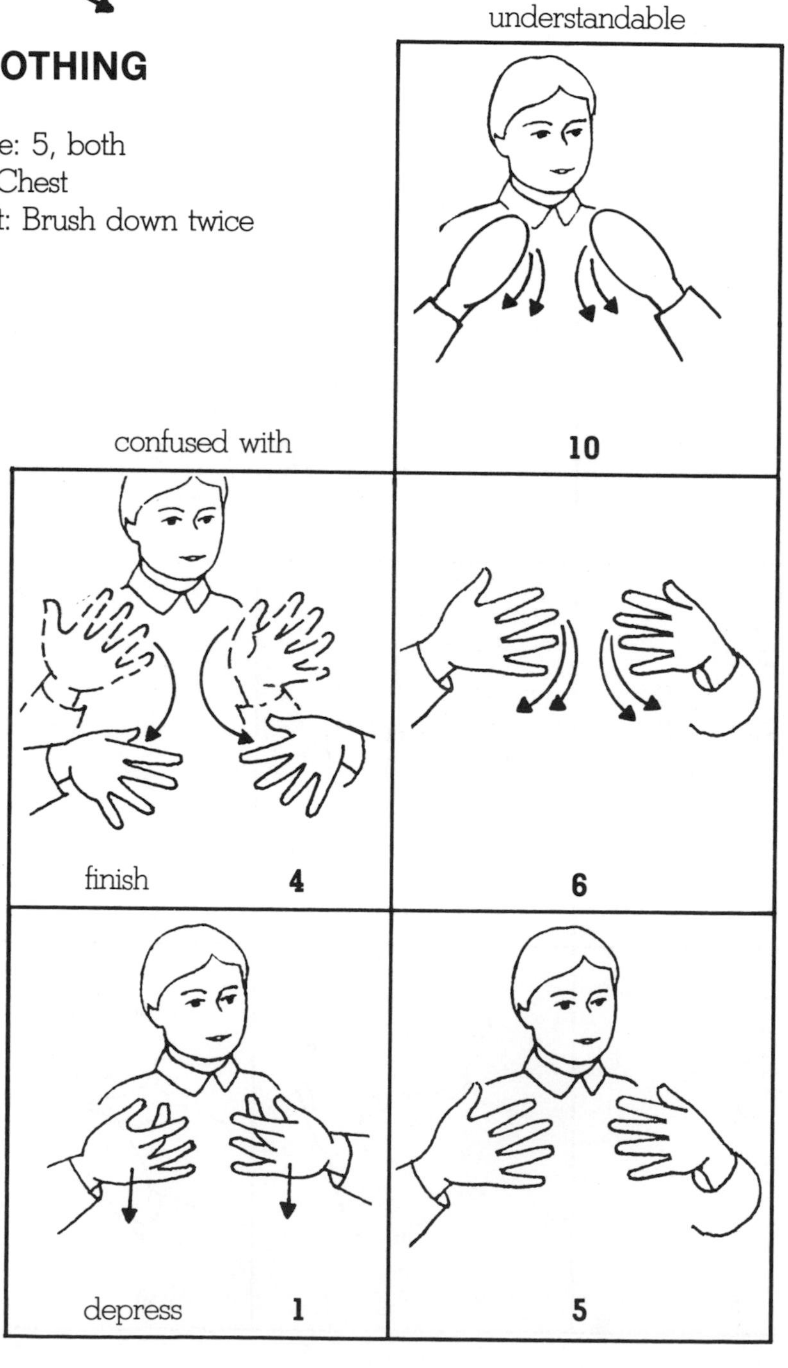

WON'T

Handshape: A, thumb extended
Location: Shoulder
Movement: Jerk back

understandable

9

6

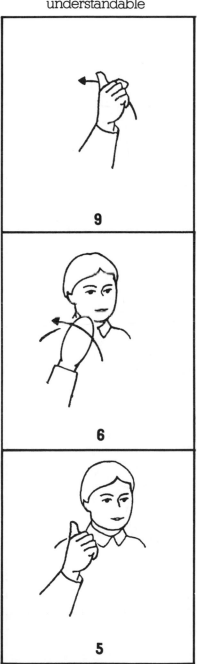

5

BROWN

Handshape: B
Location: Cheek
Movement: Rub thumb down

understandable

9

confused with

Boston **3**

6

beer **3**

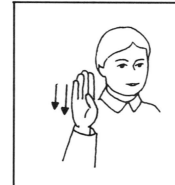

6

RIDE

Handshape: H; C
Location: In front of body
Movement: Move forward

understandable

9

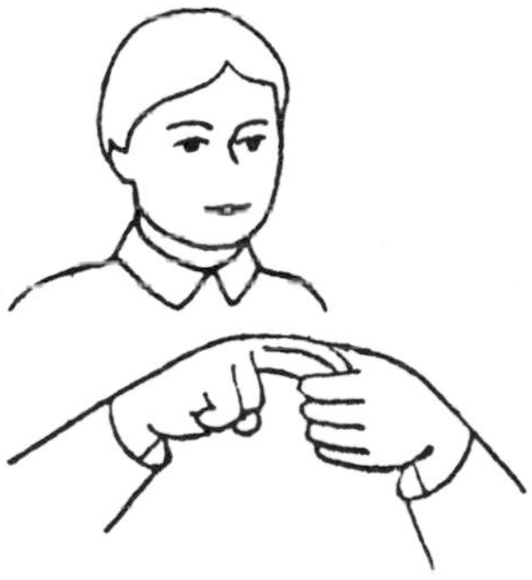

8

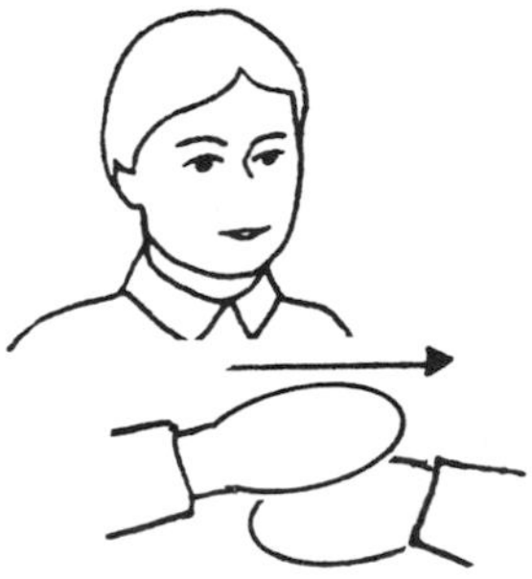

3

confused with

chair 1

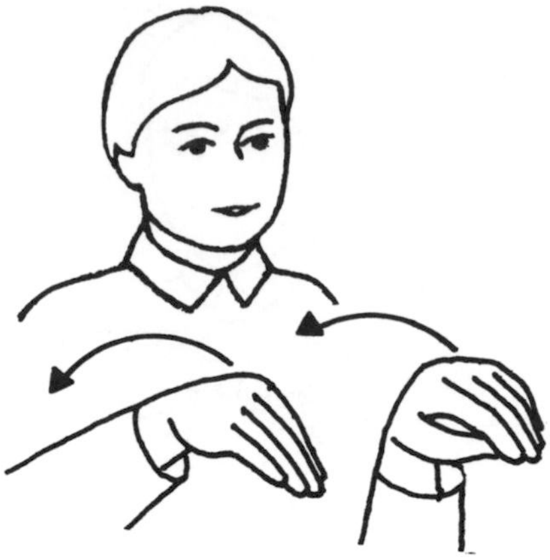

MOVE

Handshape: Flat O, both
Location: In front of body
Movement: From one side to other

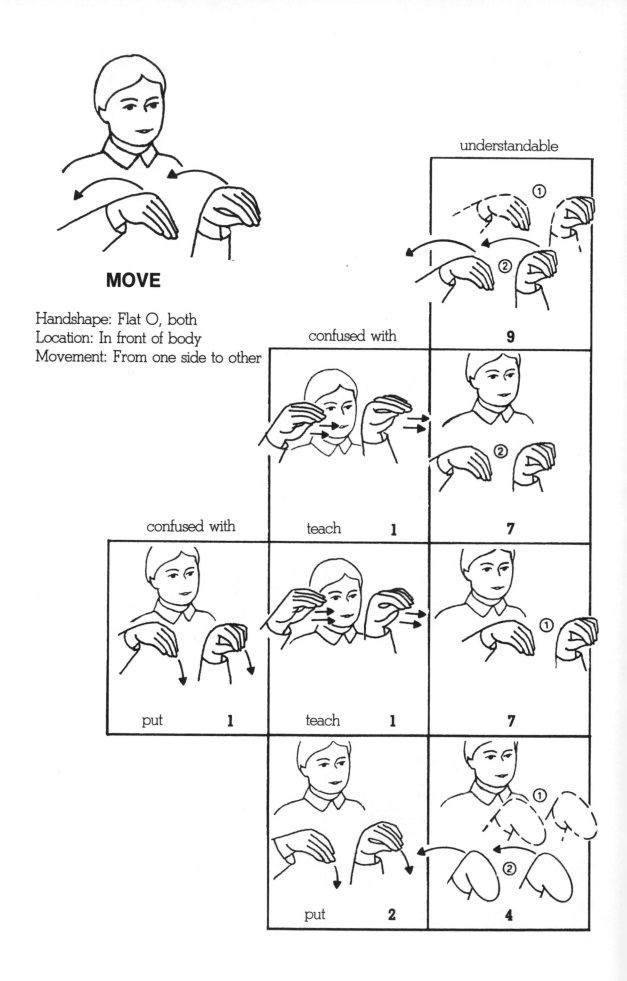

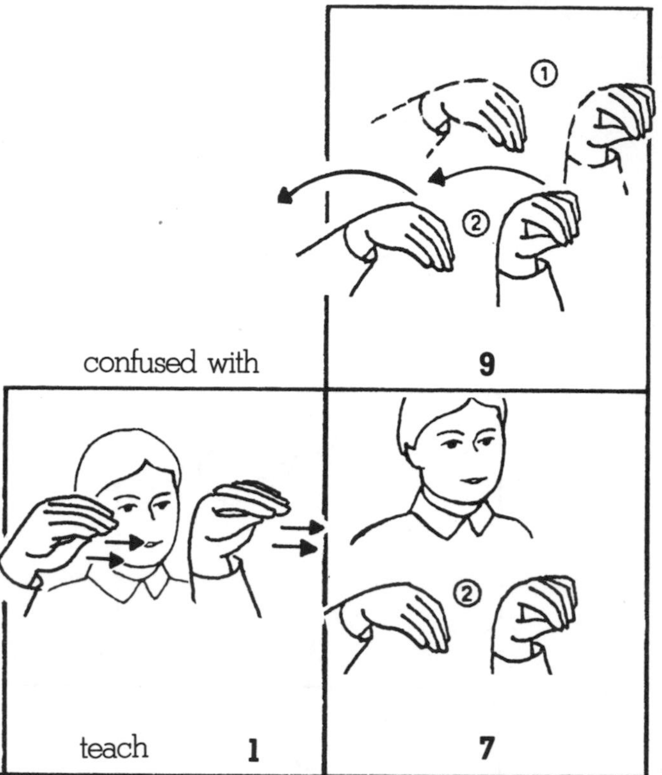

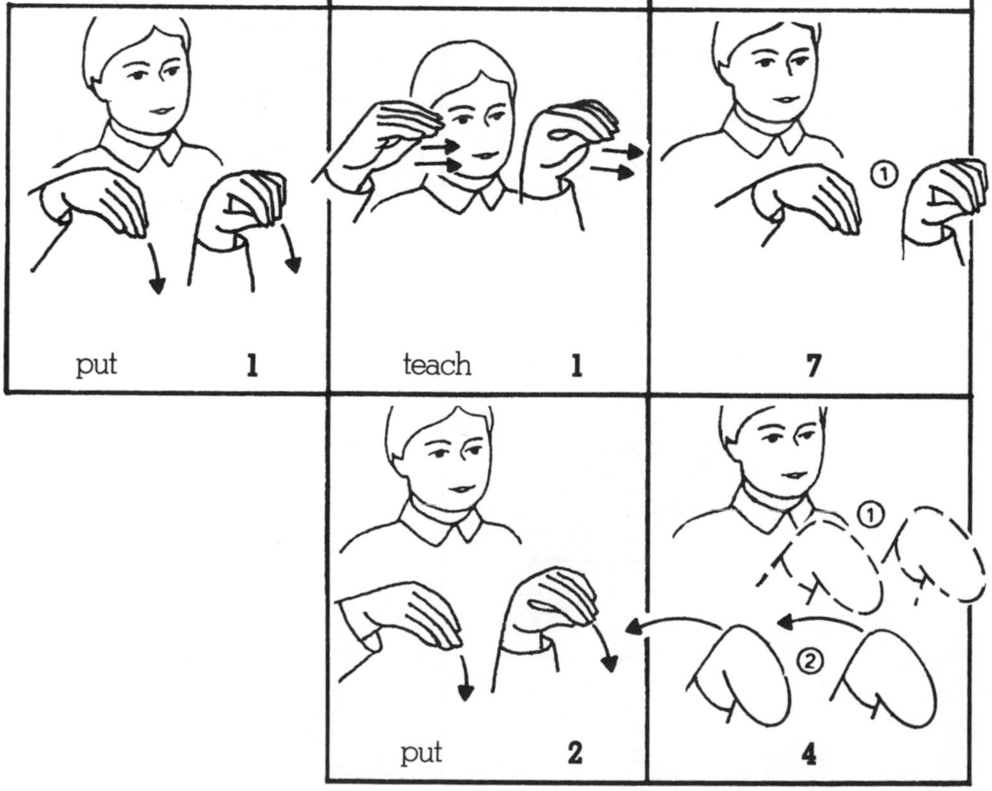

EAT

Handshape: Flat O
Location: Mouth
Movement: Touch lips once

understandable

8

8

4

confused with

good **2**

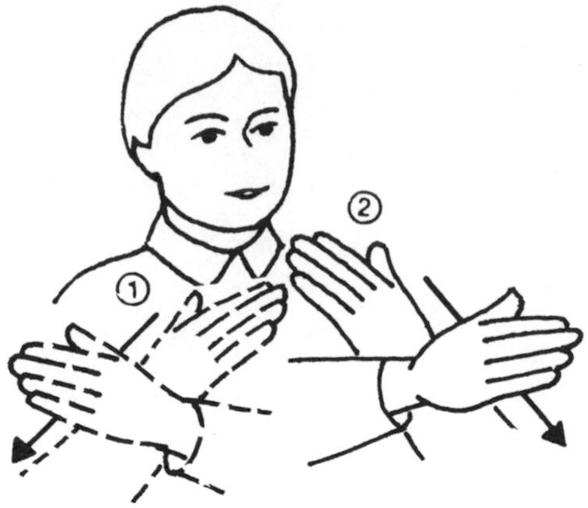

HAND

Handshape: Open B, both
Location: In front of body
Movement: Alternately trace at wrists

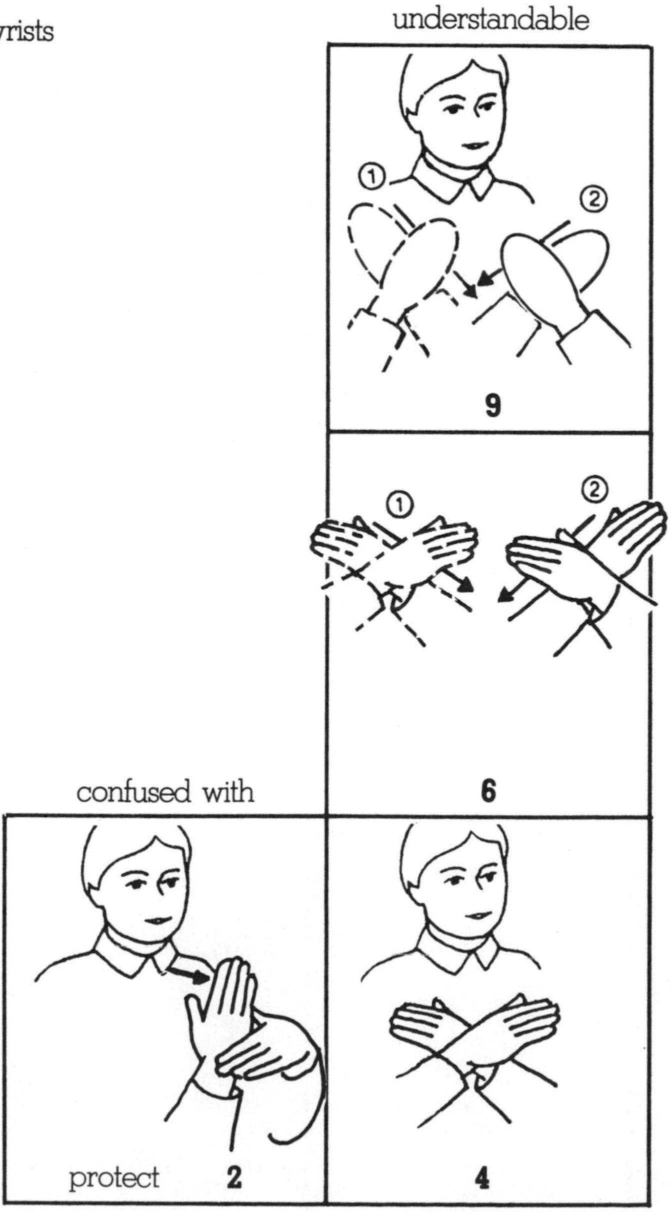

WINDOW

Handshape: Open B, both
Location: In front of body
Movement: Up an down

understandable

10

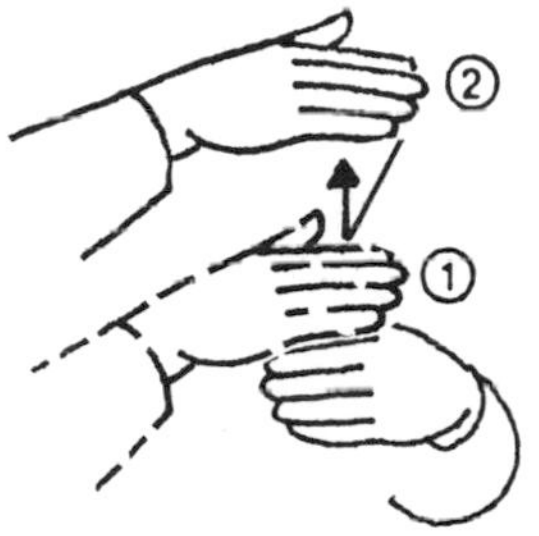

6

confused with

open **4**

5

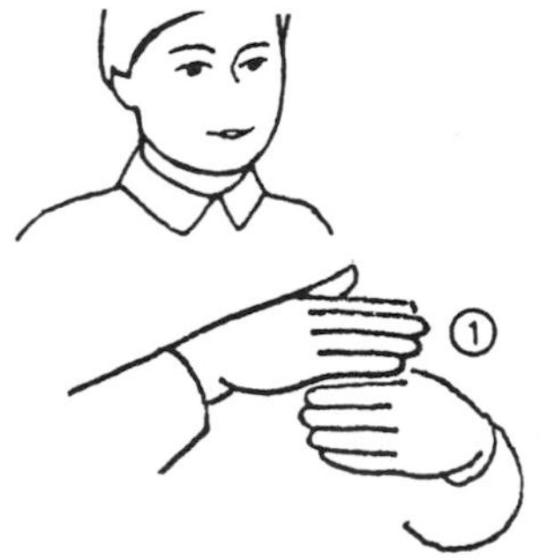

5

confused with

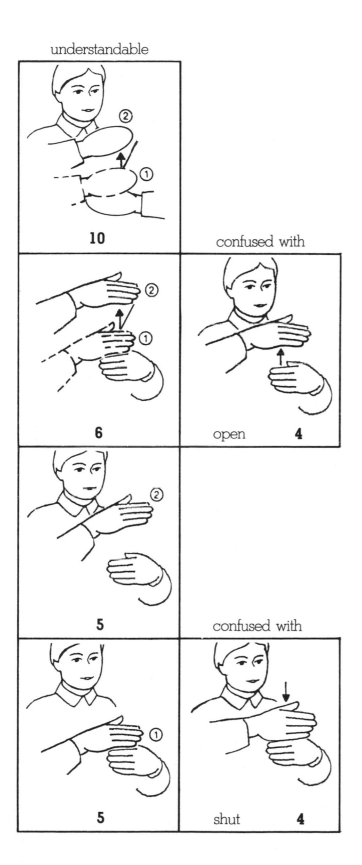

shut **4**

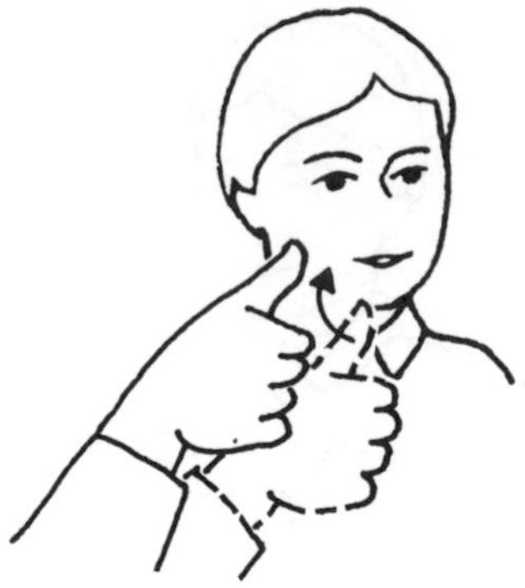

YESTERDAY

Handshape: A, thumb extended
Location: Cheek
Movement: Tap, move back, tap again

understandable

9

confused with

home **2**

6

refuse **2**

6

pipe **2**

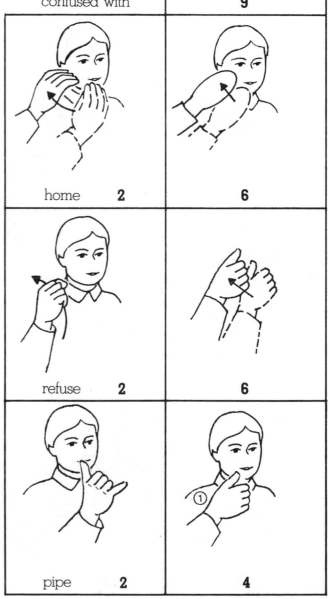

4

ON

Handshape: Open B, both
Location: In front of body
Movement: Place one hand on other

understandable

8

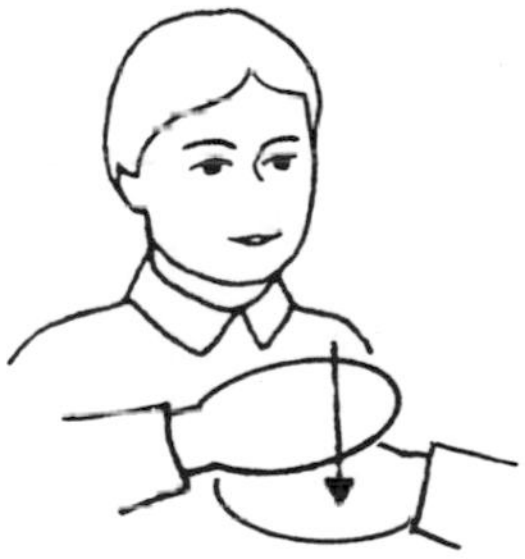

8

3

confused with

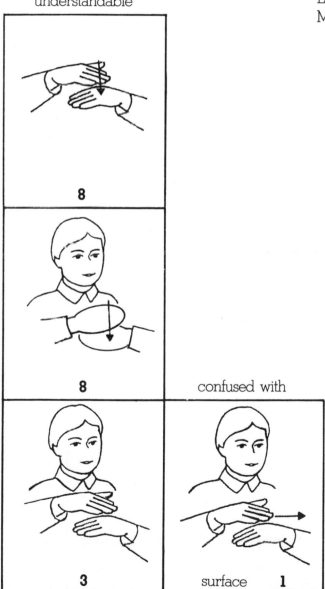

surface 1

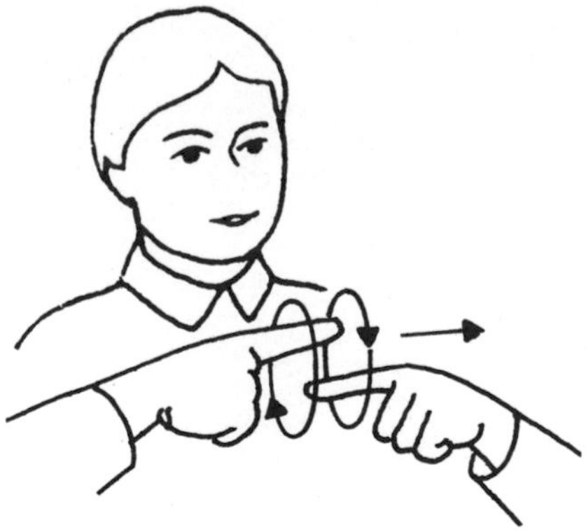

GO

Handshape: 1, both
Location: In front of body
Movement: From one side to other

understandable

9

confused with

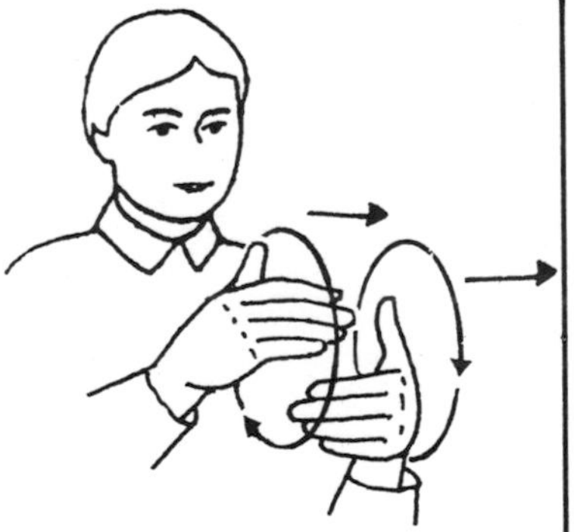

process 1

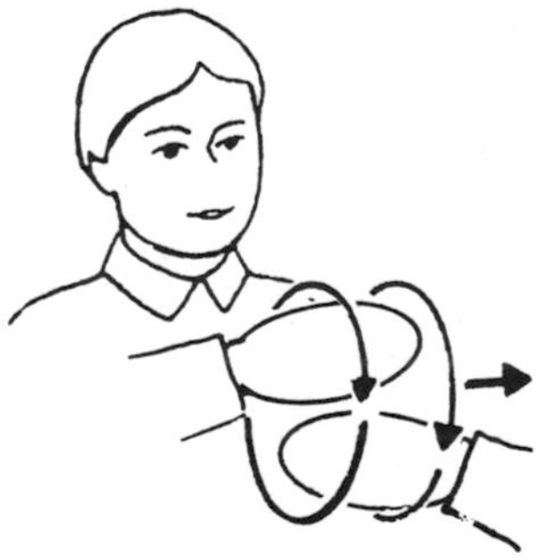

6

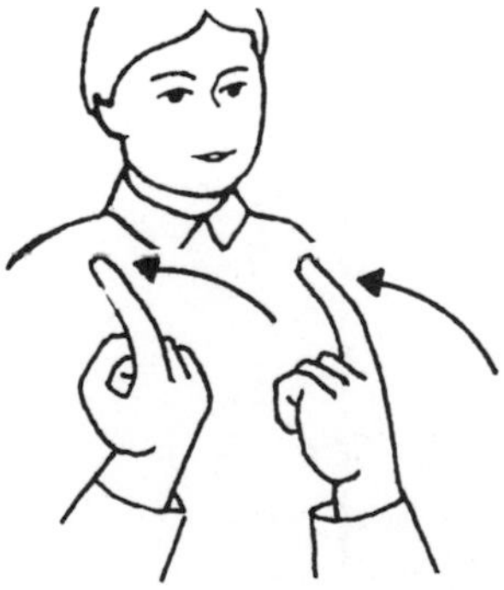

come 2

4

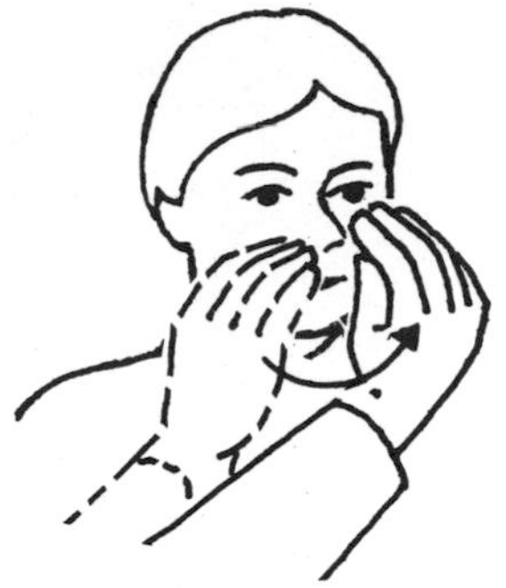

FLOWER

Handshape: Flat O
Location: Nose
Movement: Circle outward

understandable

8

confused with

6

smell **1**

5

our **2**

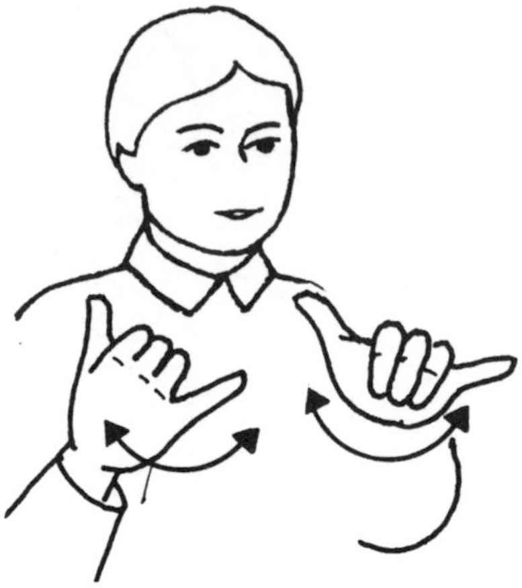

PLAY

Handshape: Y, both
Location: In front of body
Movement: Shake from wrists

understandable

10

confused with

wow **1**

4

now **5**

4

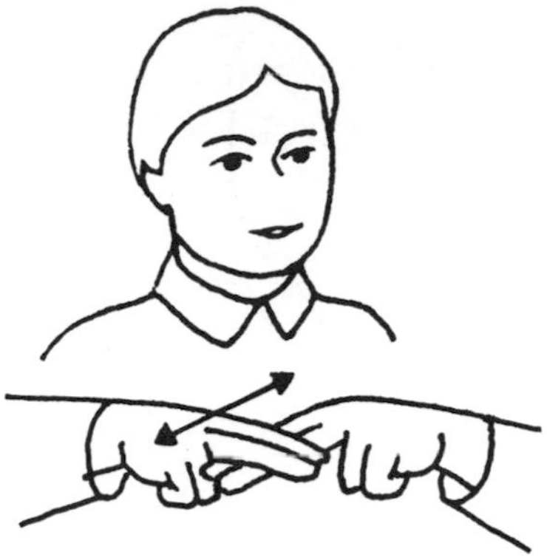

TRAIN

Handshape: H, both
Location: In front of body
Movement: Back and forth

understandable

8

6

4

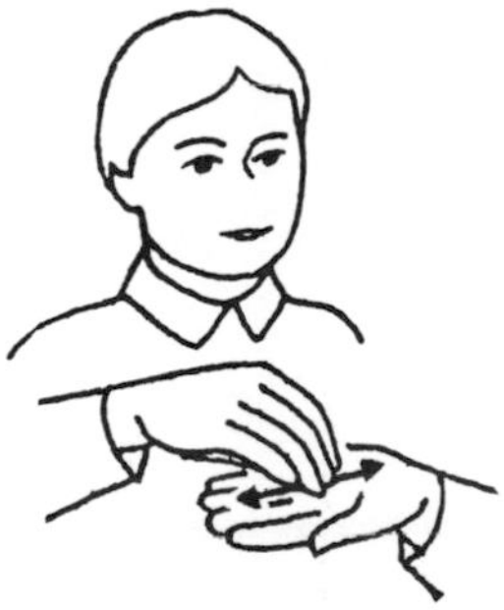

excuse **2**

confused with

surface **1**

confused with

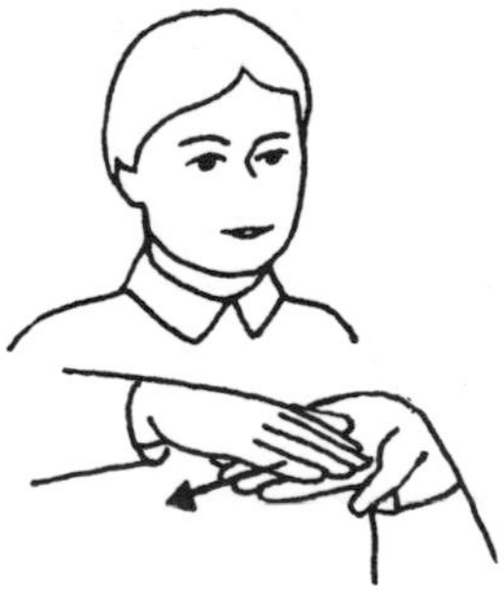

clean **1**

PERSON

Handshape: P, both
Location: At sides
Movement: Move down

understandable

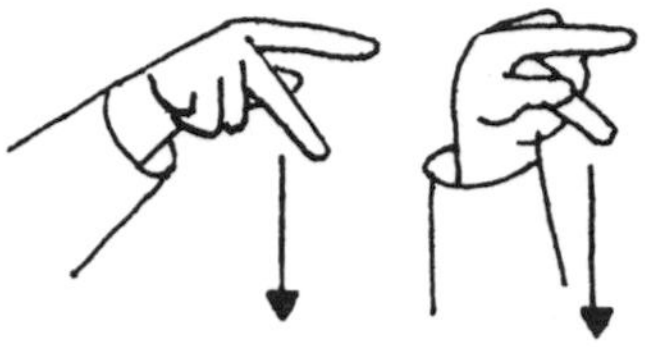

10

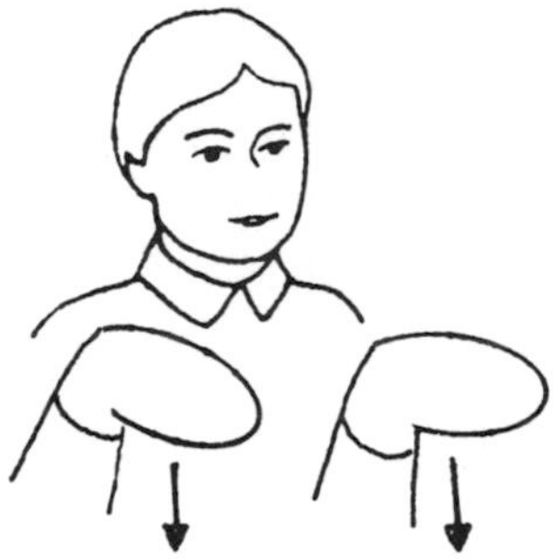

6

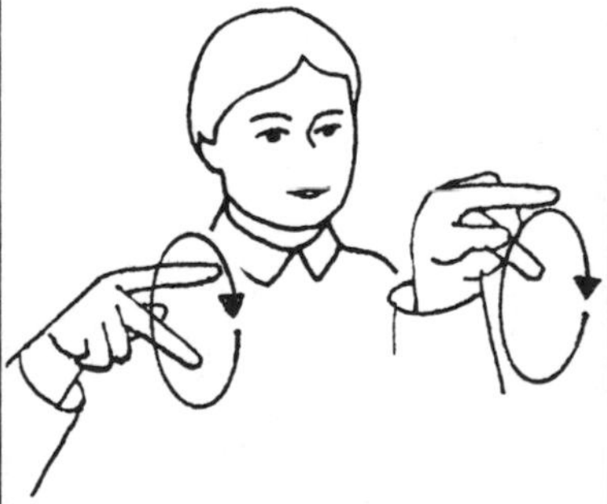

people 3

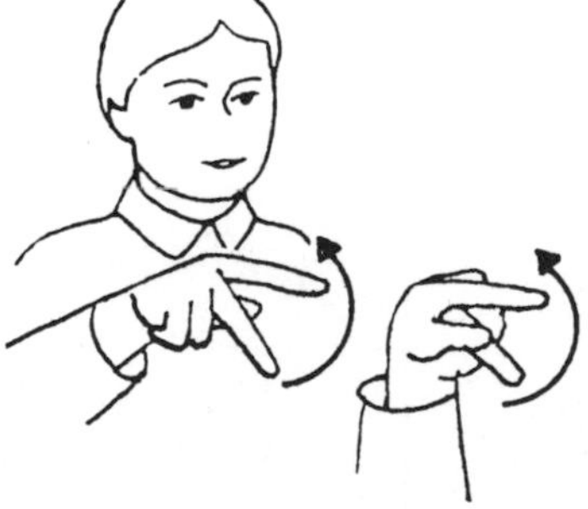

permit 3

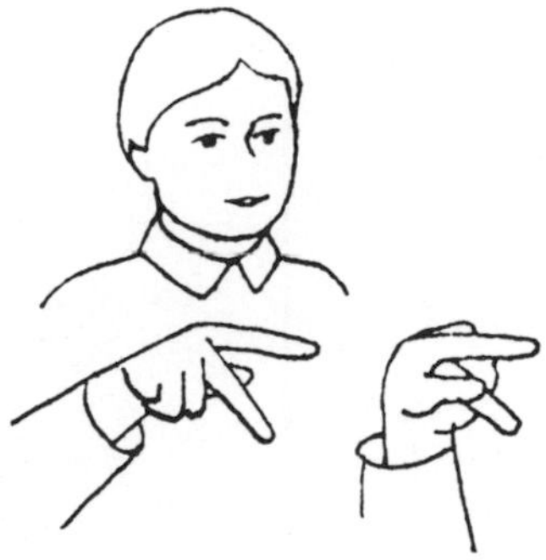

2

LOOK

Handshape: V
Location: Eyes
Movement: Point and move forward

understandable

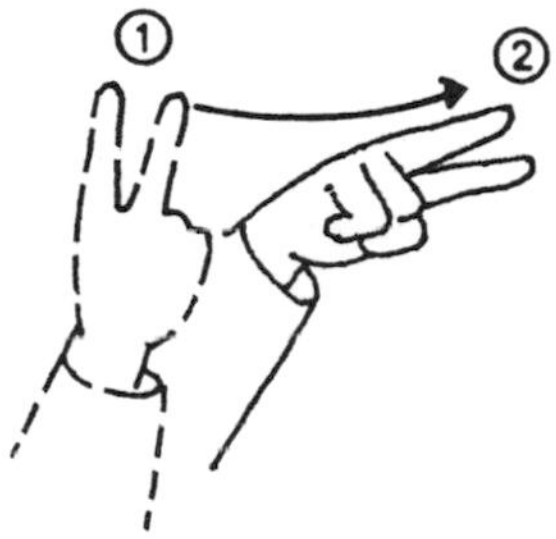

10

confused with

4

two **3**

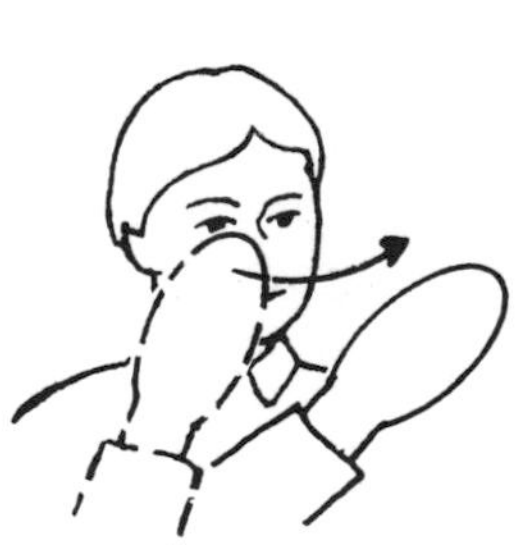

4

ignore **3**

RAIN

Handshape: Bent 5, both
Location: In front of body
Movement: Up and down

understandable

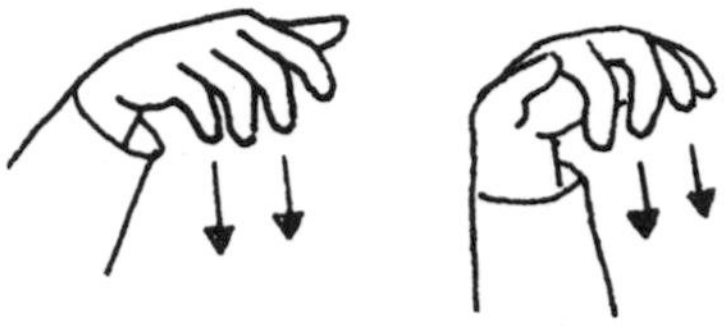

8

confused with

expect **2**

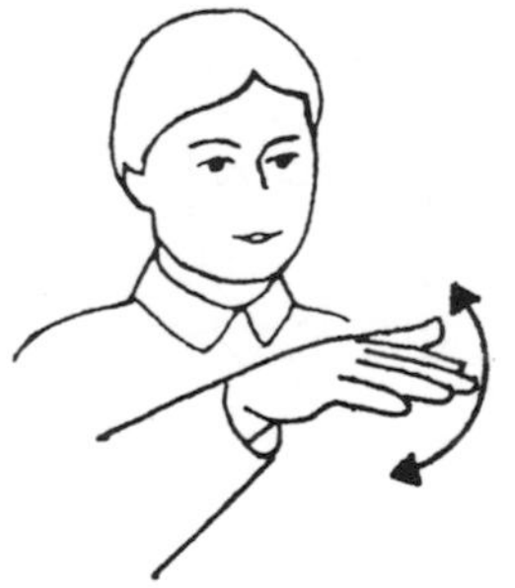

bye-bye **2**

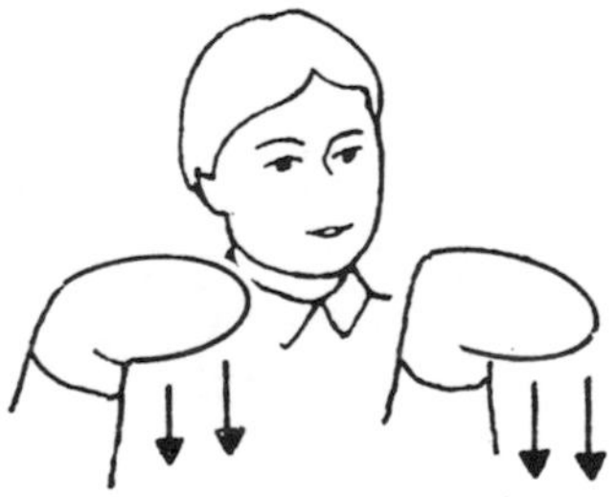

5

monster **1**

4

UGLY

understandable

Handshape: 1 to X
Location: Nose
Movement: Across face, changing to X

8

confused with

5

from **2**

puzzle **1**

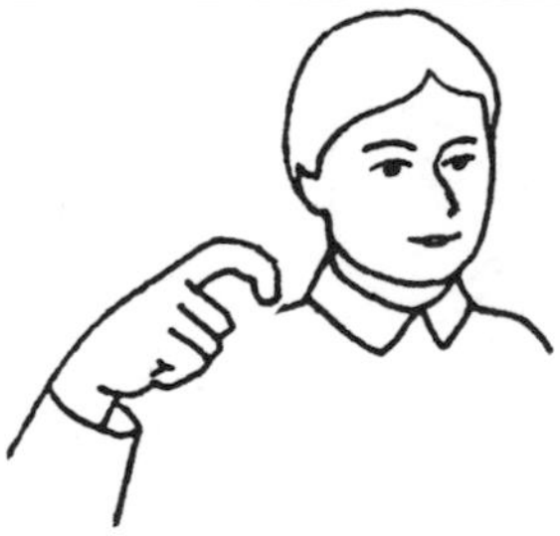

4

boring / dry **1**

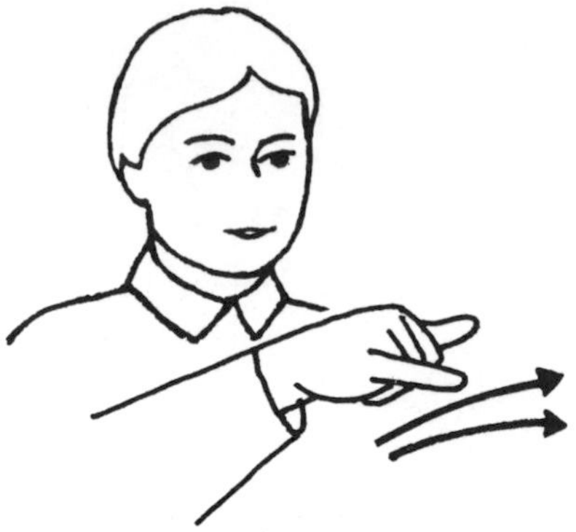

AIRPLANE

Handshape: Y, index extended
Location: In front of body
Movement: In and out twice

confused with | understandable

fly **1**

8

love **3**

5

confused with

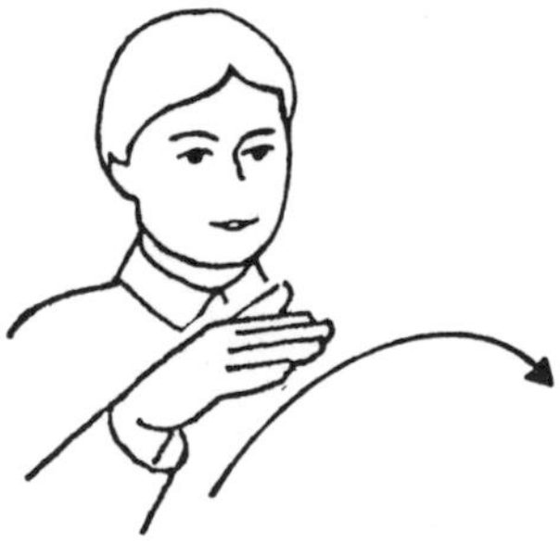

hill **2**

fly **3**

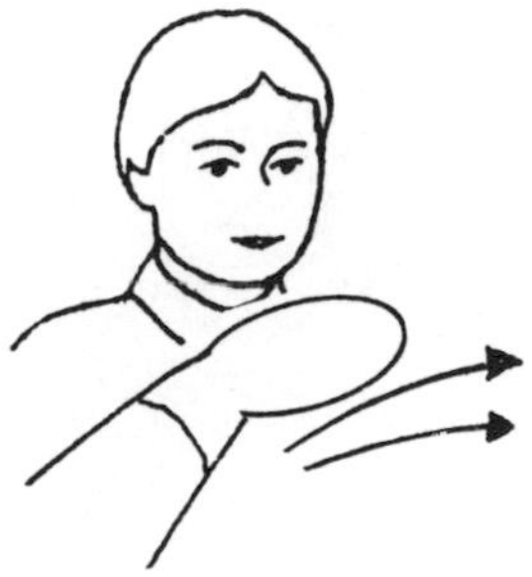

4

BATHE

Handshape: A, thumb extended, both
Location: Chest
Movement: Scrub up and down

understandable

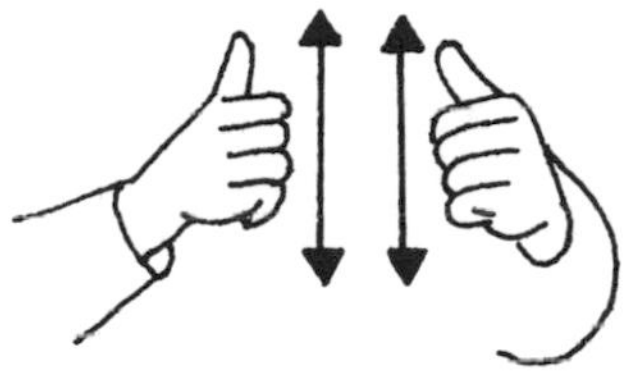

9

confused with

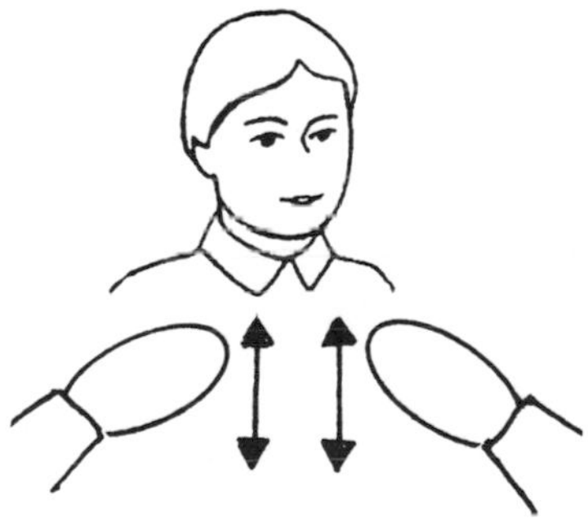

5

dress 3

happy 1

3

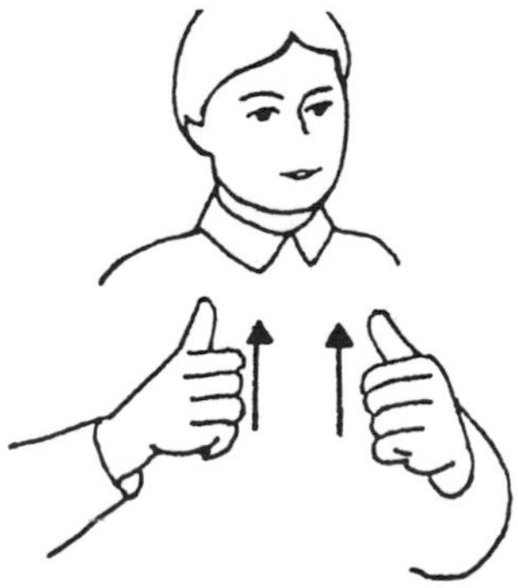

address 2

have 2

DRAW

Handshape: I
Location: In front of body
Movement: Trace down palm

understandable

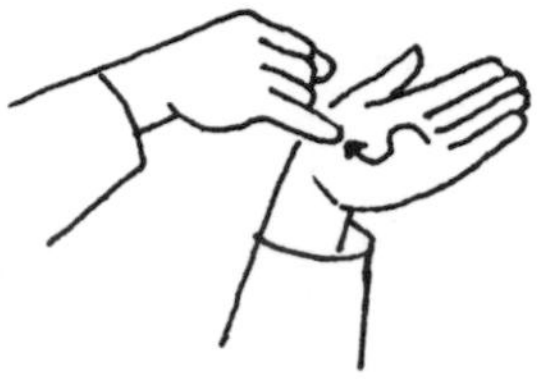

9

confused with

some 2

crack 3

3

just 1

it 3

3

BOX

Handshape: Open B, both
Location: In front of body
Movement: Outline sides of box

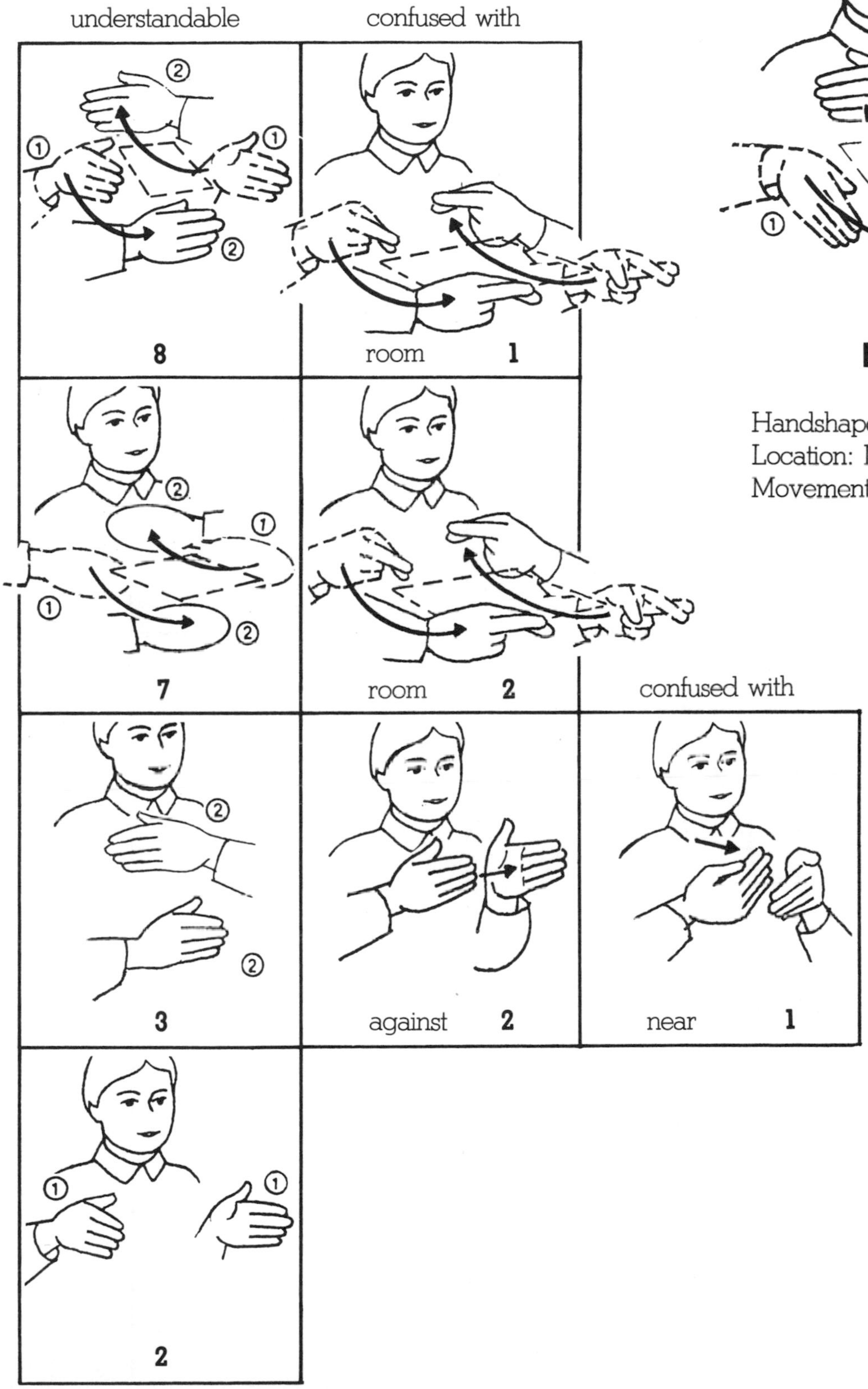

COMB

Handshape: Bent 5
Location: Head
Movement: Brush down twice

understandable

8

confused with

weak-minded **1**

4

strange **1**

rain **2**

2

BIRD

Handshape: G
Location: Mouth
Movement: Close fingers twice

understandable

8

4

2

confused with

twenty **3**

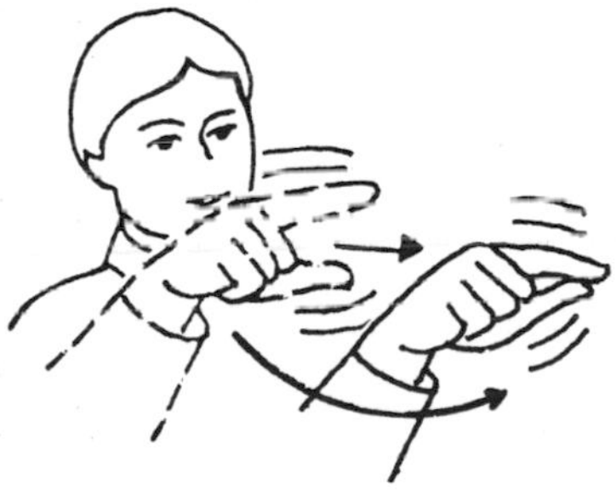

gossip **1**

pig **3**

WHERE

Handshape: 1
Location: In front of body
Movement: Wave back and forth

SINGLE ROBUST

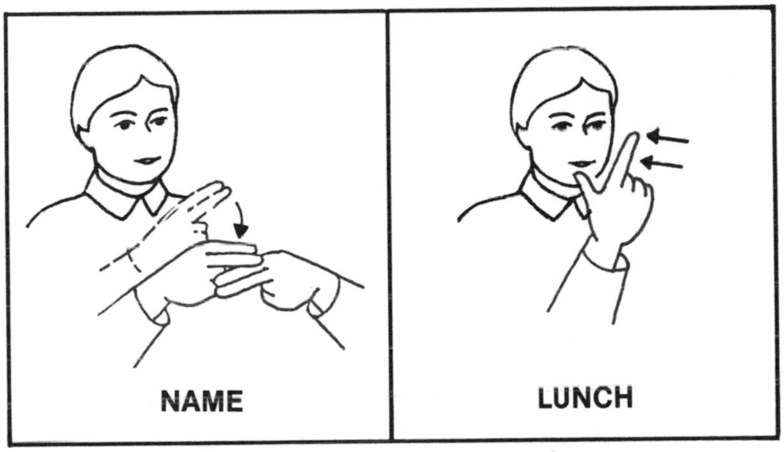

SINGLE FRAGILE

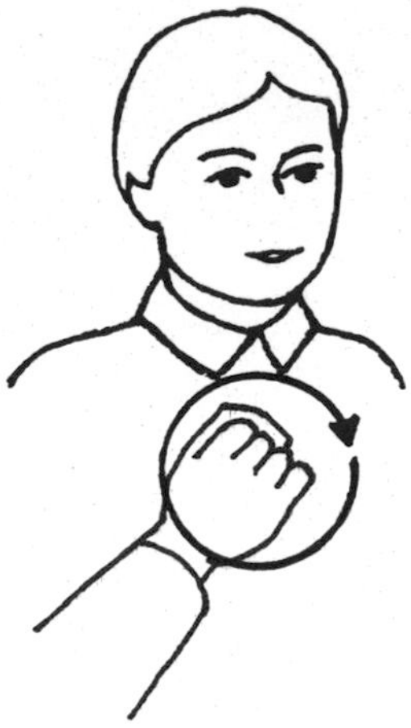

SORRY

Handshape: S
Location: Chest
Movement: Small circle

understandable

10

7

1

confused with

please **7**

enjoy **1**

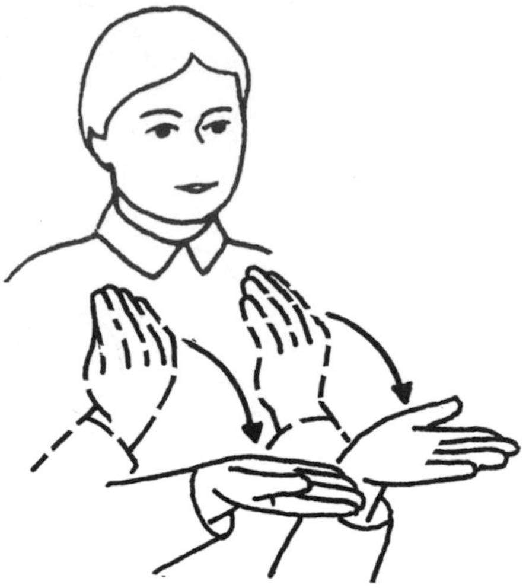

GIVE

Handshape: Flat O, both
Location: In front of body
Movement: Flip hands outward

understandable

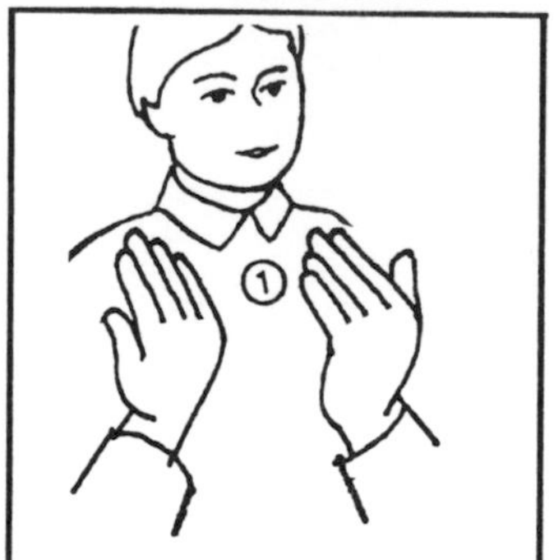

9

8

6

confused with

carry **6**

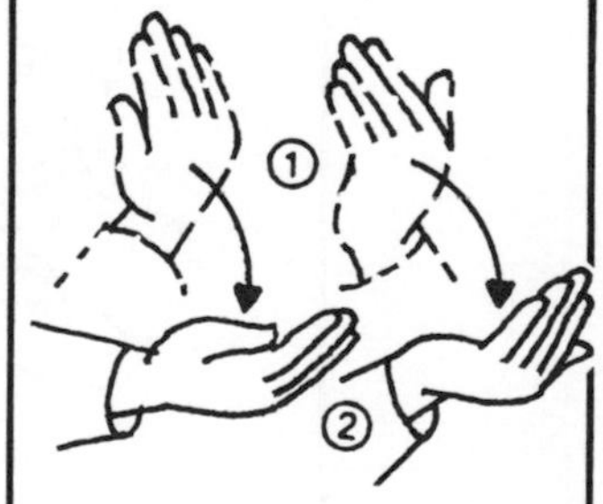

1

NO

Handshape: H, thumb extended
Location: In front of body
Movement: Snap closed

understandable

10

7

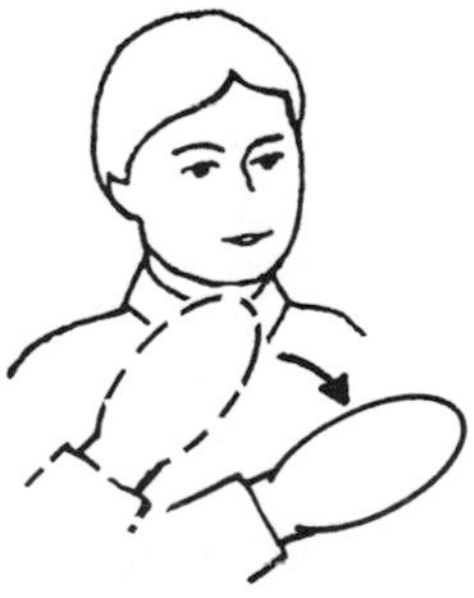

0

confused with

stubborn **2**

bye-bye **2**

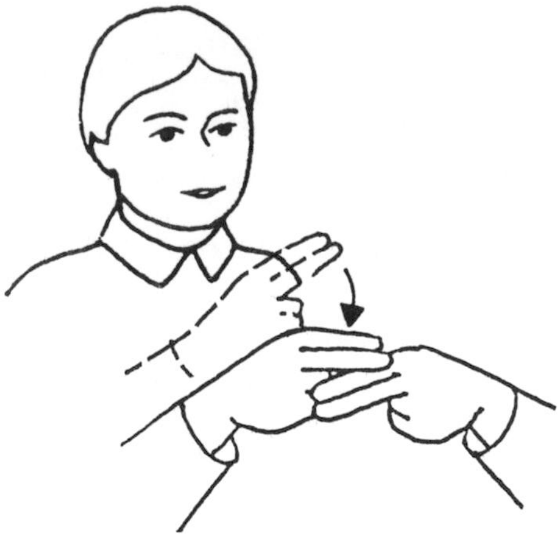

NAME

Handshape: H, both
Location: In front of body
Movement: Tap twice

understandable

10

confused with

short **1**

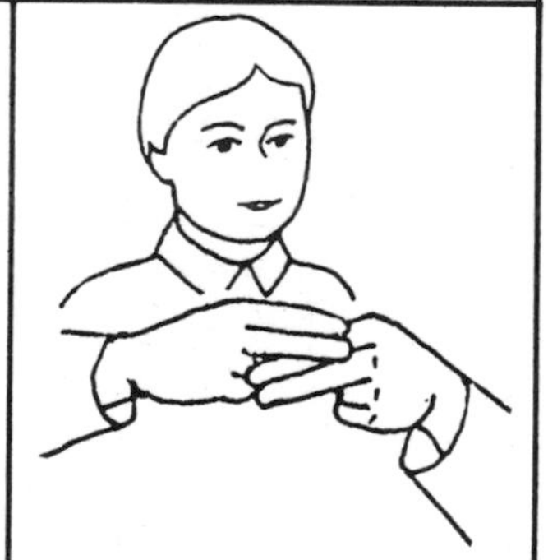

7

confused with

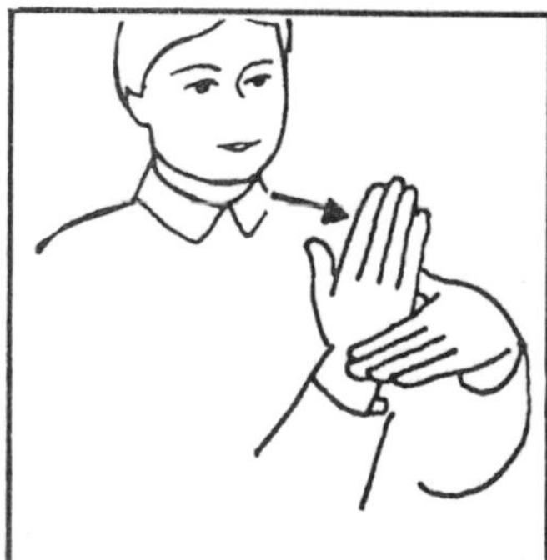

protect **1**

medium **1**

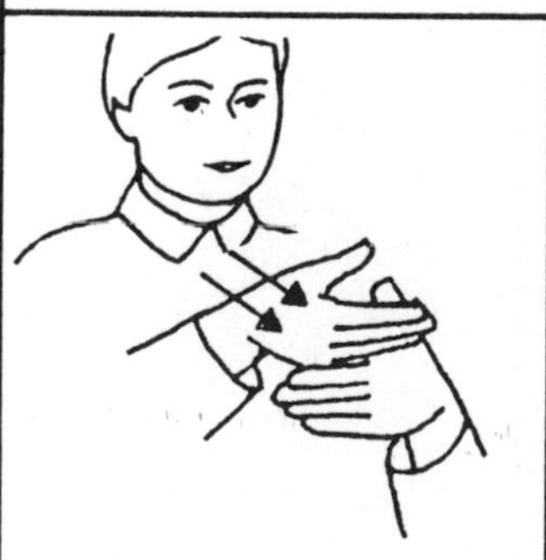

bother **3**

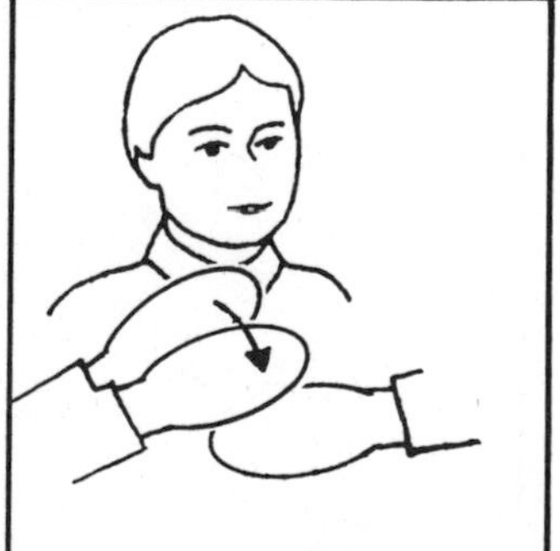

1

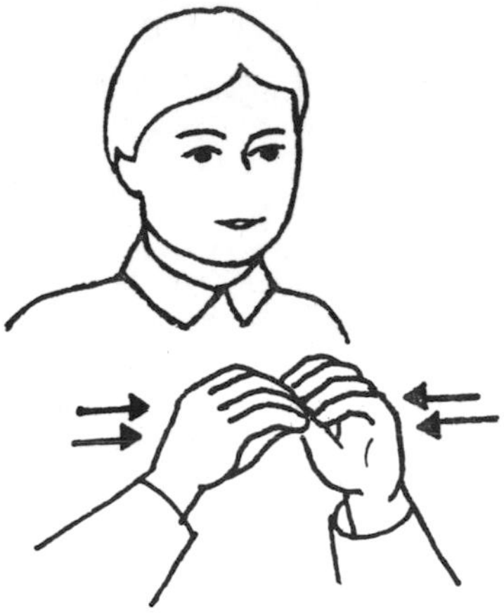

MORE

Handshape: Flat O, both
Location: In front of body
Movement: Tap fingertips twice

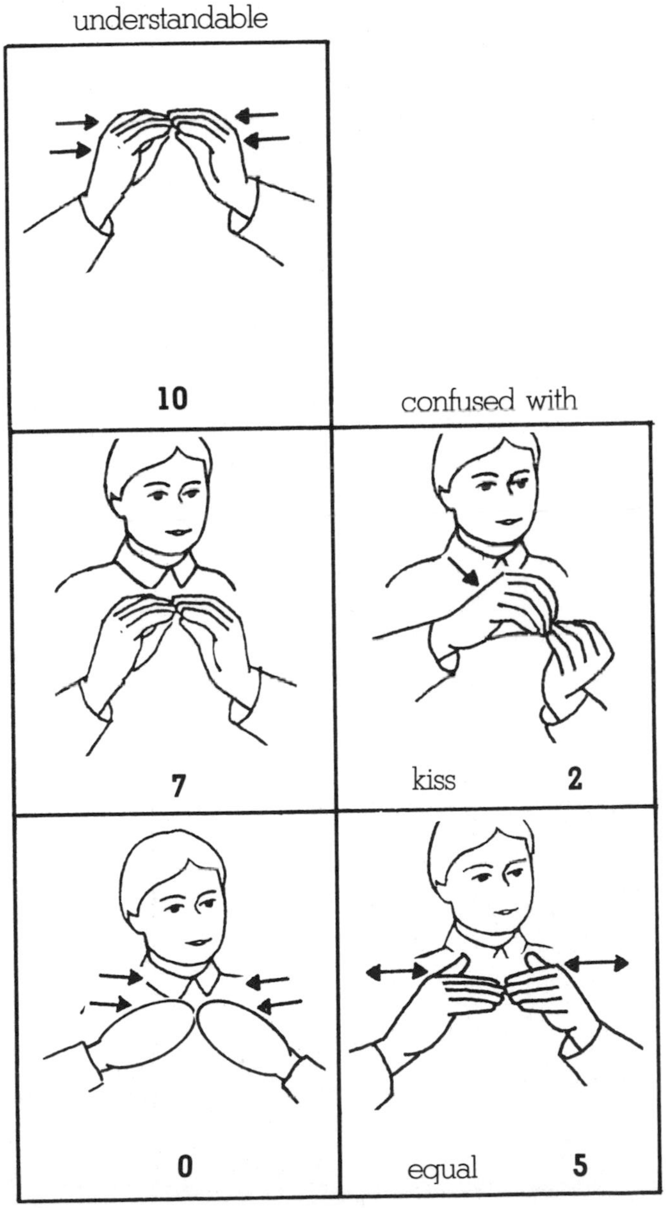

TEA

Handshape: F; O
Location: In front of body
Movement: Stir

understandable

10

6

confused with

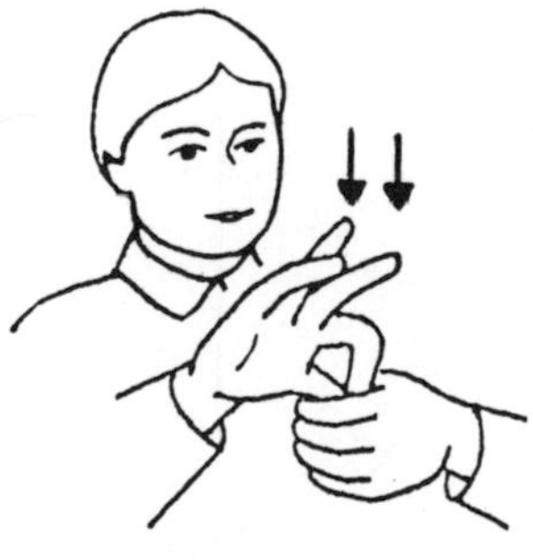

vote **2**

2

FROM

Handshape: 1, both
Location: In front of body
Movement: Toward body, bending finger

understandable

confused with

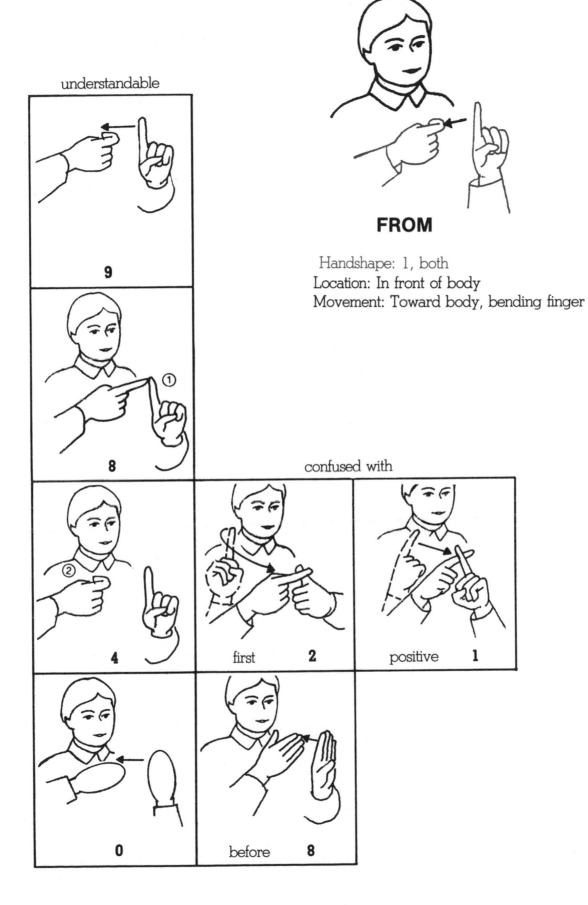

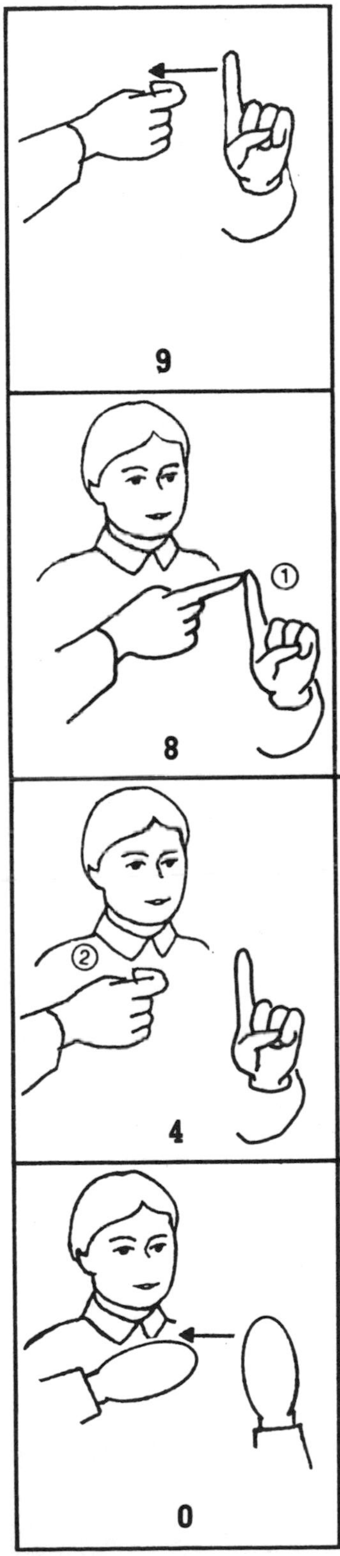

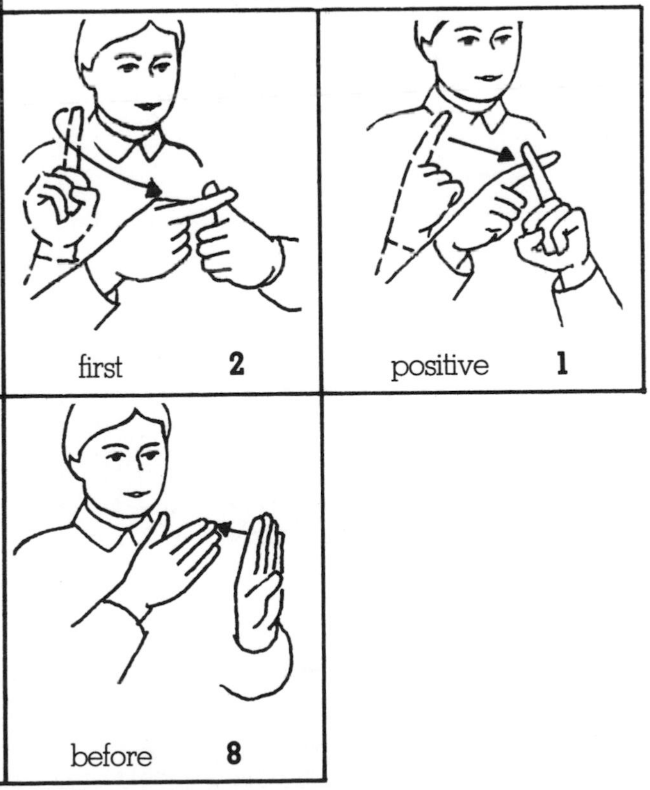

MONDAY

Handshape: M
Location: In front of body
Movement: Small circle

understandable

9

confused with

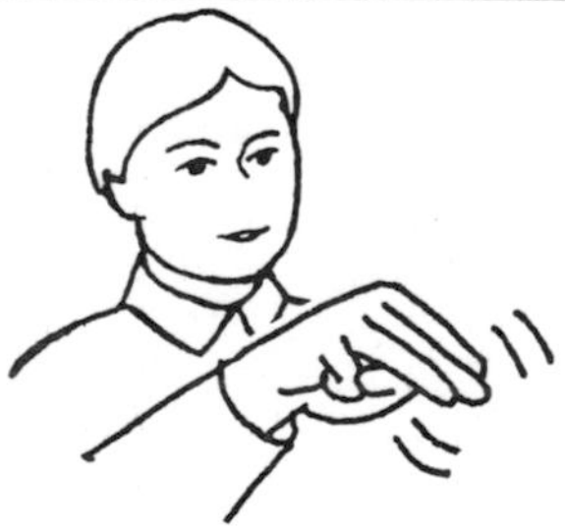

"M" **2**

8

confused with

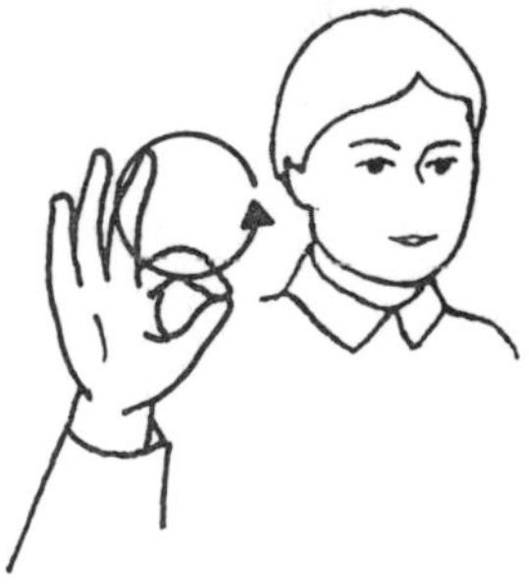

Friday **1**

Sunday **6**

0

TELEPHONE

Handshape: Y
Location: Side of face
Movement: None

understandable

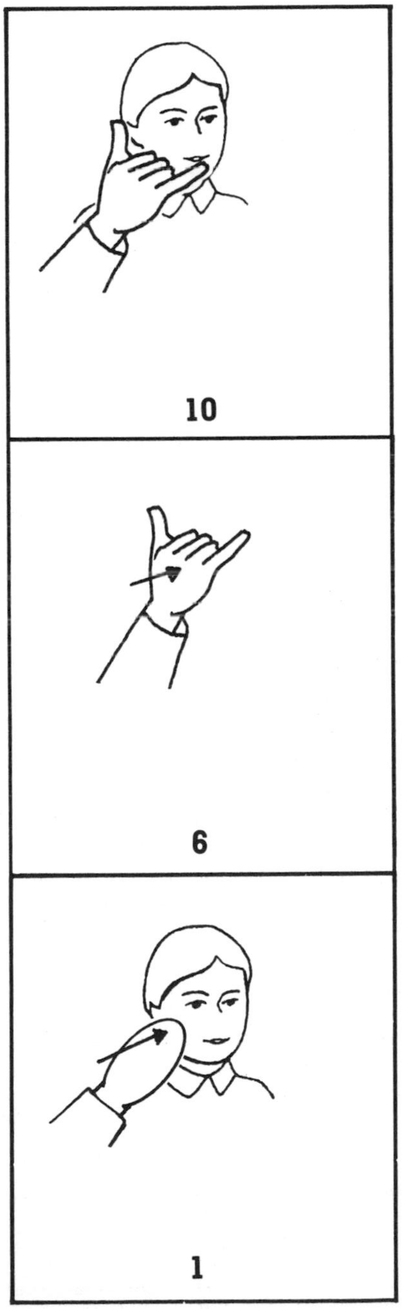

SIGN

Handshape: 1, both
Location: In front of body
Movement: Circular, inward

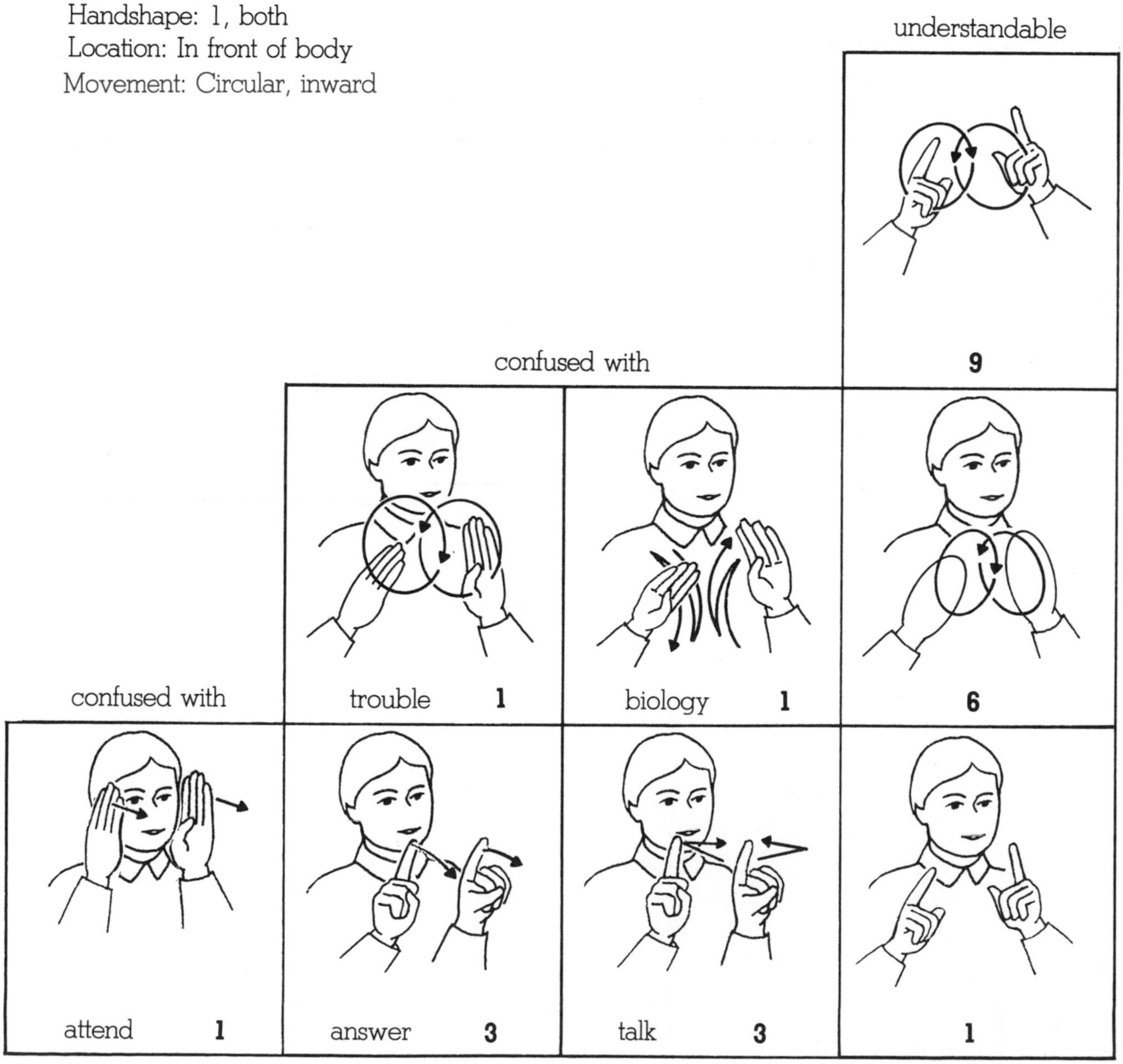

COOKIE

Handshape: C; open B
Location: In front of body
Movement: Tap, twist, tap again

understandable

9

8

confused with

cake 1

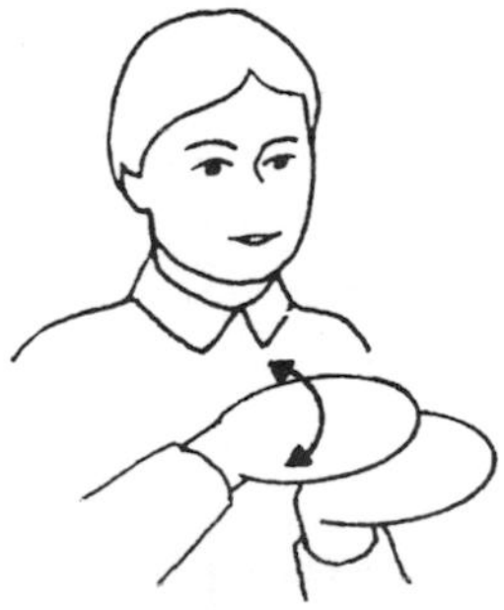

0

hamburger 4

confused with

become 1

school 1

FAST

Handshape: L, both
Location: In front of body
Movement: Close fingers quickly

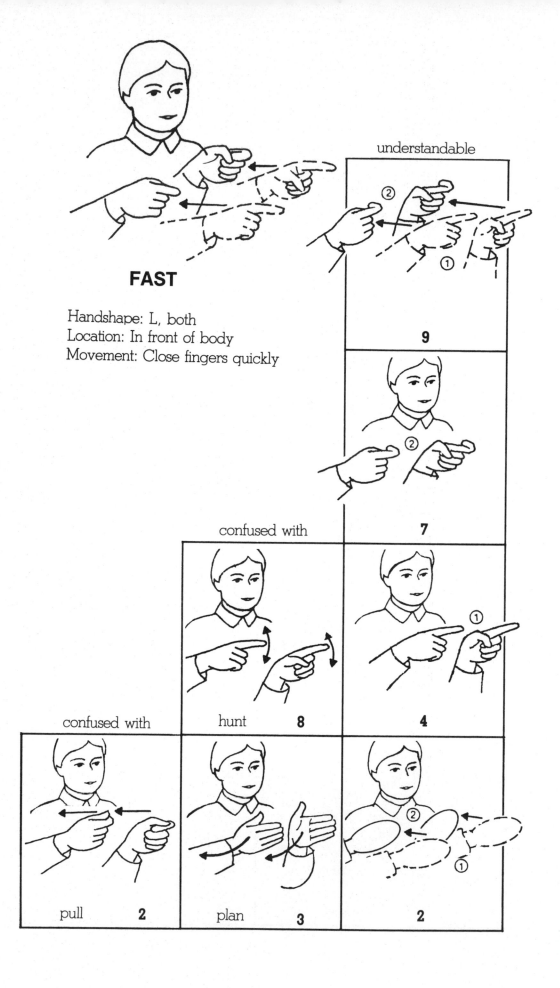

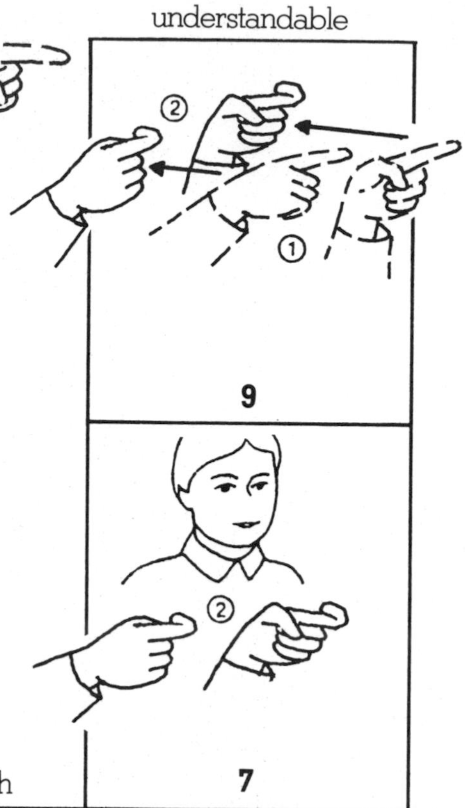

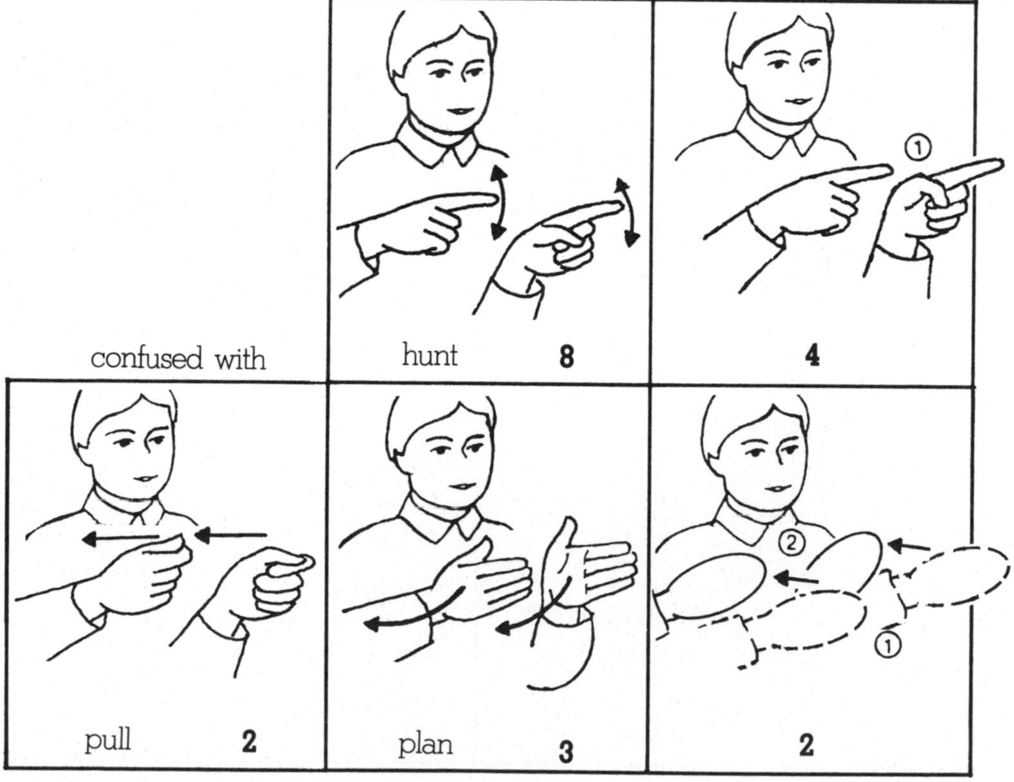

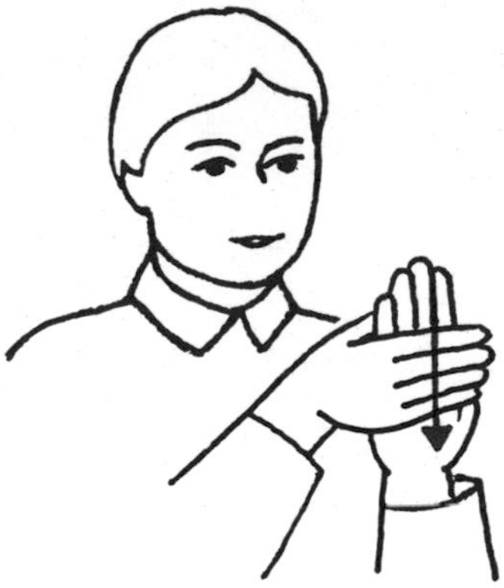

GLOVE

Handshape: C, both
Location: In front of body
Movement: Alternately slide over fingers

understandable

9

5

2

confused with

sandwich **2**

magazine **1**

EGG

Handshape: H, both
Location: In front of body
Movement: Down and apart

understandable

10

confused with

quiet 1

divide 3

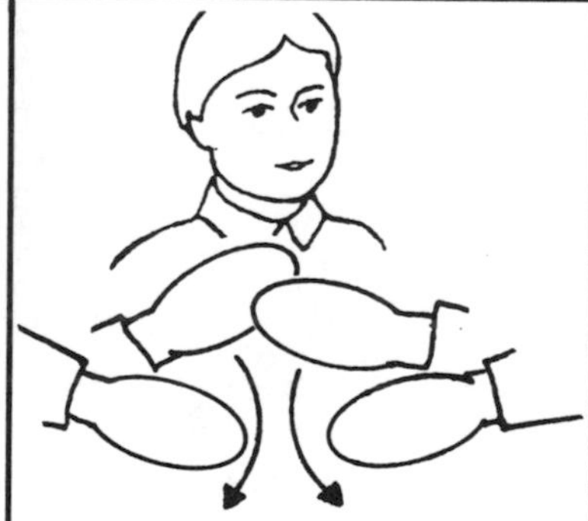

5

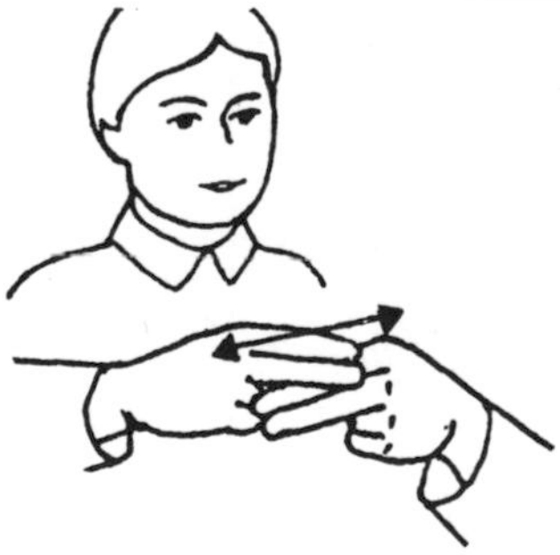

short 3

name 5

1

PIG

Handshape: B
Location: Chin
Movement: Flap fingers

understandable

8

confused with

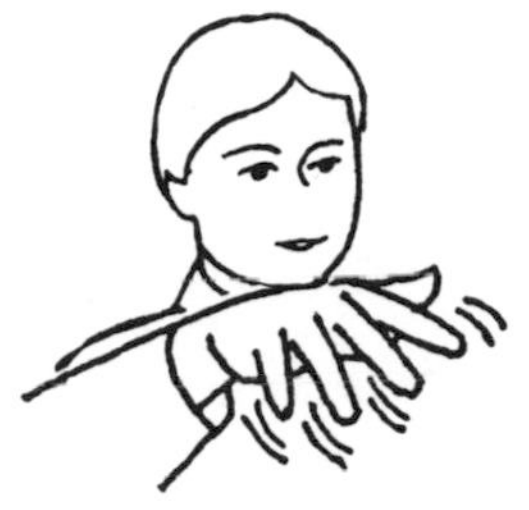

dirty 1

8

0

confused with

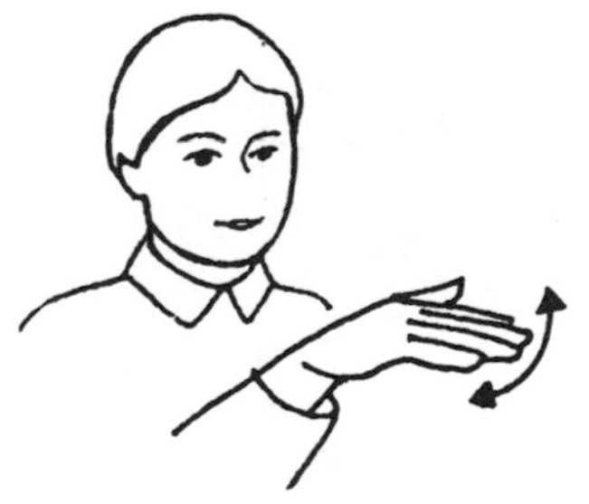

bye-bye 7

FIVE

Handshape: 5
Location: In front of body
Movement: None

understandable

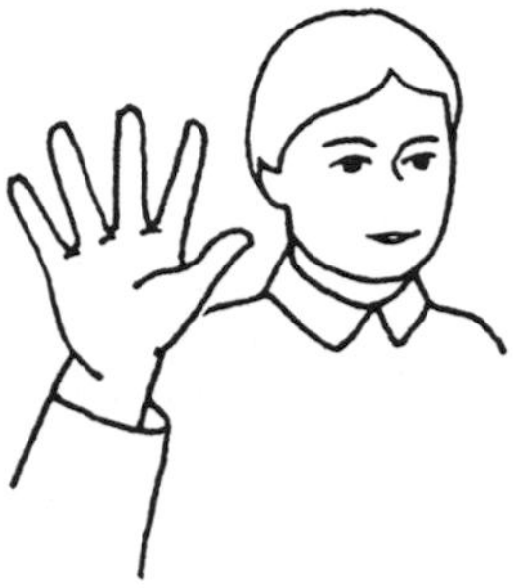

8

7

confused with

your 7

1

CAN

Handshape: S, both
Location: In front of body
Movement: Down forcefully

understandable

9

confused with

5

possible 1

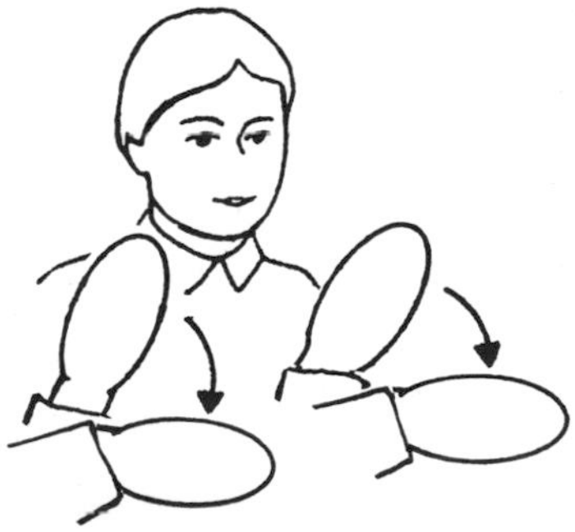

1

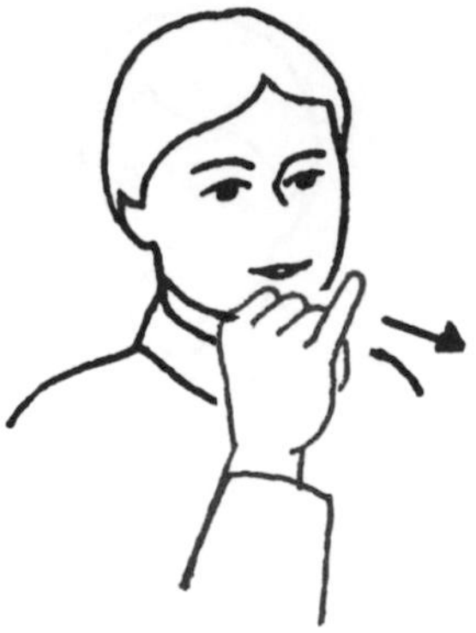

IS

Handshape: I
Location: Mouth
Movement: Move out

understandable

10

confused with

idiot **1**

I **1**

5

be **4**

1

WRONG

Handshape: Y
Location: Chin
Movement: Tap chin

understandable

10

5

confused with

now **3**

0

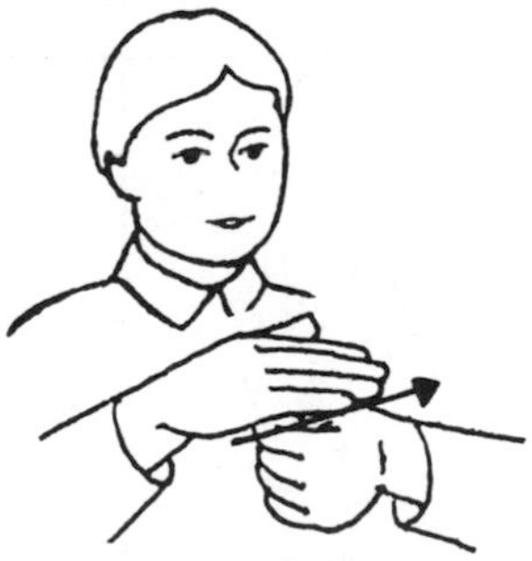

full **1**

confused with

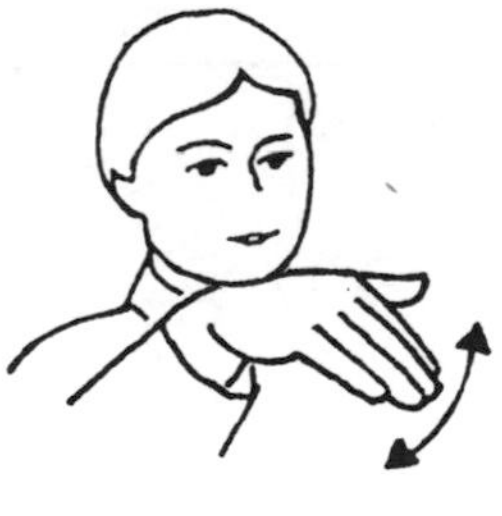

pig **1**

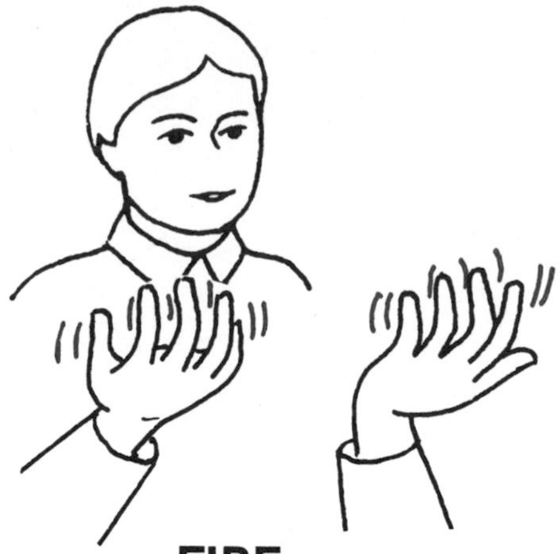

FIRE

Handshape: 5
Location: In front of body
Movement: Fluttering upward

understandable

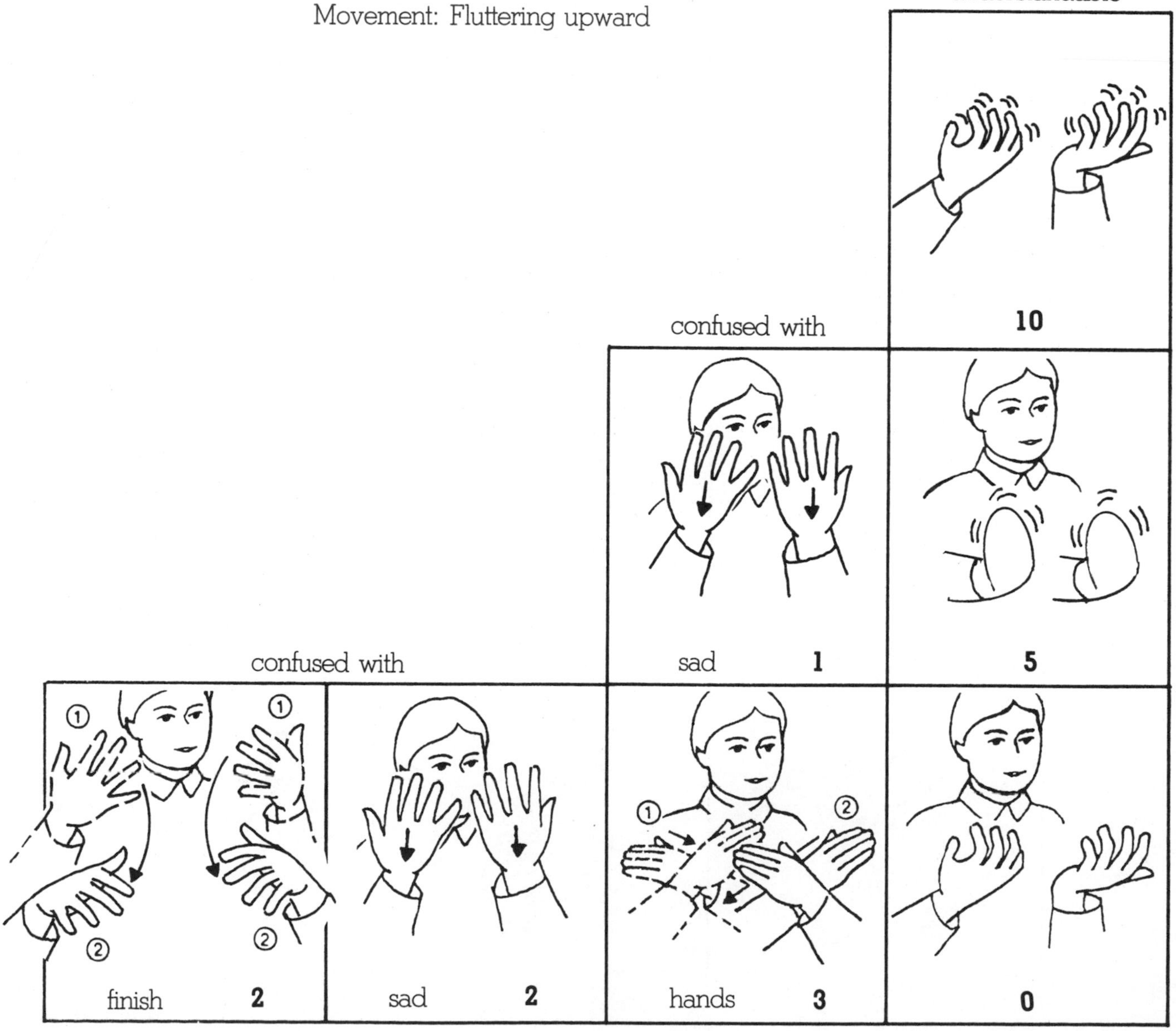

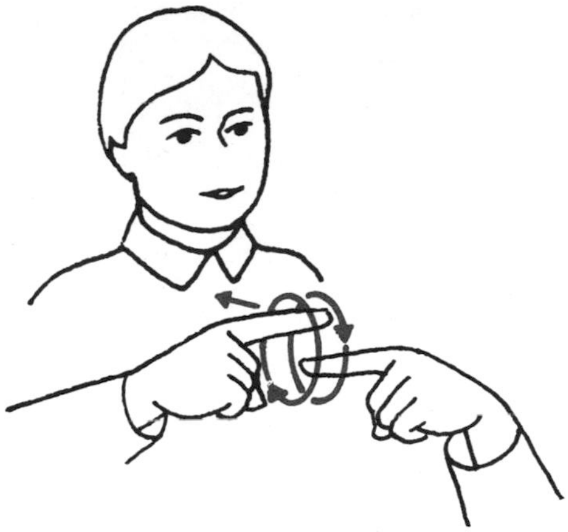

COME

Handshape: 1, both
Location: In front of body
Movement: Circle toward self

understandable

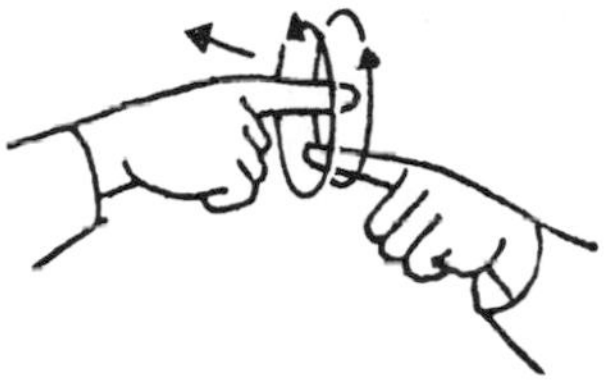

9

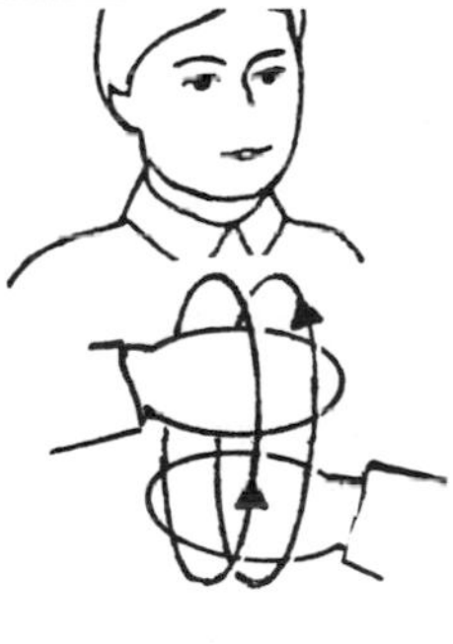

5

confused with

past 2

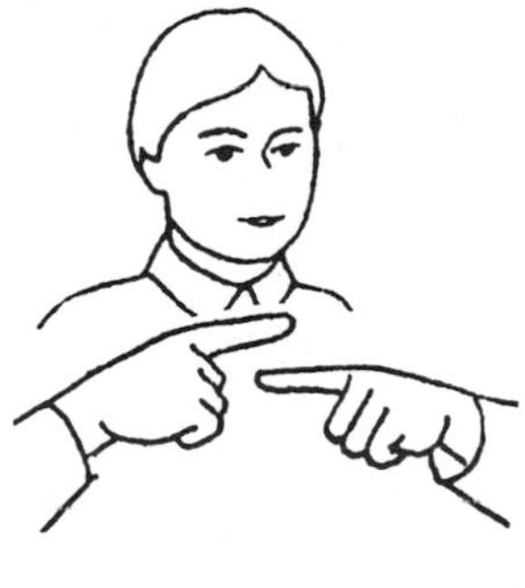

1

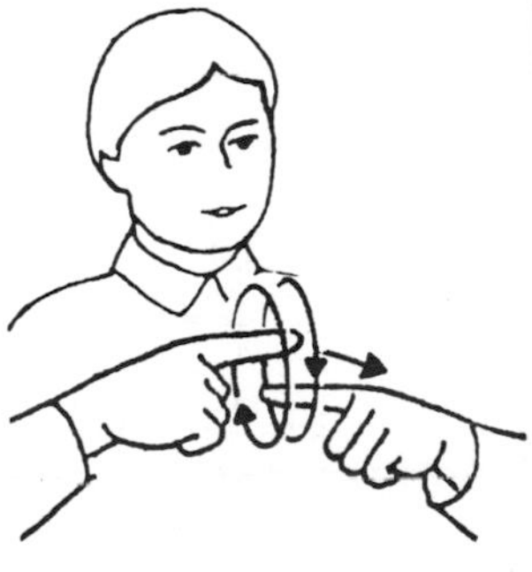

go 2

confused with

sign 2

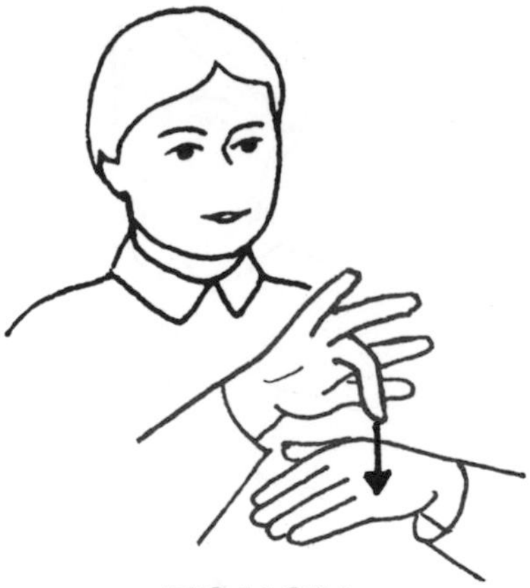

TOUCH

Handshape: 5, middle finger bent
Location: In front of body
Movement: Touch back of hand

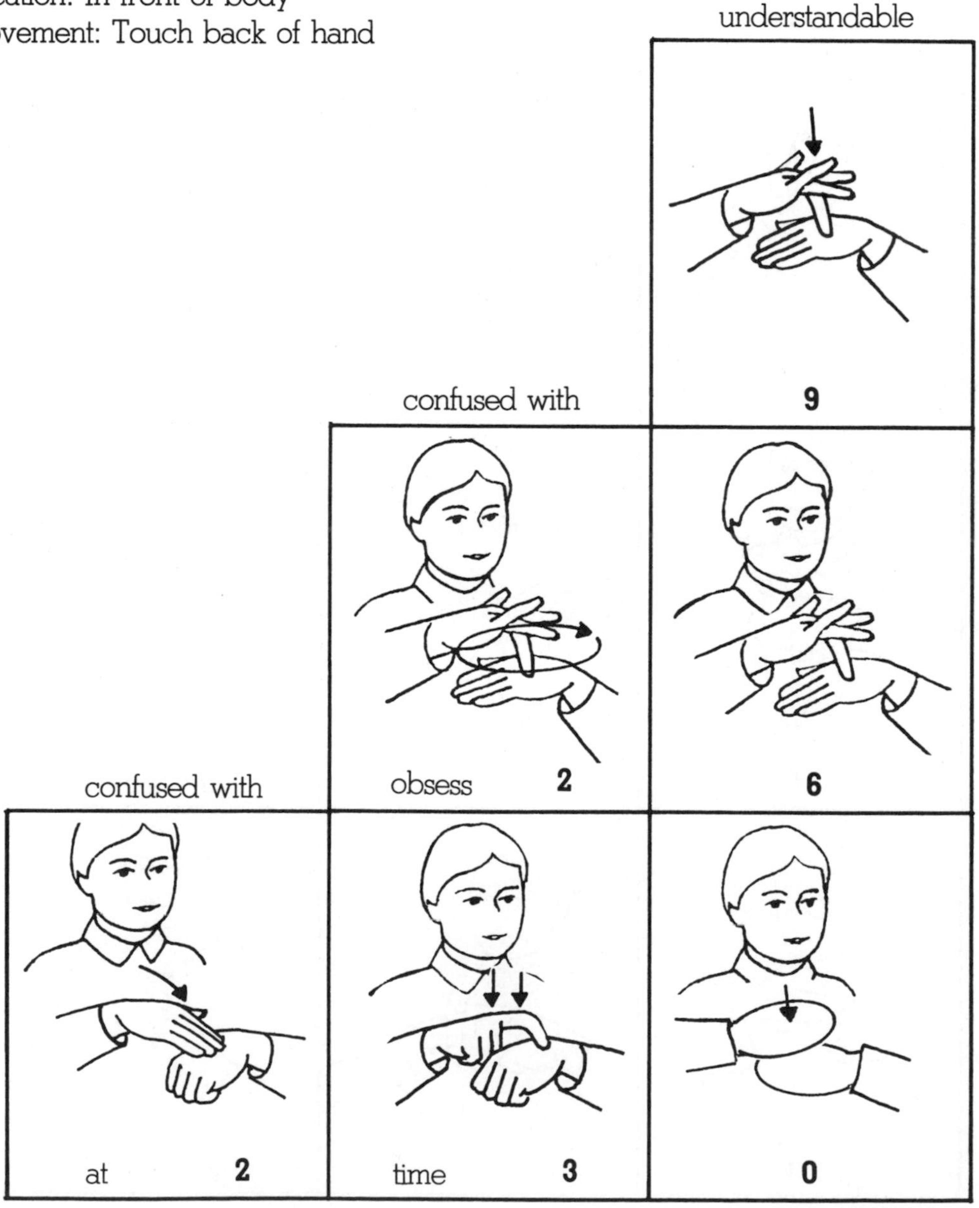

US

Handshape: U
Location: Chest
Movement: From one side to the other

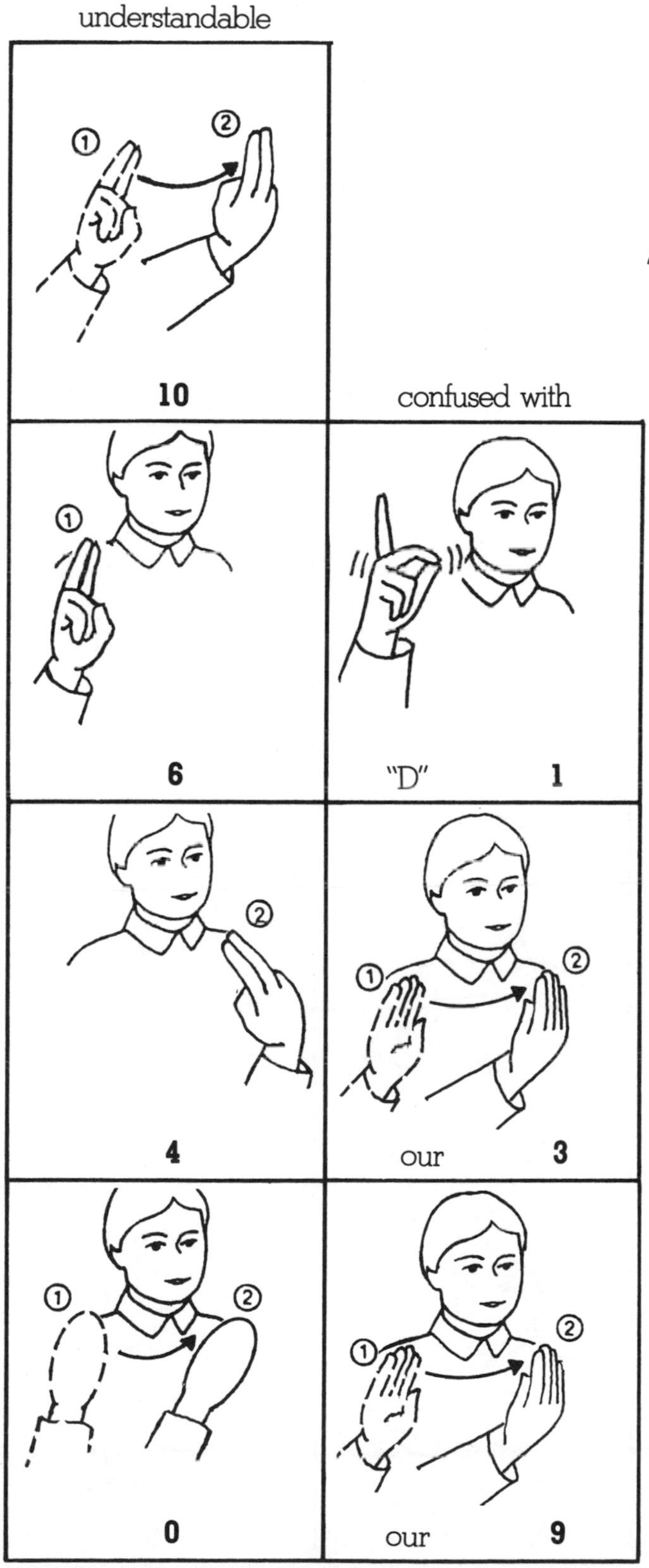

THAT

Handshape: Y
Location: In front of body
Movement: Tap palm

understandable

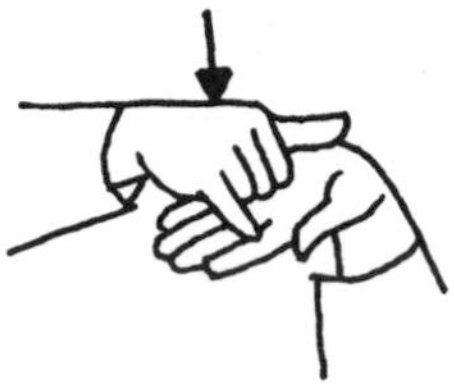

9

confused with

New York 2

6

school 2

0

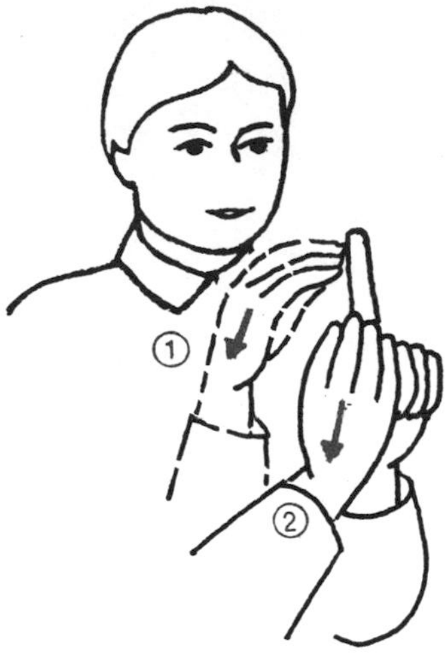

BANANA

Handshape: Flat O; 1
Location: In front of body
Movement: Peel a banana

understandable

9

confused with

6

mitten **1**

3

popular **1**

confused with

1

minute **2**

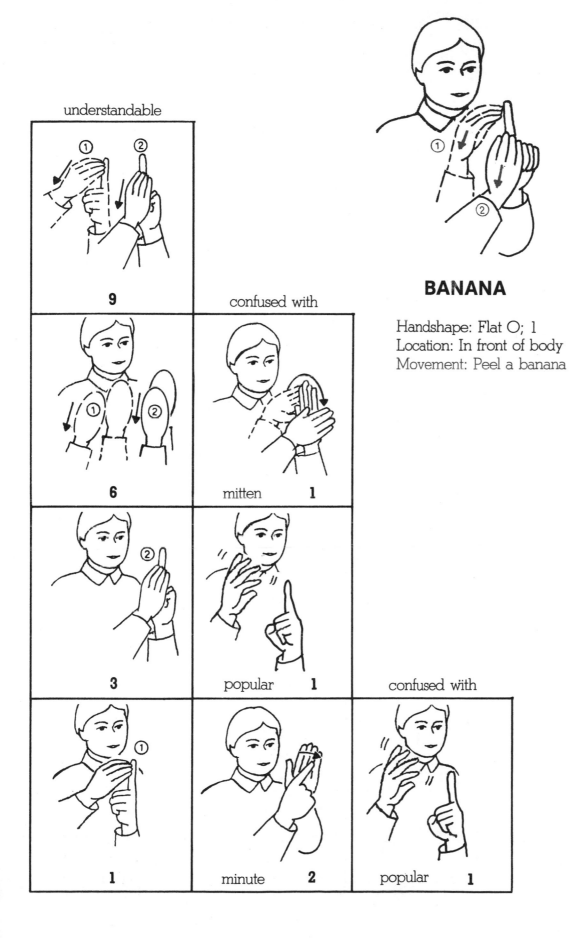

popular **1**

SIT

Handshape: H, both
Location: In front of body
Movement: Tap once

understandable

10

confused with

push down **3**

4

confused with

train **2**

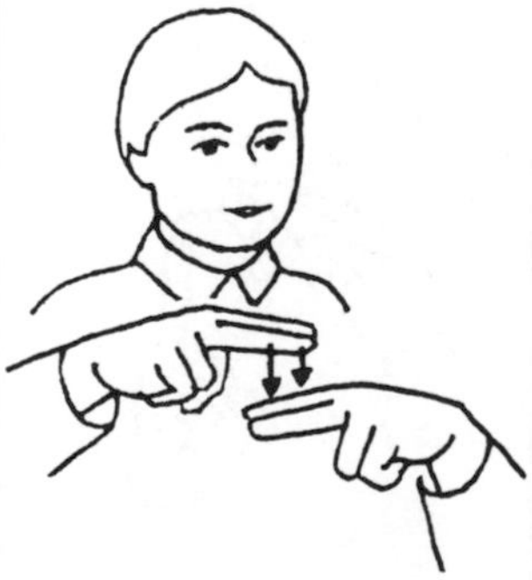

chair **2**

bedtime **4**

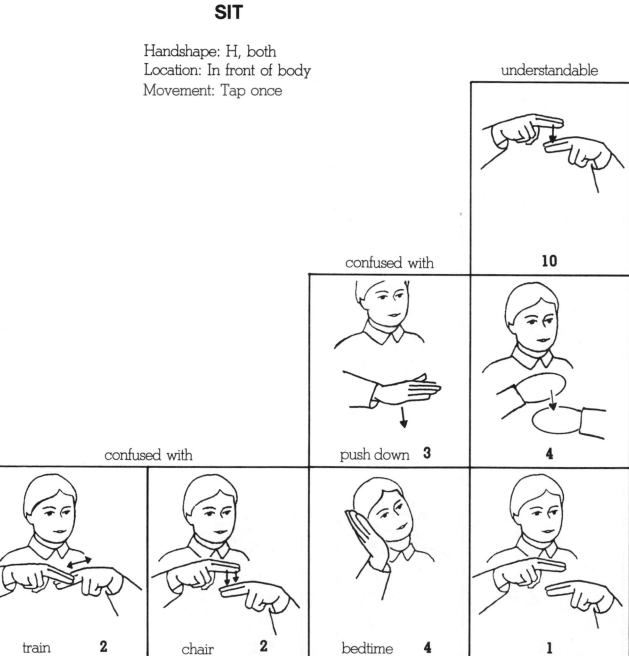

1

LUNCH

Handshape: L
Location: Chin
Movement: Tap twice with thumb

understandable

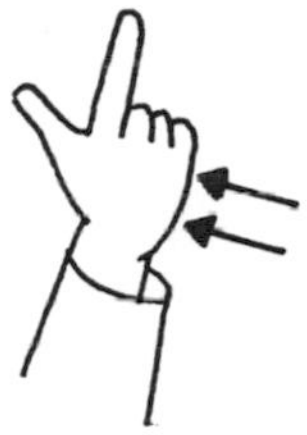

8

confused with

5

who 1

confused with

1

mother 4

breakfast 3

PURPLE

Handshape: P
Location: In front of body
Movement: Shake back and forth

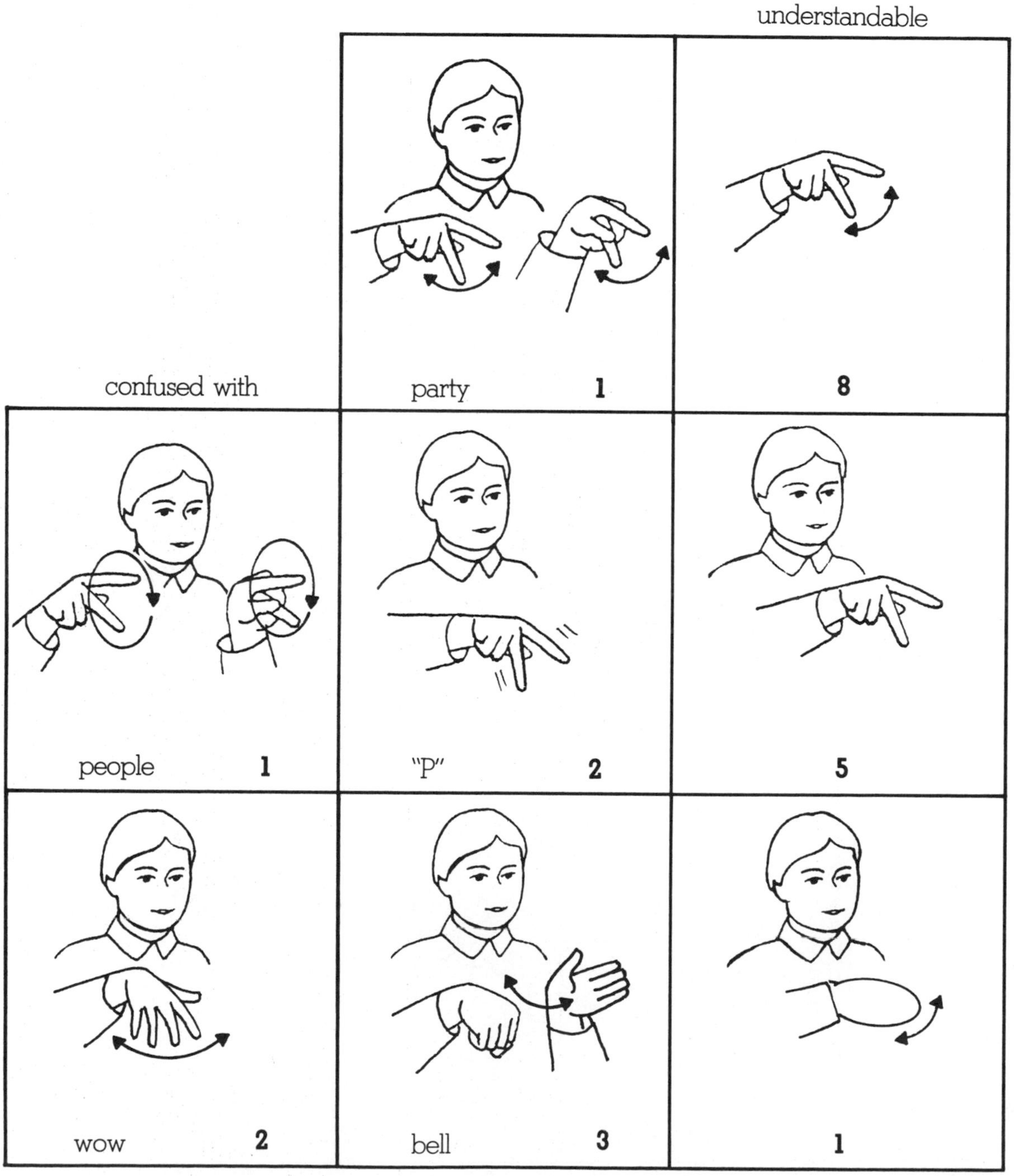

CAT

Handshape: F
Location: Side of mouth
Movement: Pull away twice

understandable

9

confused with

3

laugh **2**

makeup **1**

2

French **1**

family **1**

ELEPHANT

Handshape: Open B
Location: Face
Movement: Trace down trunk

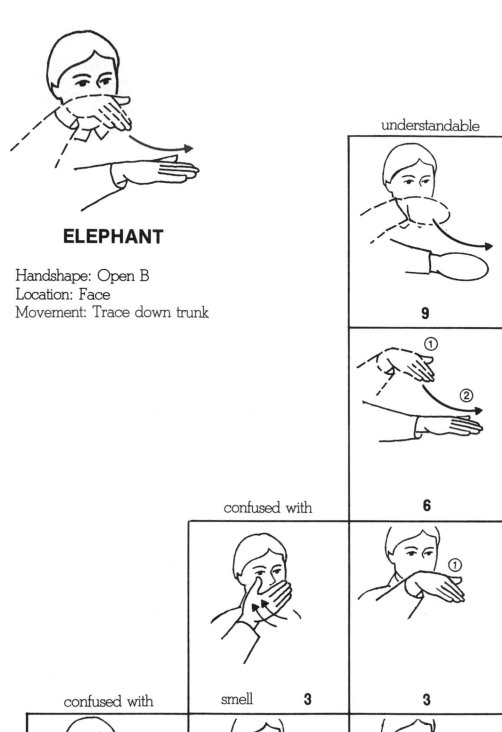

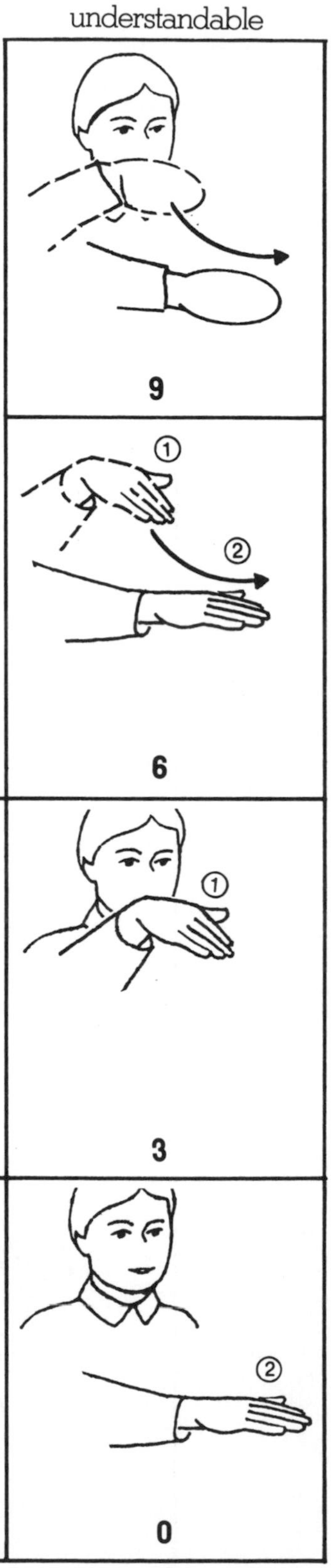

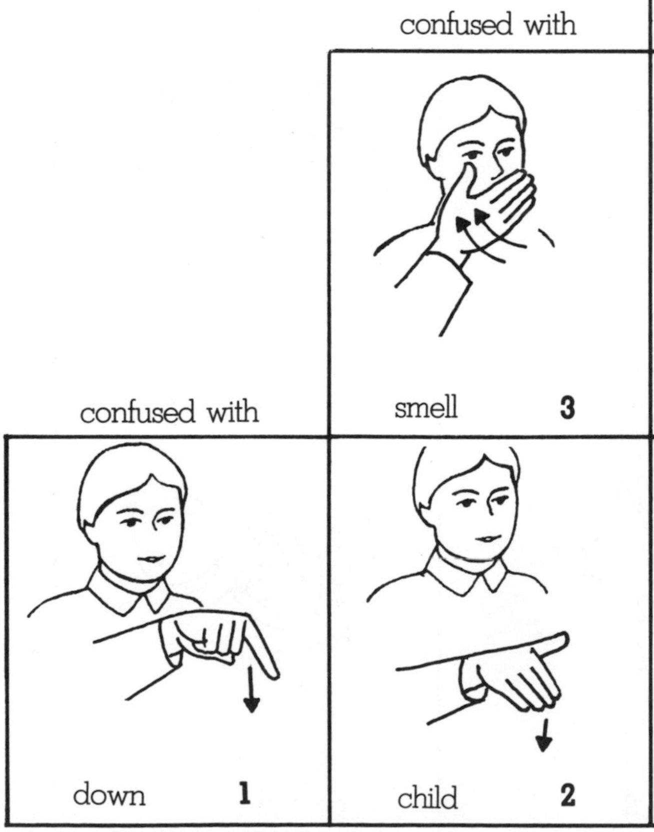

FATHER

Handshape: 5
Location: Forehead
Movement: Tap twice

understandable

10

confused with

4

cabbage **2**

dumb **1**

confused with

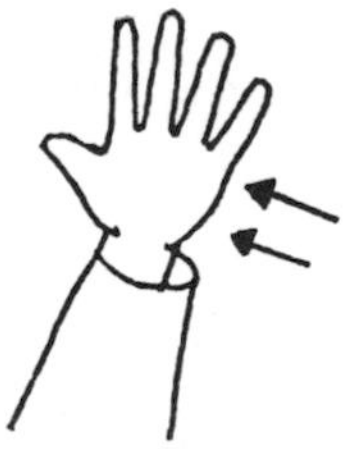

0

fine **5**

mother **2**

polite **1**

SWIM

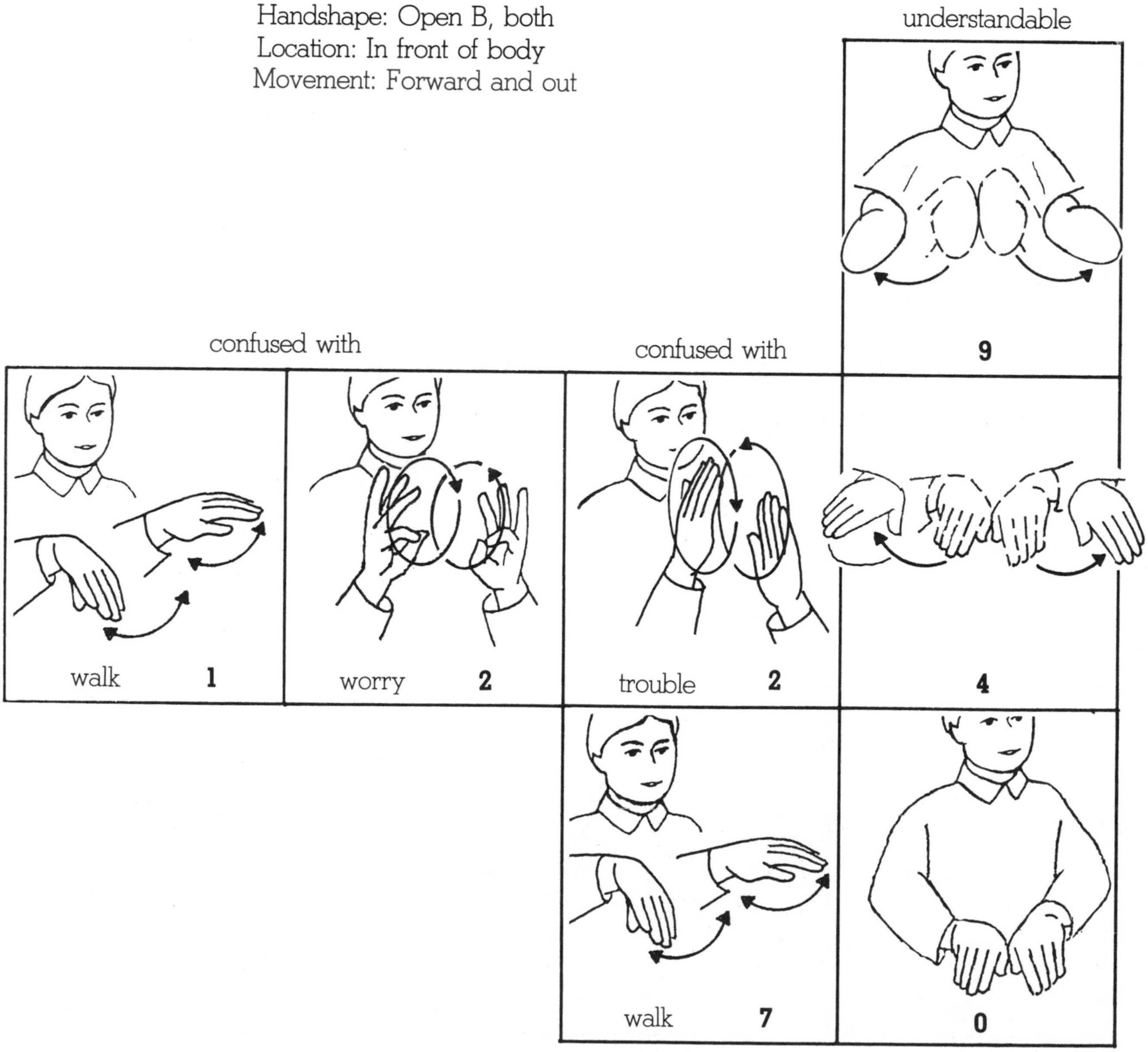

Handshape: Open B, both
Location: In front of body
Movement: Forward and out

confused with

walk 1

worry 2

confused with

trouble 2

walk 7

understandable

9

4

0

SKIRT

Handshape: 5, both
Location: Waist
Movement: Brush down and out

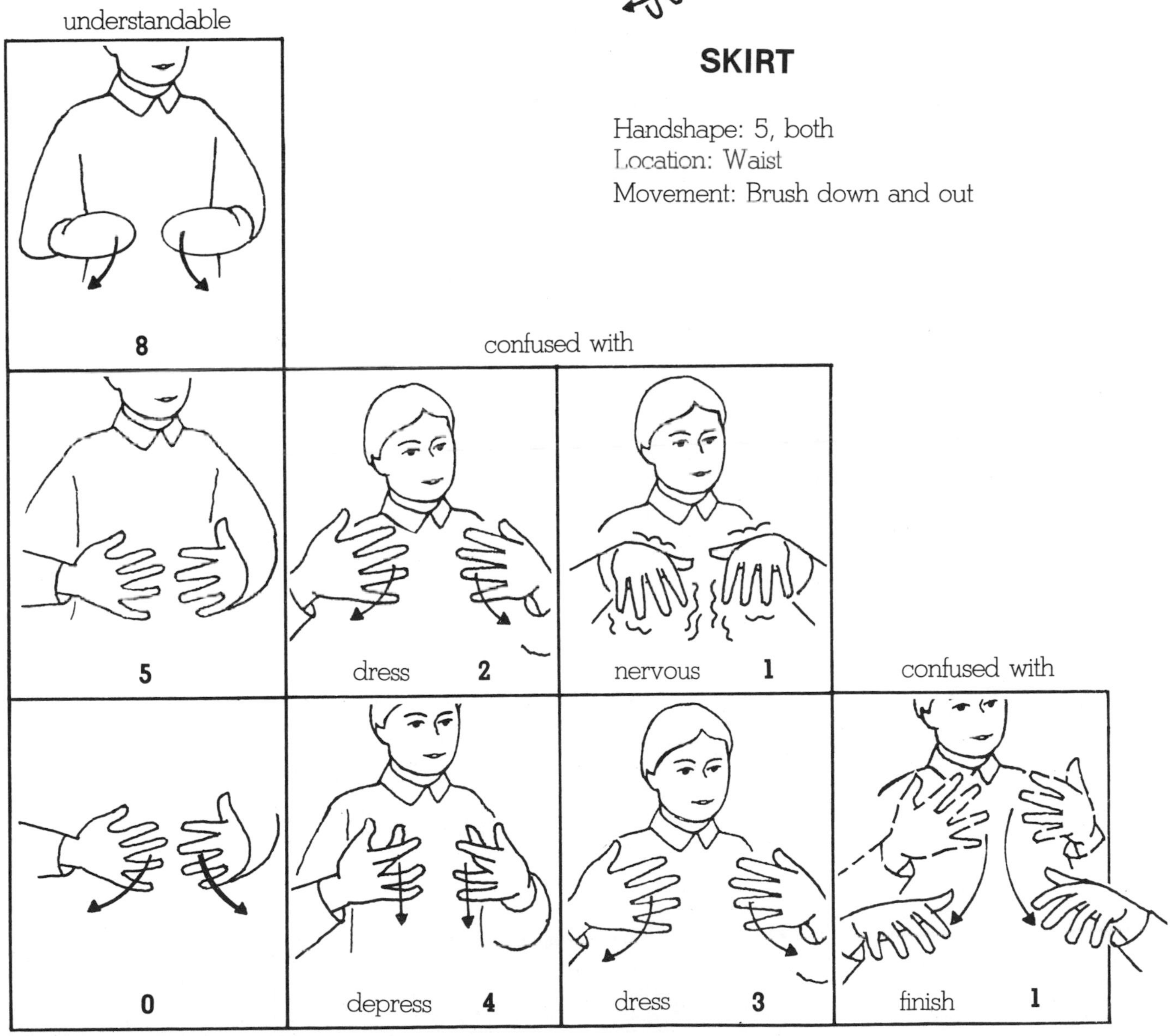

MOTHER

Handshape: 5
Location: Chin
Movement: Tap twice

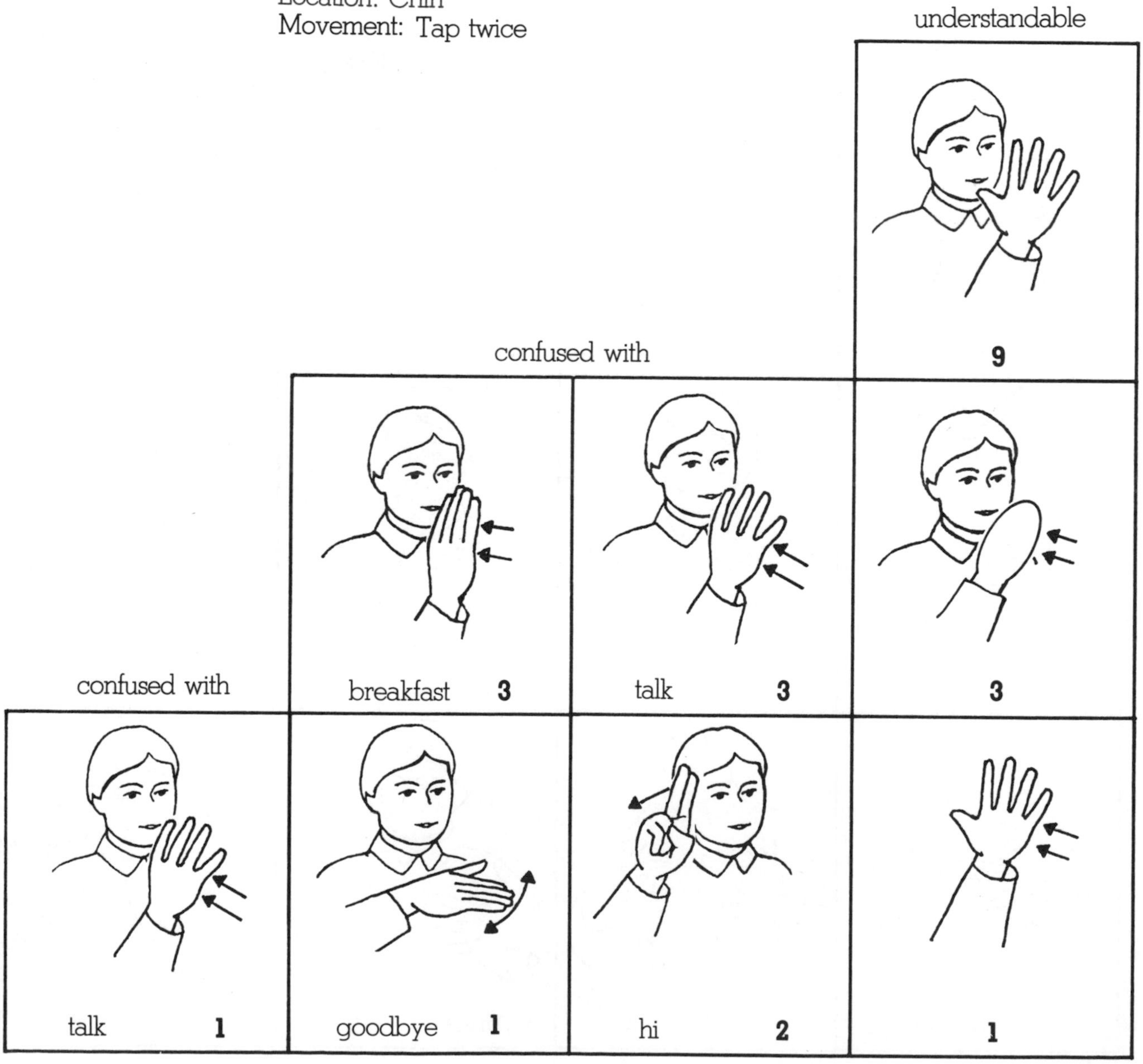

DINNER

Handshape: D
Location: Chin
Movement: Tap Twice

understandable

8

4

0

confused with

dentist **1**

confused with

smell **2**

flower **1**

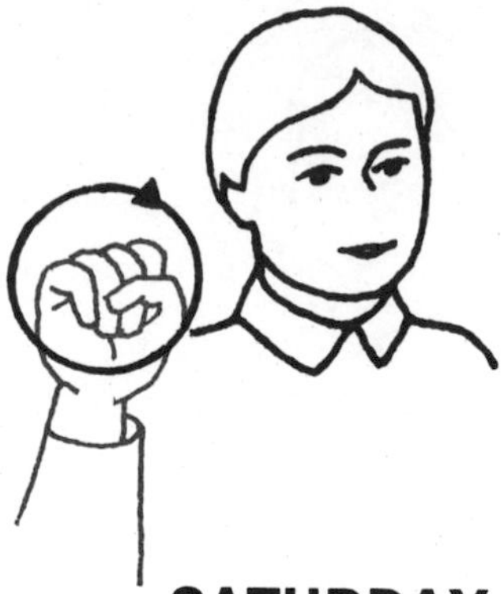

SATURDAY

Handshape: S
Location: In front of body
Movement: Small circle

understandable

10

confused with

"S" **2**

yes **3**

2

Sunday **1**

always **1**

0

NOSE

Handshape: 1
Location: Nose
Movement: Touch

understandable

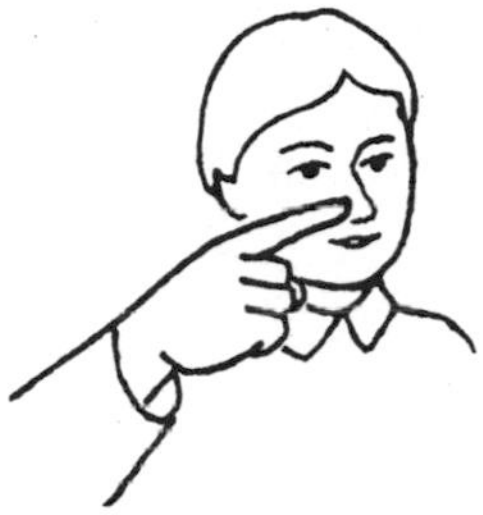

8

confused with

bored 1

3

smell 2

0

me 4

confused with

say 2

HAIR

Handshape: F
Location: Head
Movement: Pull out slightly

understandable

9

2

confused with

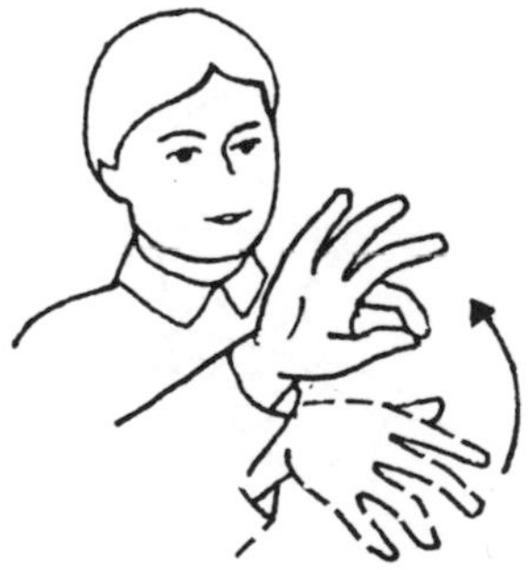

find 1

cat 3

0

SINGLE ROBUST

DOUBLE FRAGILE

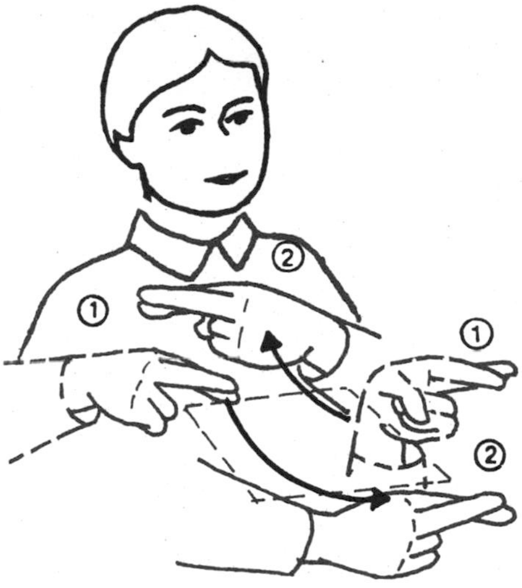

ROOM

Handshape: R, both
Location: In front of body
Movement: Outline sides of room

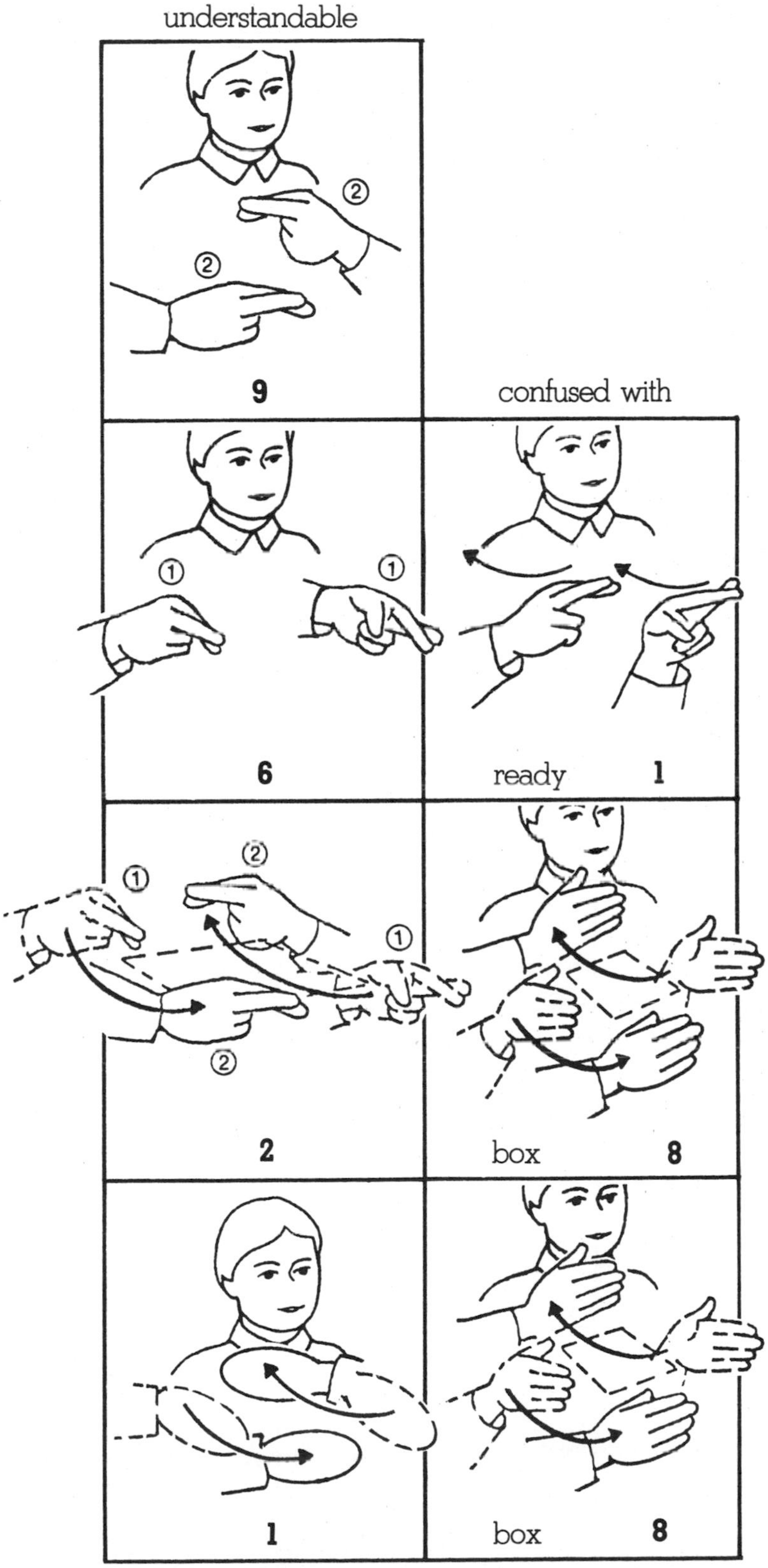

WHAT

Handshape: 1; open B
Location: In front of body
Movement: Down across fingers

understandable

9

confused with

half 1

cut 1

confused with

some 5

2

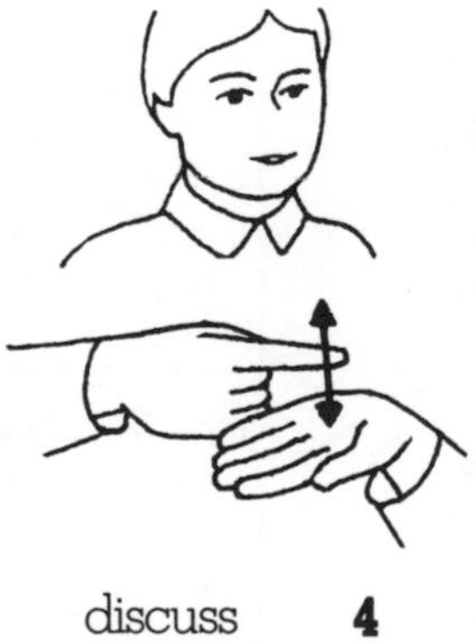

discuss 4

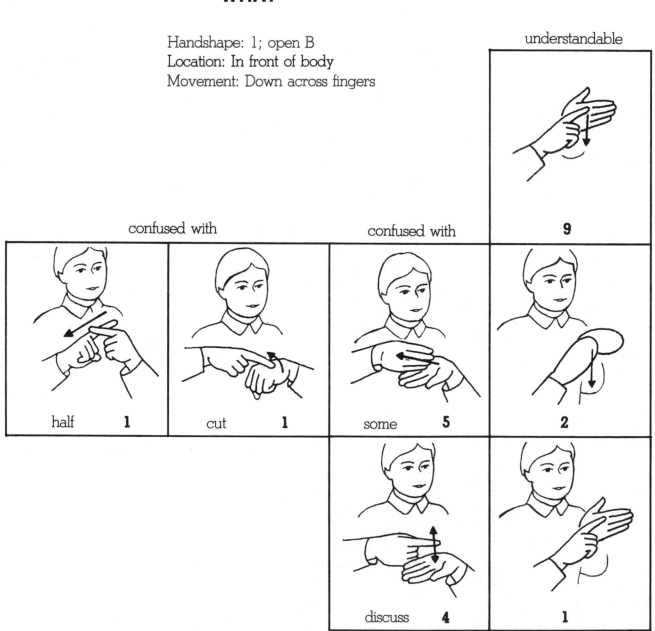

1

YES

Handshape: S
Location: In front of body
Movement: Shake up and down

understandable

10

confused with

2

"S" **4**

Saturday **1**

0

bye-bye **7**

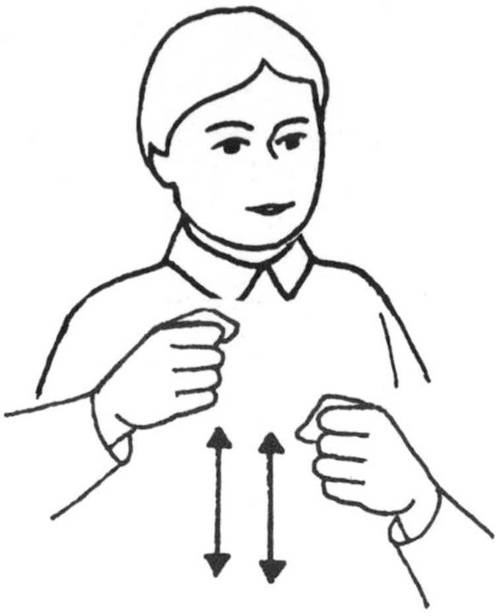

MILK

Handshape: S, both
Location: In front of body
Movement: Squeeze up and down

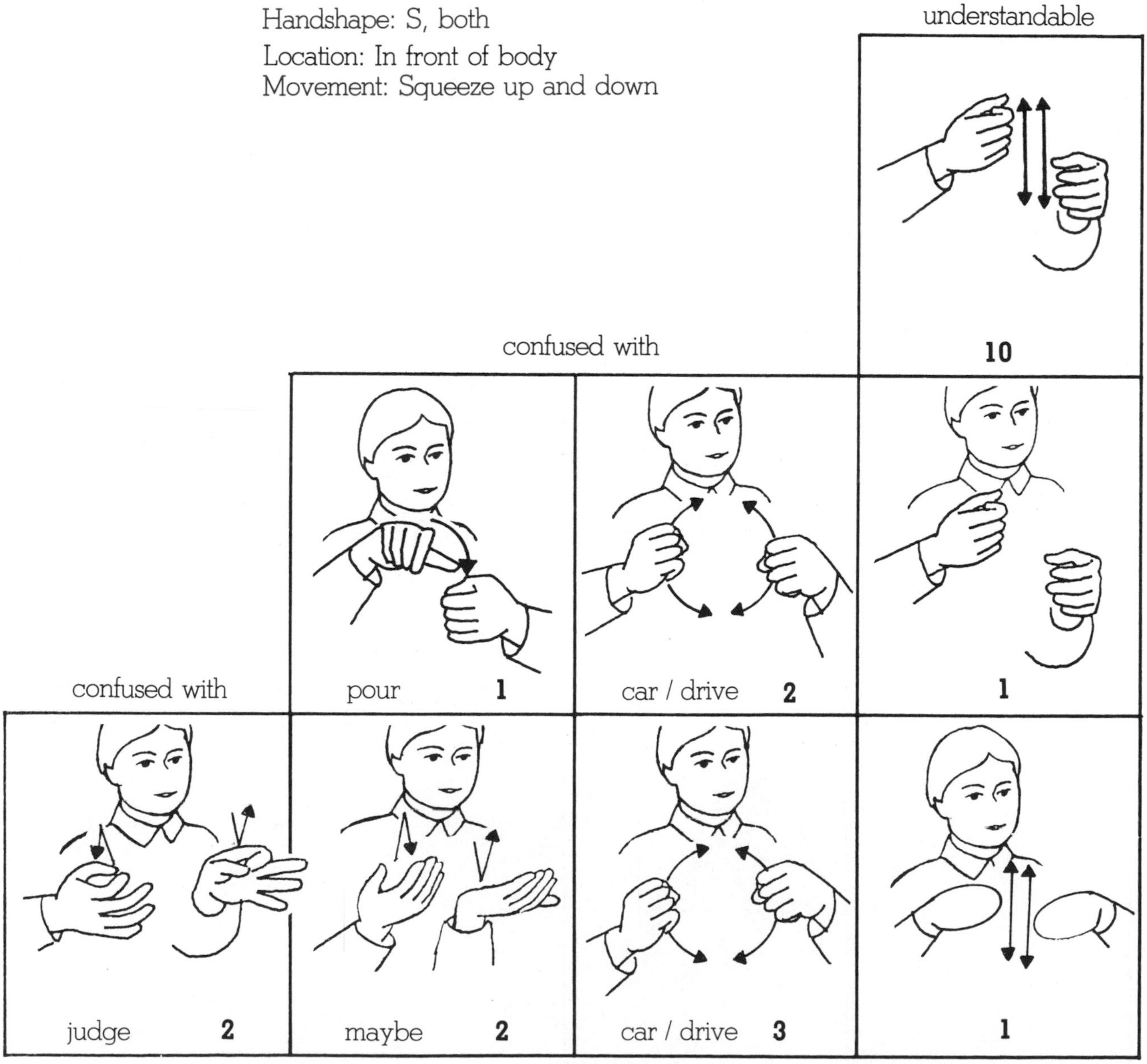

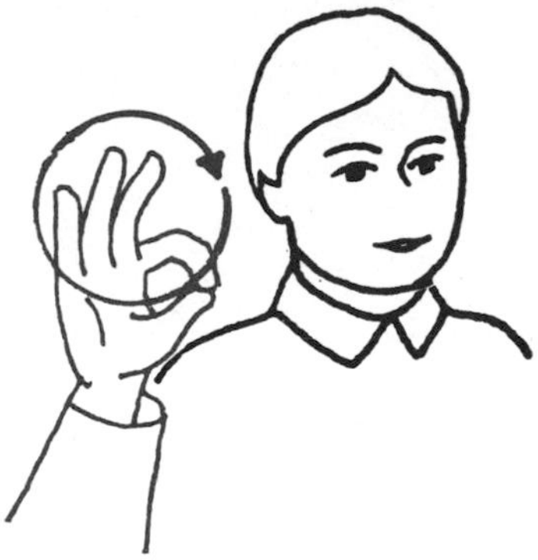

FRIDAY

Handshape: F
Location: In front of body
Movement: Small circle

understandable

9

1

confused with

Sunday **6**

1

"9" **2**

confused with

preach **2**

"F" **1**

FIGHT

Handshape: S, both
Location: In front of body
Movement: Flex at wrist twice

understandable

9

confused with

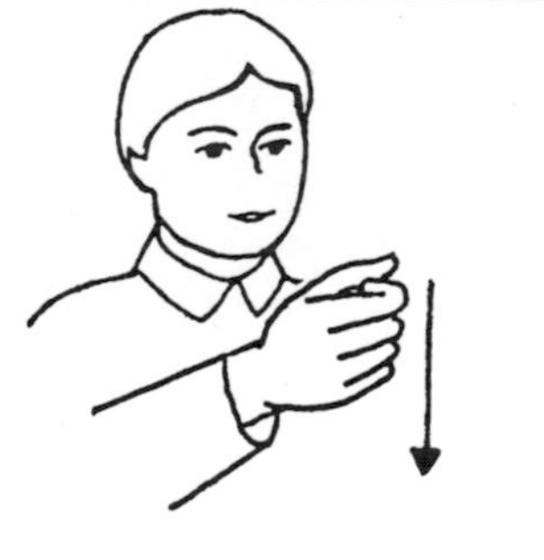

pull down **1**

advertise **4**

1

trouble **9**

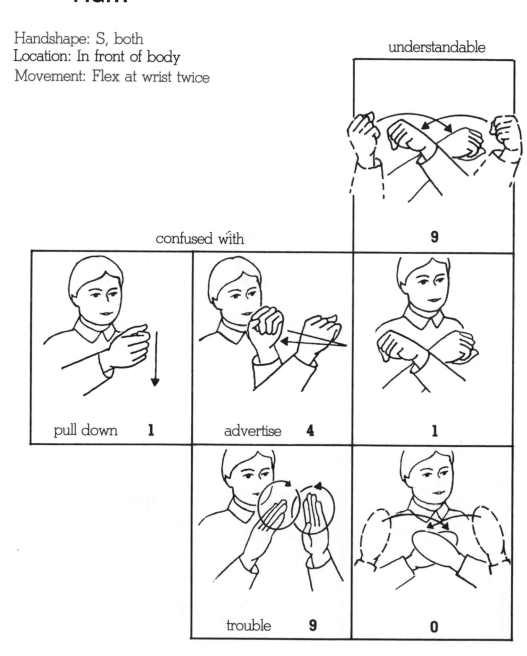

0

NOT

Handshape: A, thumb extended
Location: Chin
Movement: Out forcefully

understandable

8

2

confused with

tomorrow 4

self 1

confused with

ten 1

0

sure / true 4

be 2

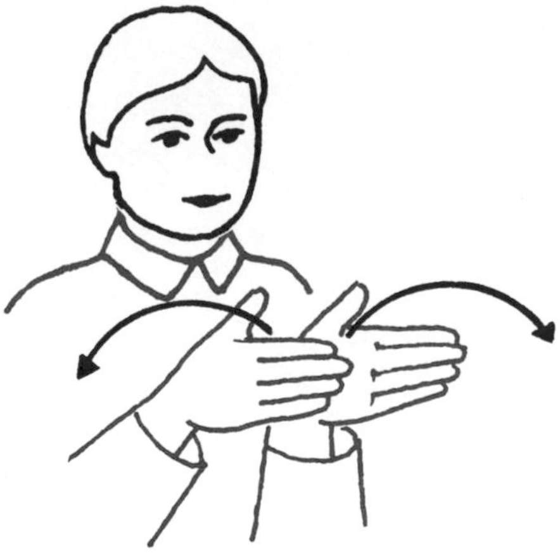

BIG

Handshape: Open B, both
Location: In front of body
Movement: Separate hands

understandable

9

confused with

book 2

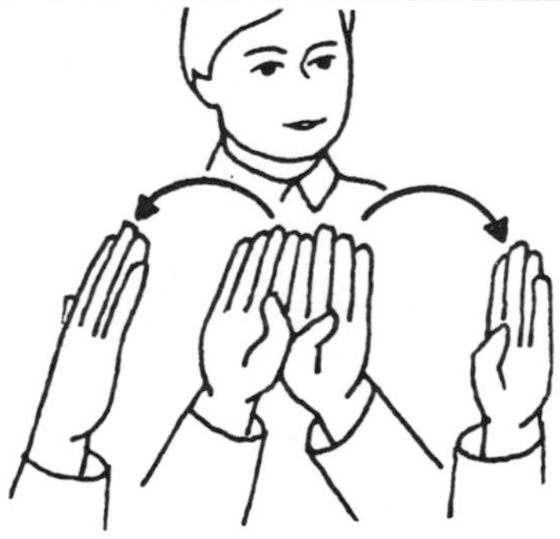

open 7

1

book 4

open 5

0

ICE CREAM

Handshape: S
Location: Mouth
Movement: Down twice

understandable

8

confused with

1

mouse **2**

confused with

0

cough **2**

shut up **1**

HEAD

Handshape: Open B
Location: Temple
Movement: Lower to chin

understandable

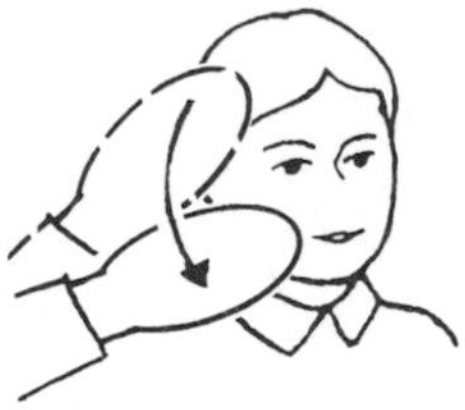

8

confused with

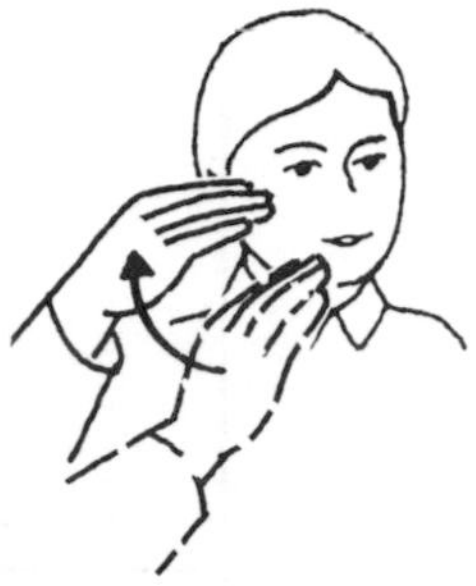

home **4**

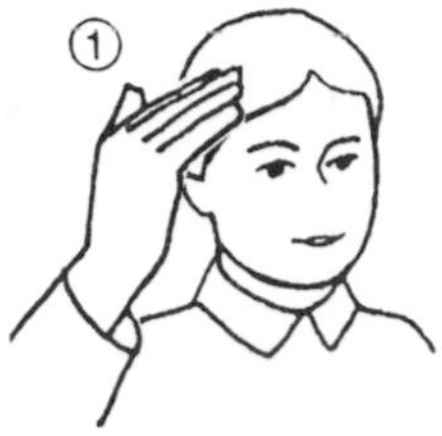

1

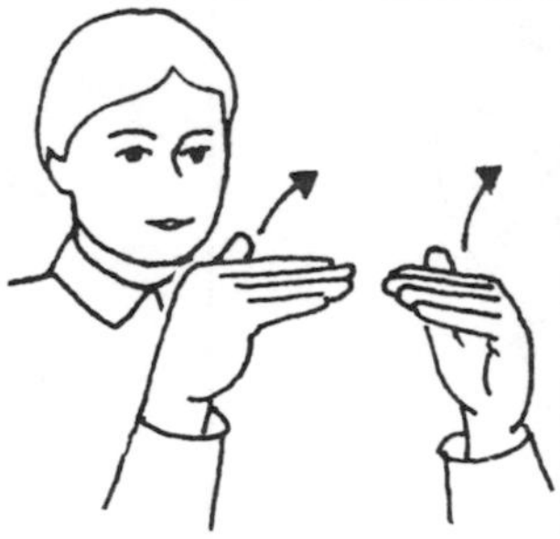

advance **1**

1

confused with

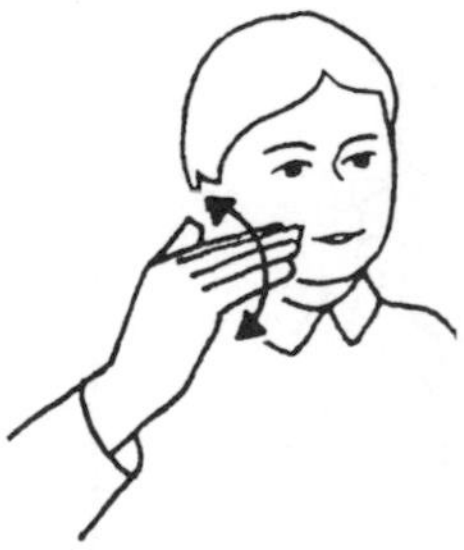

tobacco **1**

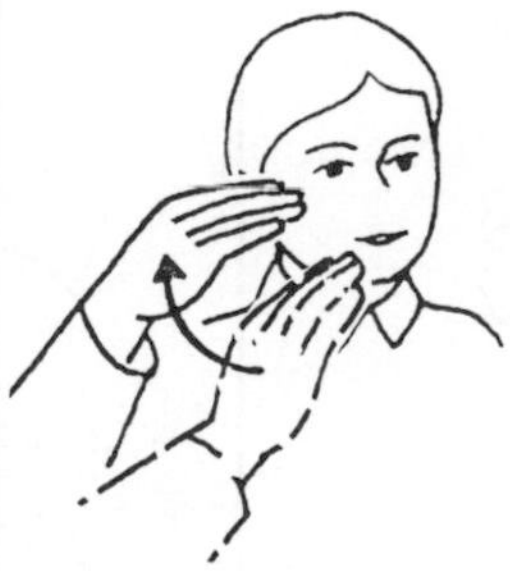

home **4**

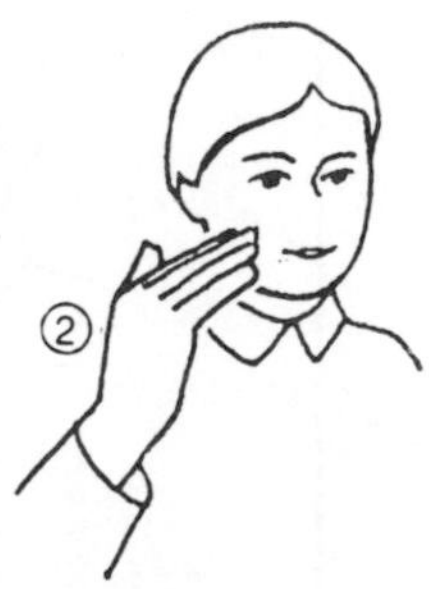

1

BREAK

Handshape: S, both
Location: In front of body
Movement: Snap apart

understandable

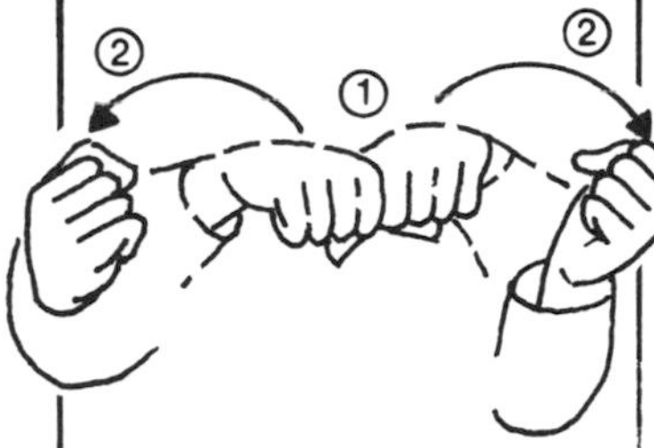

10

confused with

cold 1

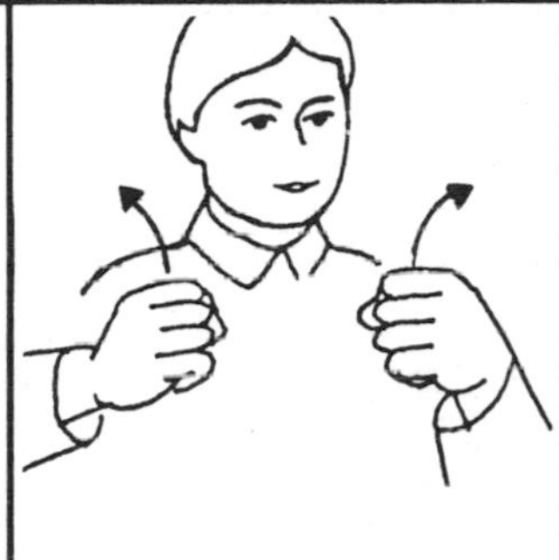

strong 2

confused with

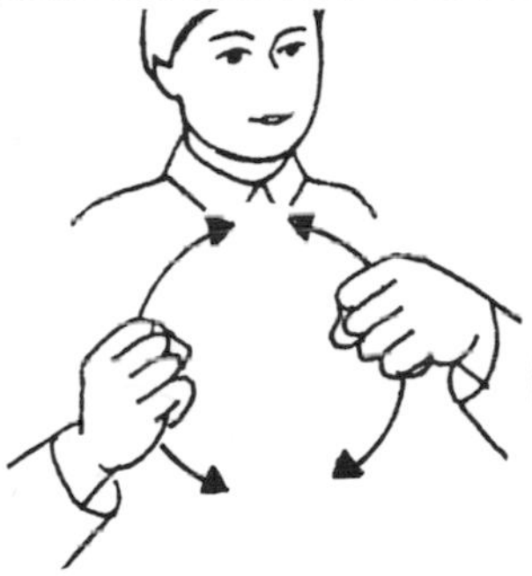

drive / car 3

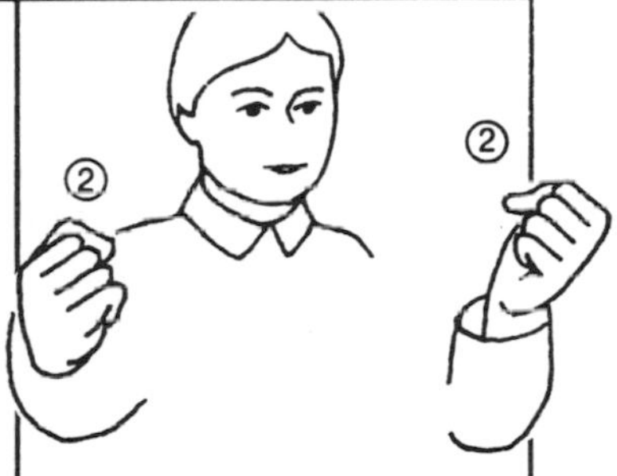

1

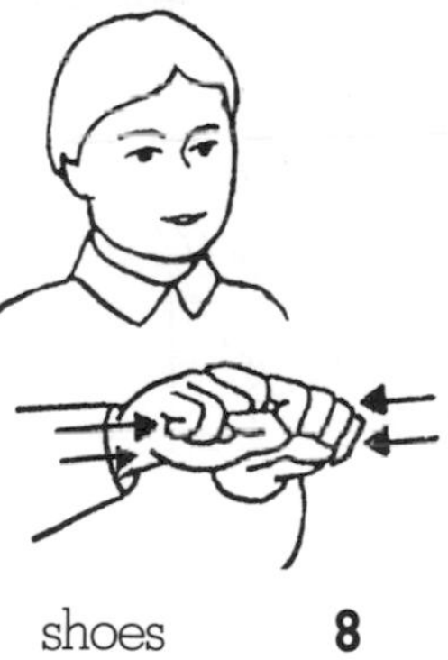

shoes 8

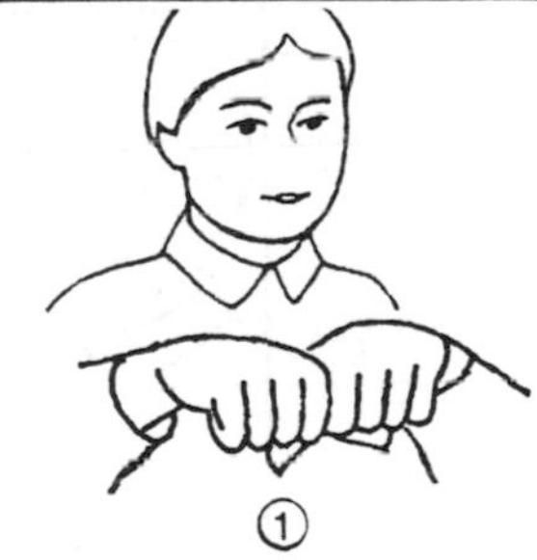

0

open 9

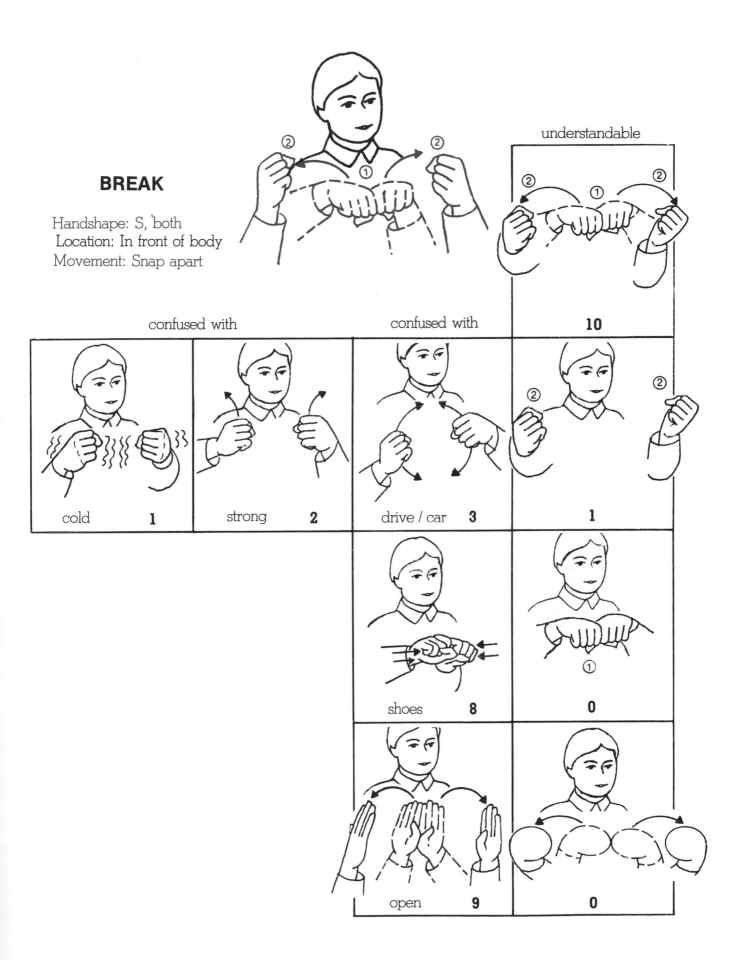

0

MAN

Handshape: A to five
Location: Forehead
Movement: Lower to chest

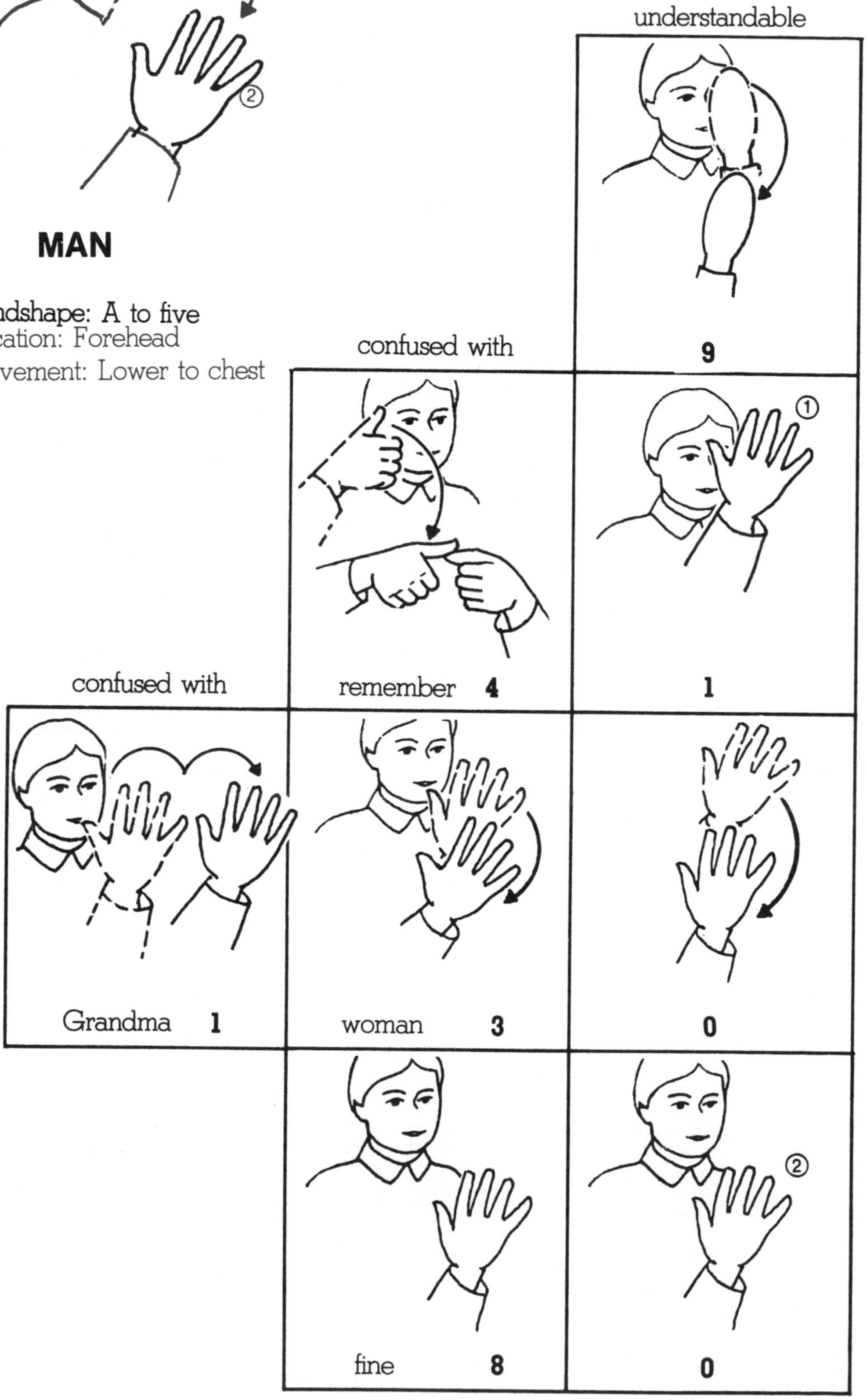

ZERO ROBUST

ZERO FRAGILE

THIRSTY

Handshape: 1
Location: Throat
Movement: Trace down throat

understandable

8

confused with

8

throat **1**

5

red **1**

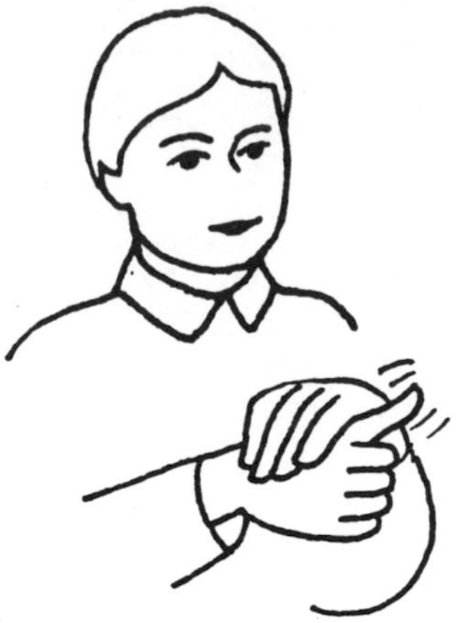

TURTLE

Handshape: A; open B
Location: In front of body
Movement: Wiggle thumb

understandable

10

confused with

hide **1**

8

fish **4**

3

WASH

Handshape: A, both
Location: In front of body
Movement: Rub in circle

understandable | confused with

7

chew 2

7

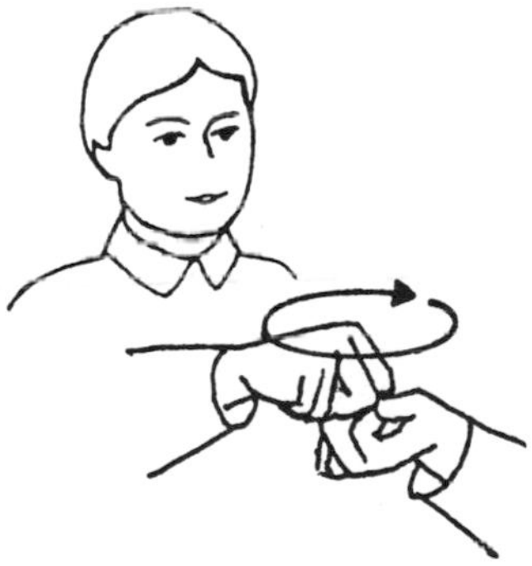

chew 1

7

DOWN

Handshape: 1
Location: In front of body
Movement: Point down

understandable

8

confused with

here 1

6

confused with

here 1

this 2

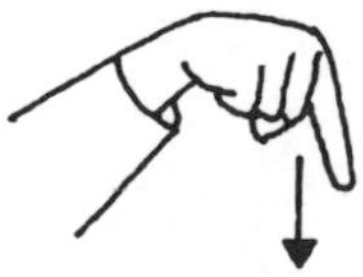

6

SOME

Handshape: Open B; both
Location: In front of body
Movement: Across palm toward self

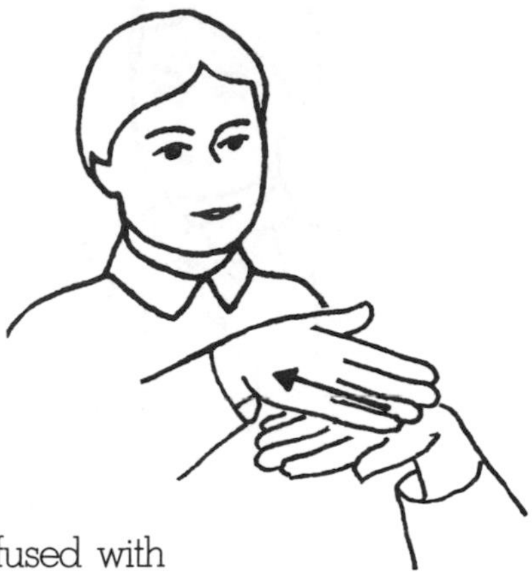

understandable	confused with
 **7**	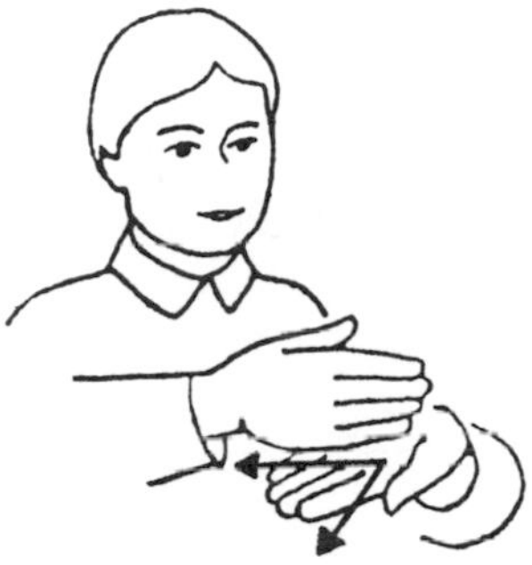 piece **3**
 **6**	 stop **2**
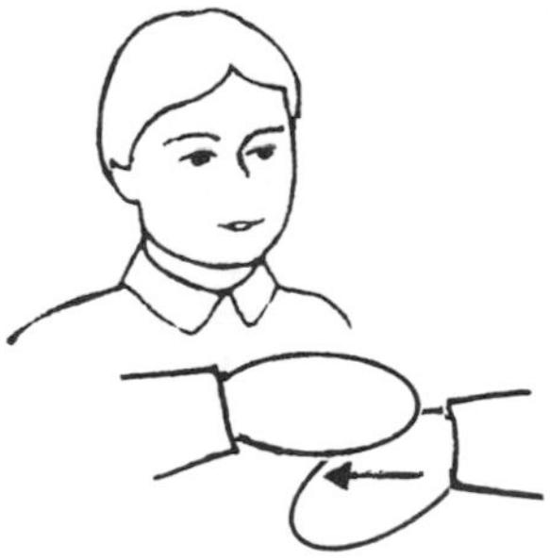 **6**	 piece **3**

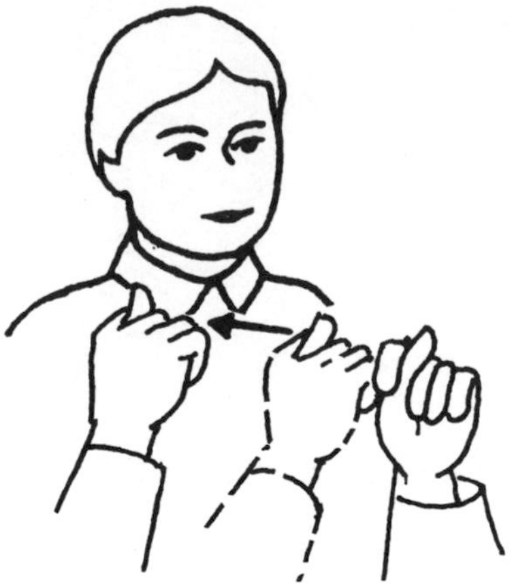

TRUCK

Handshape: T; both
Location: In front of body
Movement: Back hand toward self

understandable

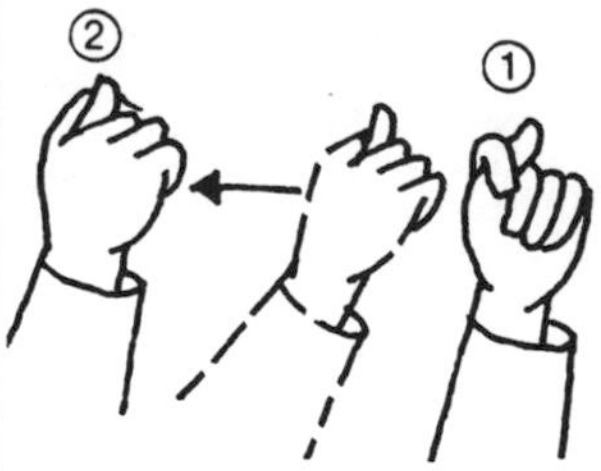

8

7

7

confused with

bus 3

3

UP

Handshape: 1
Location: In front of body
Movement: Point Up

understandable

7

confused with

6

one **2**

5

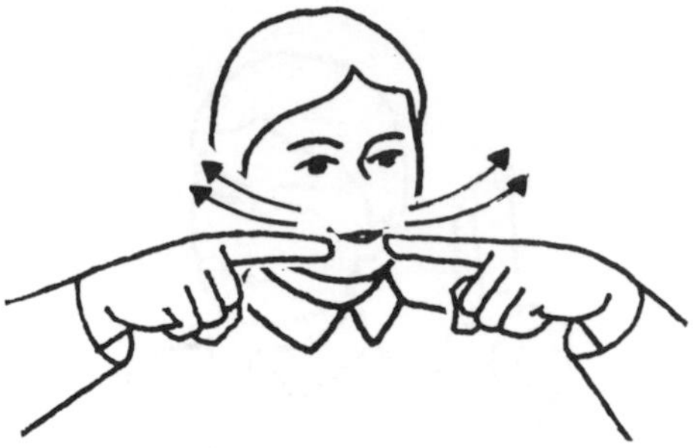

LAUGH

Handshape: L, both
Location: Sides of mouth
Movement: Up and out, twice

understandable

7

confused with

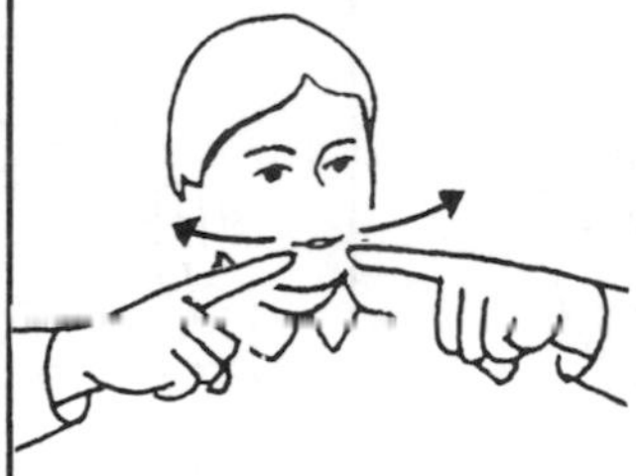

smile **3**

7

fight **5**

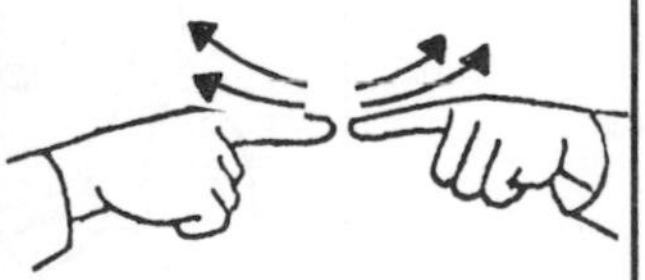

4

BUS

Handshape: B, both
Location: In front of body
Movement: Back hand toward self

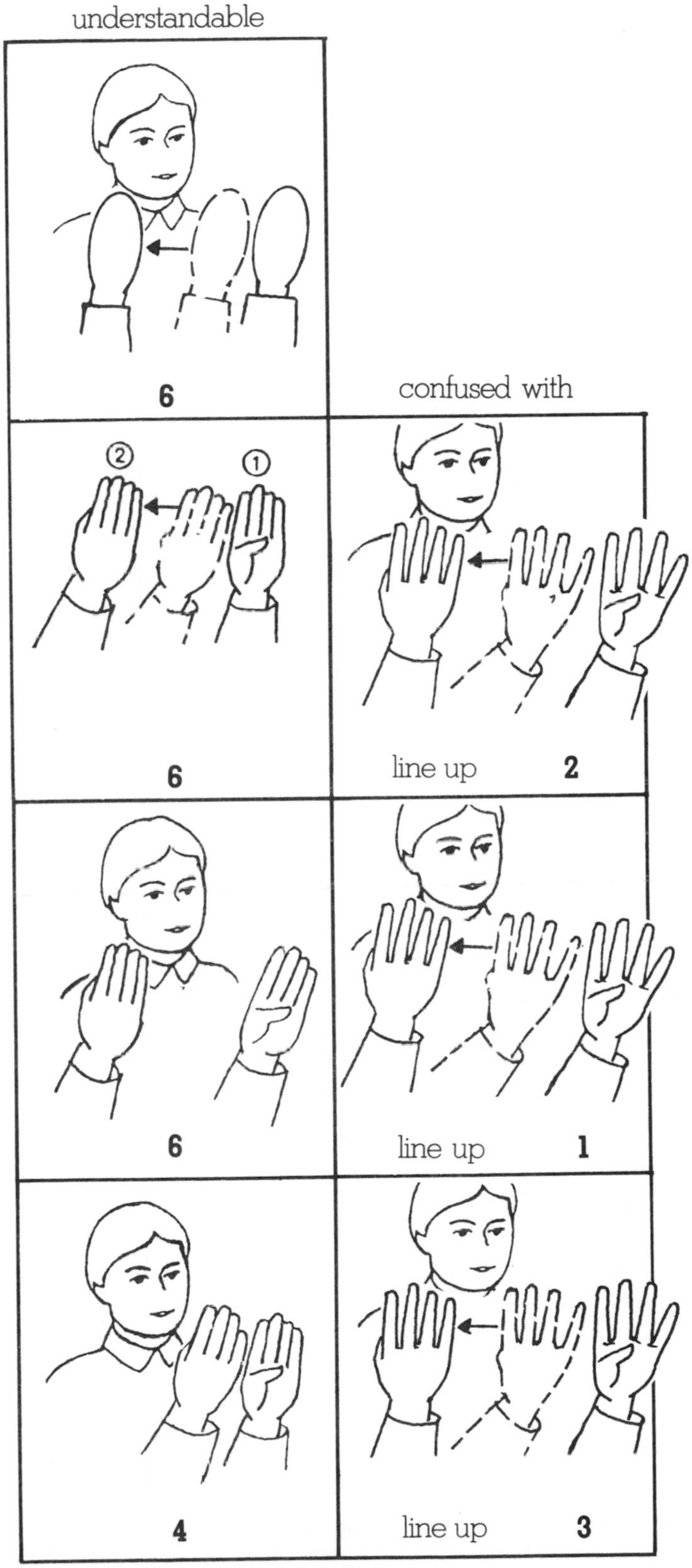

HORSE

Handshape: H, thumb extended
Location: Temple
Movement: Fingers forward twice

confused with — understandable

rabbit **2**

6

rabbit **3**

6

confused with

donkey **3**

rabbit **3**

4

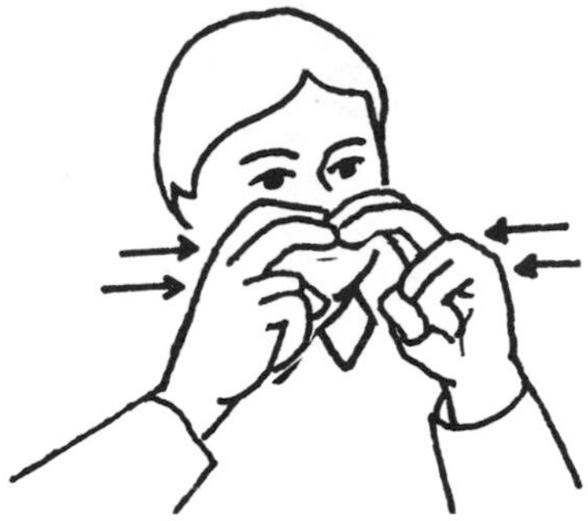

SQUIRREL

Handshape: Bent V, both
Location: In front of body
Movement: Tap together twice

understandable

8

5

3

confused with

meeting 5

PRETTY

Handshape: 5 to flat O
Location: Face
Movement: Circle and close hand

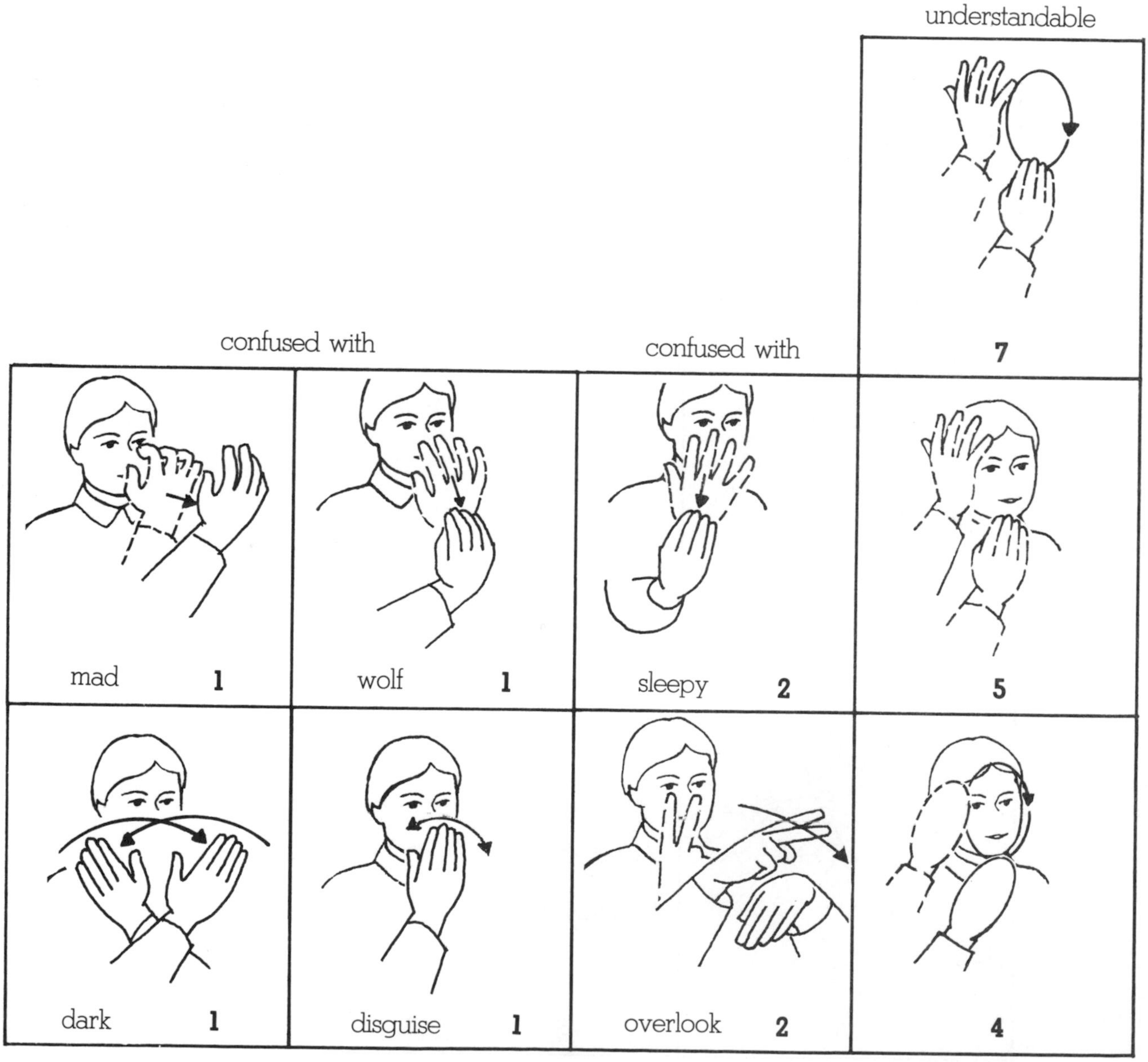

BUT

Handshape: 1, both
Location: In front of body
Movement: Uncross fingers

understandable | confused with

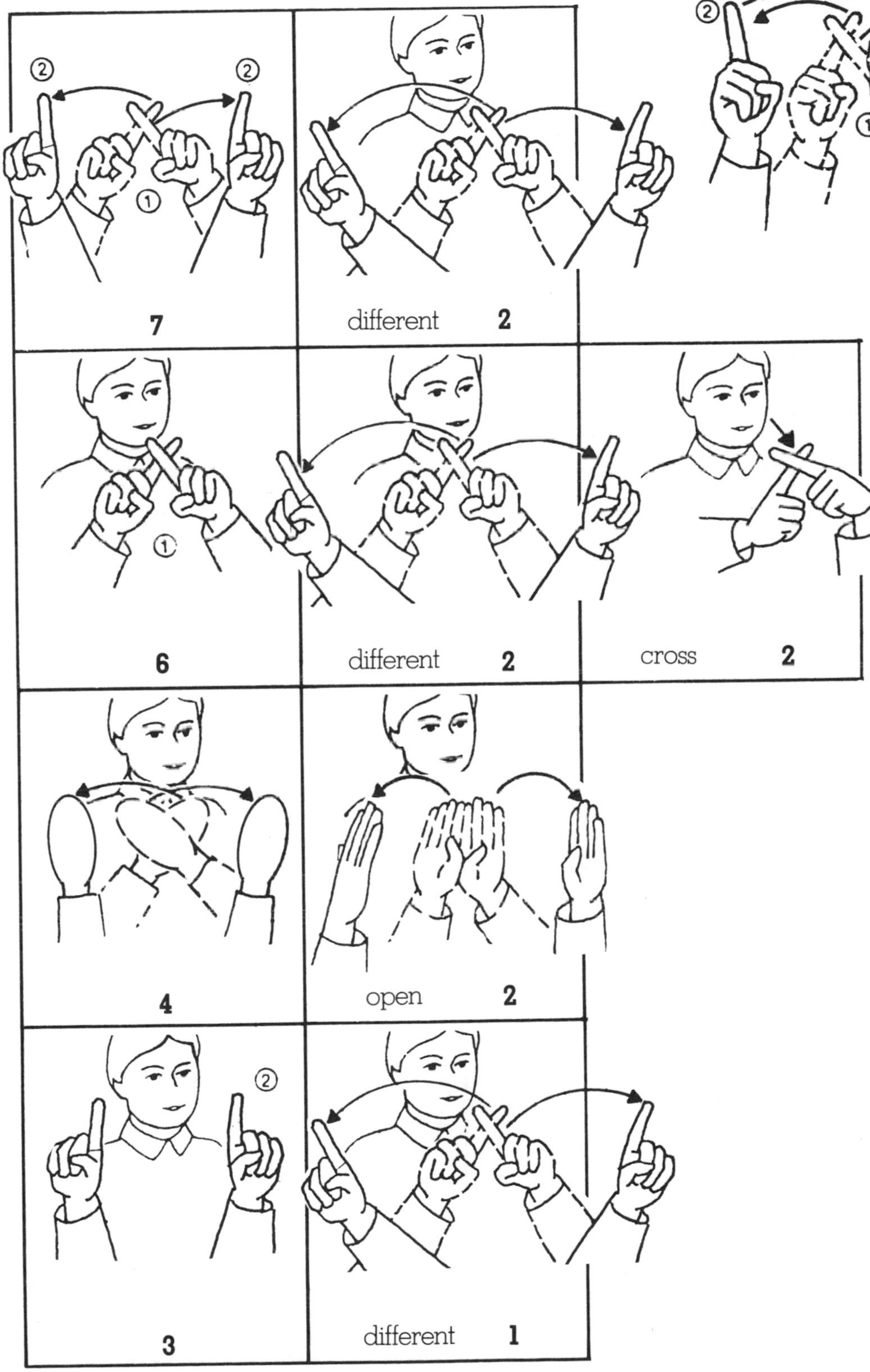

BICYCLE

Handshape: S, both
Location: In front of body
Movement: Pedal with hands

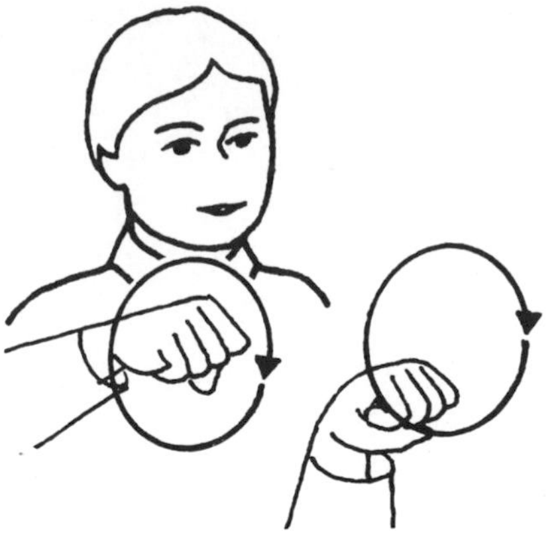

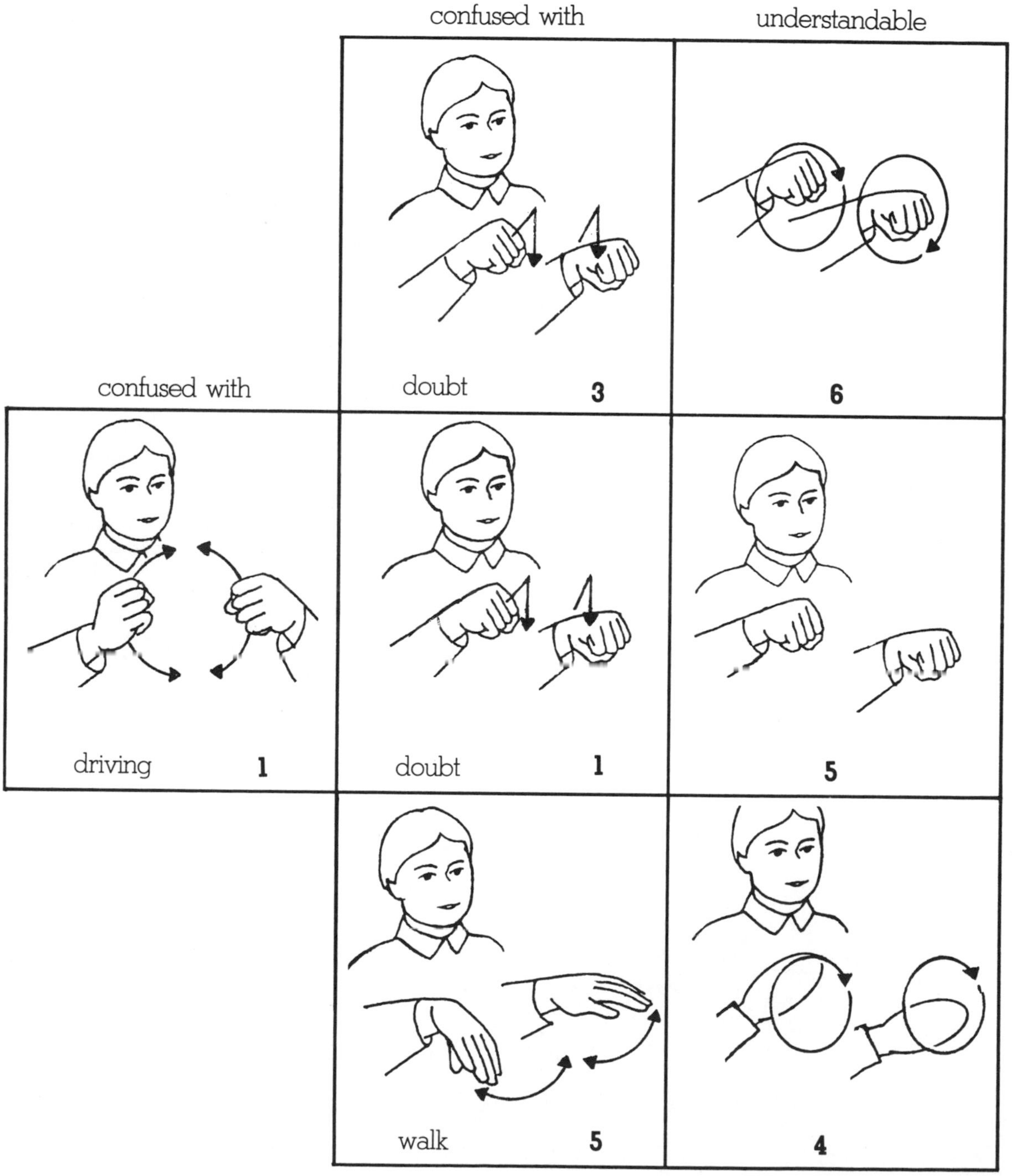

POLICE

Handshape: C
Location: Chest
Movement: Tap twice

understandable	confused with
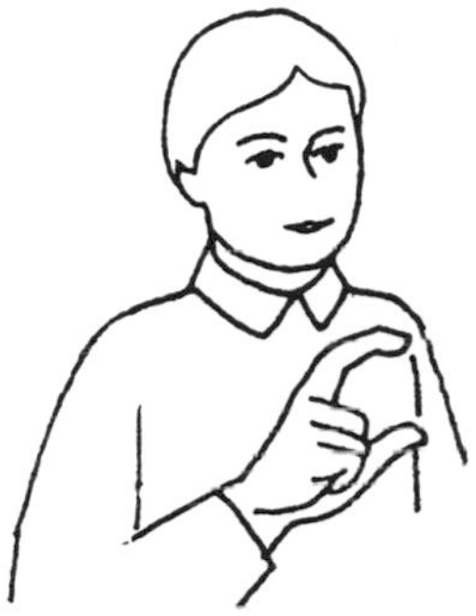 8	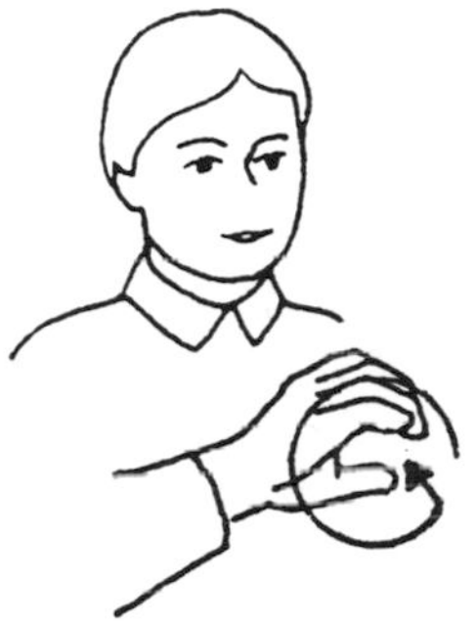 character 2
 4	
 2	

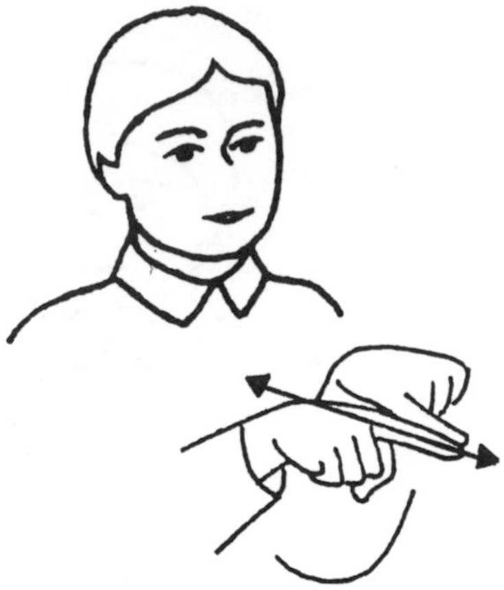

SOCK

Handshape: 1, both
Location: In front of body
Movement: Rub together

confused with	understandable
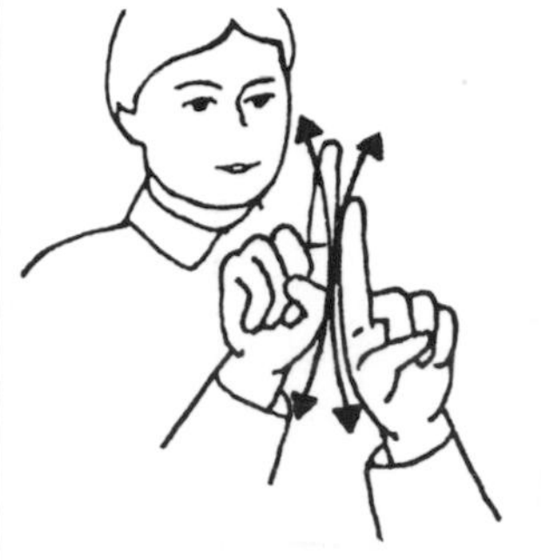 star 2	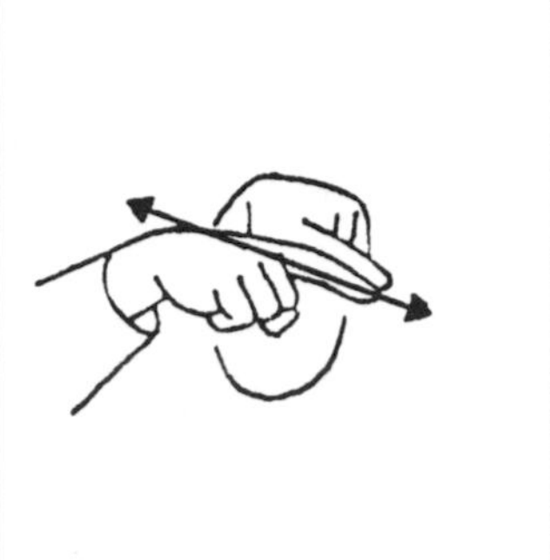 7
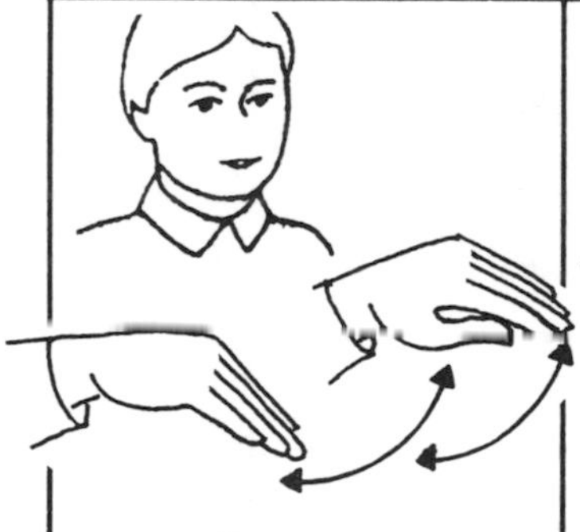 walk 1	 4
 depend 1	 3

WE

Handshape: W
Location: Chest
Movement: Arc from one side to other

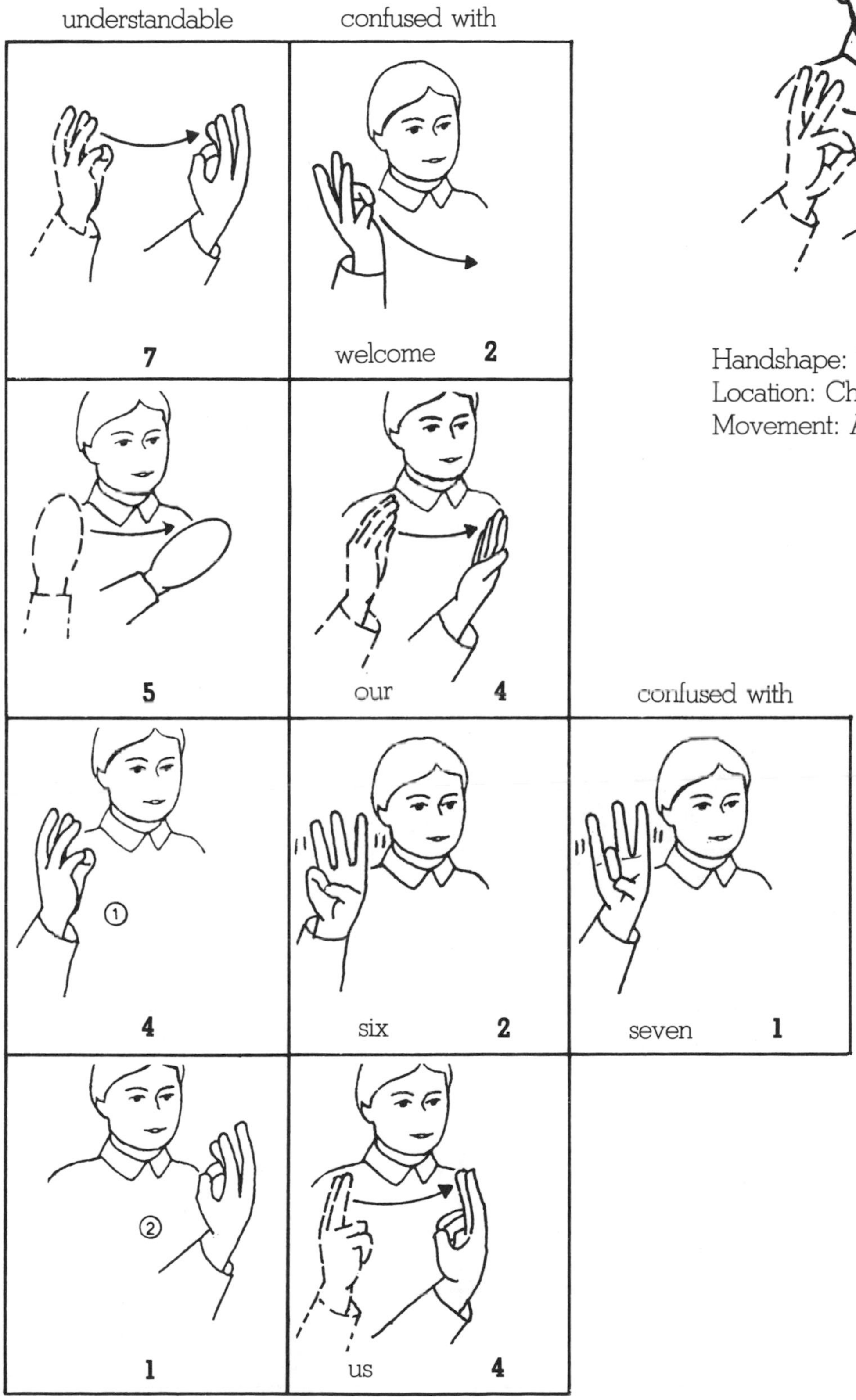

DRESS

Handshape: 5, both
Location: Chest
Movement: Down and out, twice

confused with

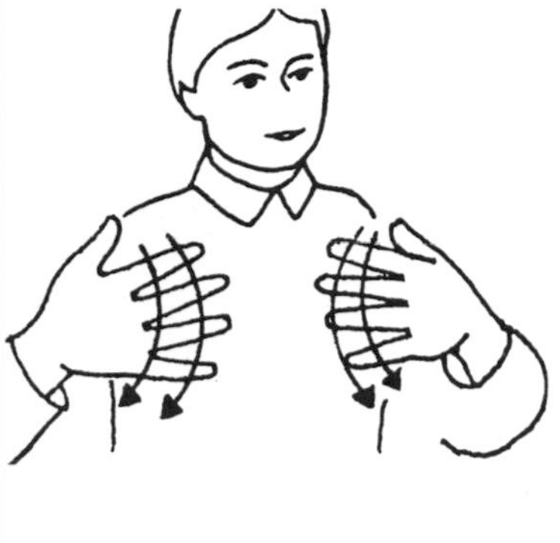

clothes **1**

understandable

6

confused with

finish **1**

clothes **3**

5

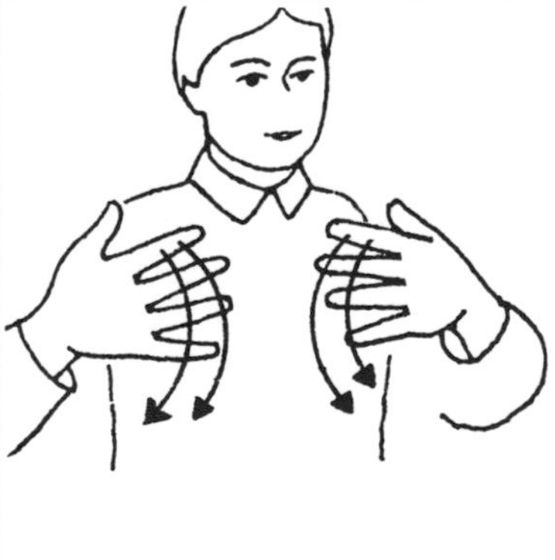

clothes **1**

finish **5**

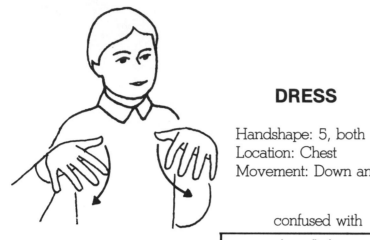

2

BLACK

Handshape: 1
Location: Forehead
Movement: Across brows

understandable | confused with

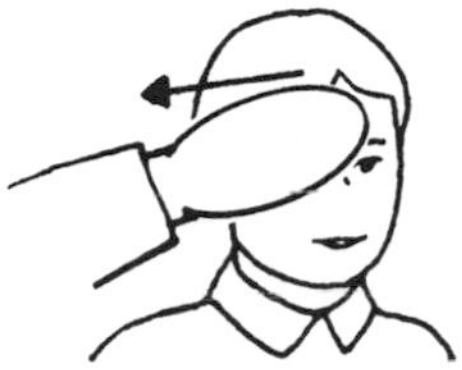

6

summer 1

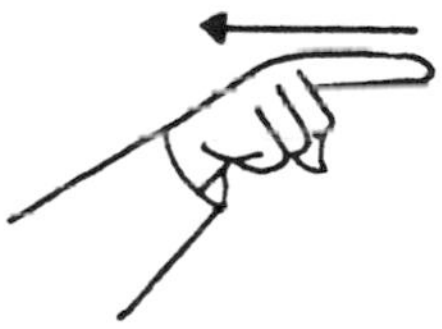

4

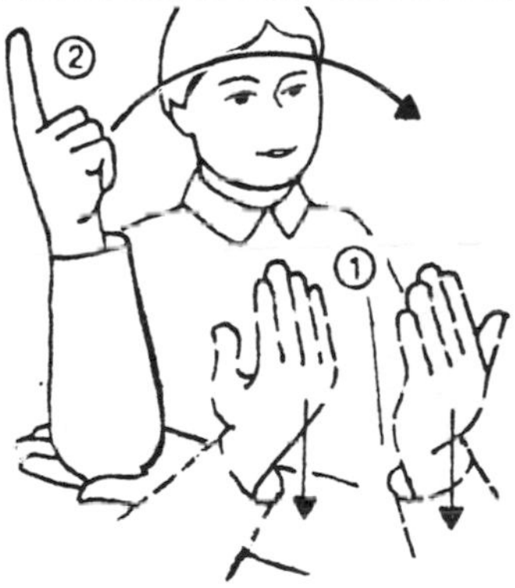

today 1

confused with

3

think 3

for 2

SOAP

Handshape: Open B, both
Location: In front of body
Movement: Scrape palm, close fingers

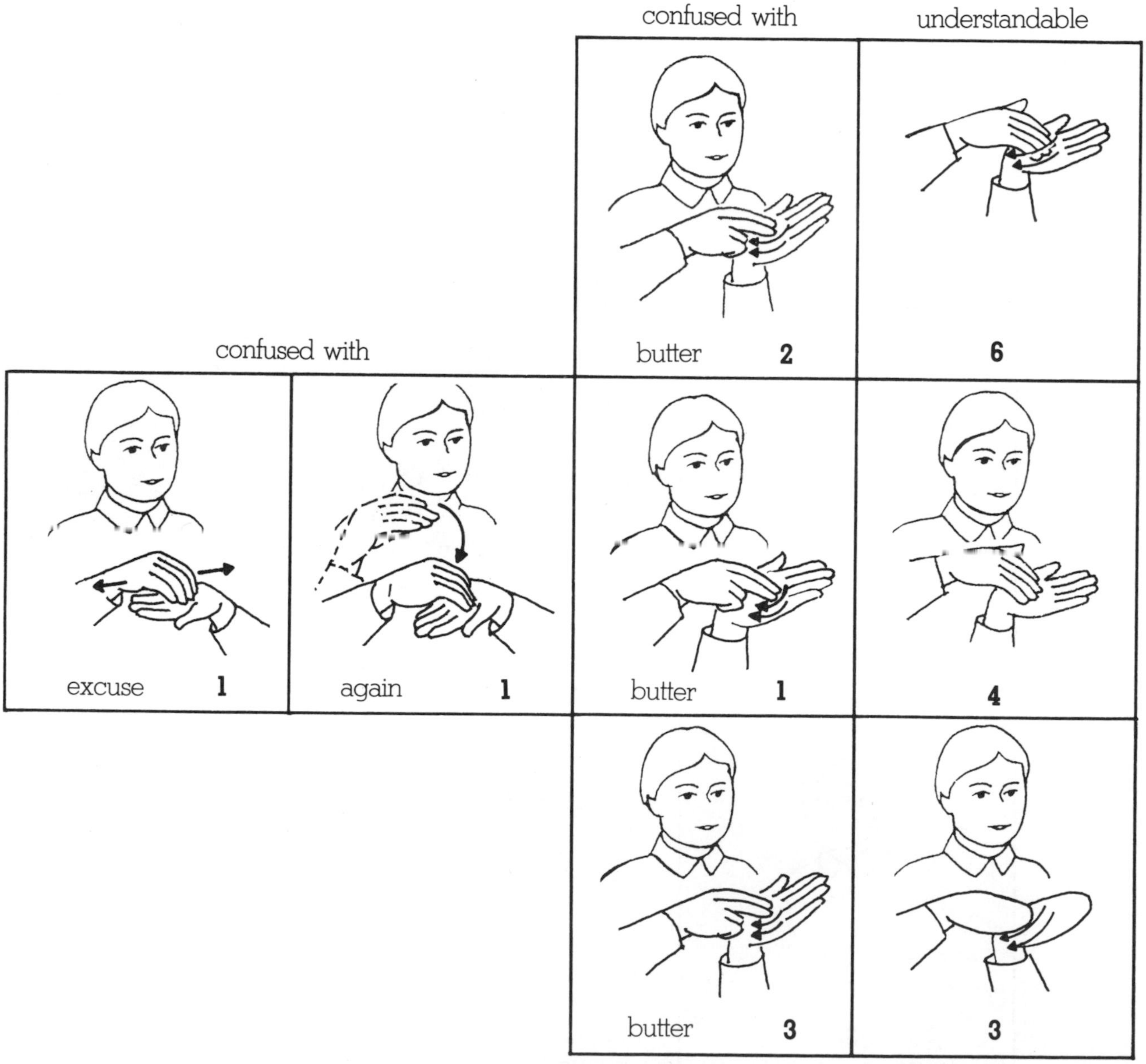

STORE

Handshape: Flat O, both
Location: In front of body
Movement: Swing out twice

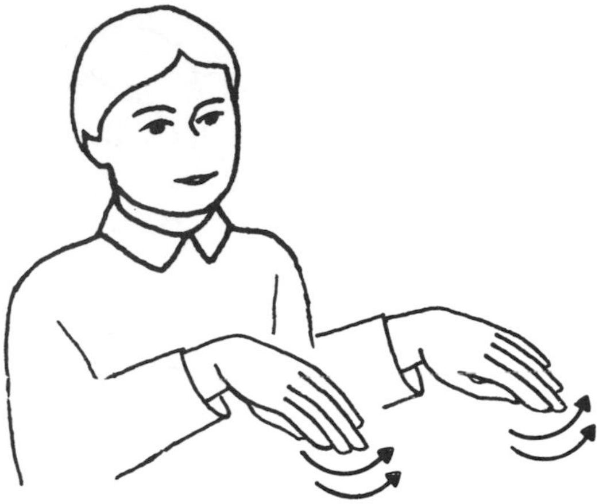

understandable | confused with

5

sell **3**

shopping **2**

4

sell **3**

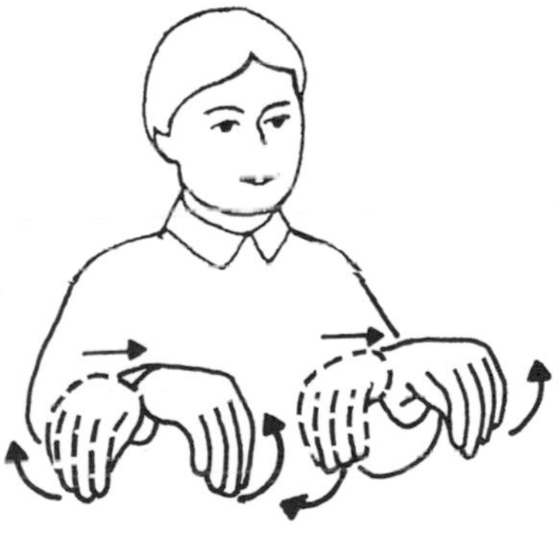

shopping **1**

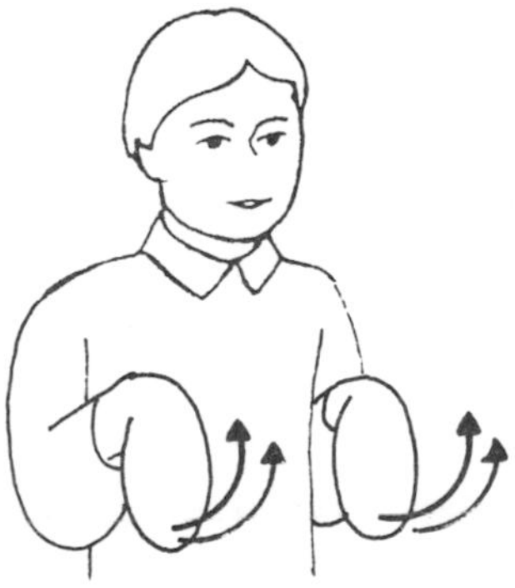

3

sell **2**

finish **1**

POTATO

Handshape: Bent V; S
Location: In front of body
Movement: Tap fist twice

understandable

7

confused with

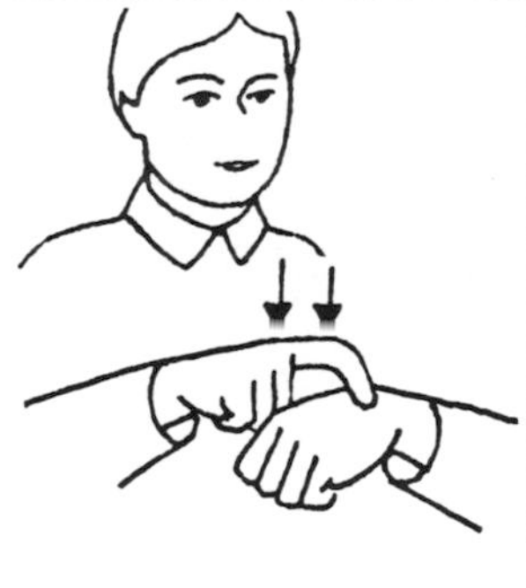

time **7**

2

confused with

bump **1**

McDonald's **1**

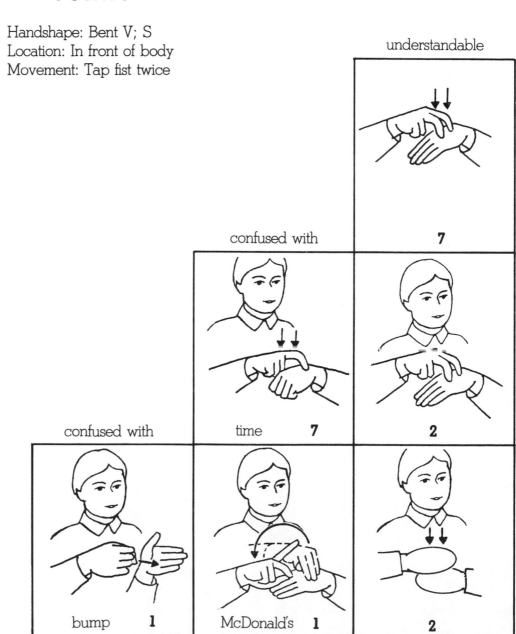

2

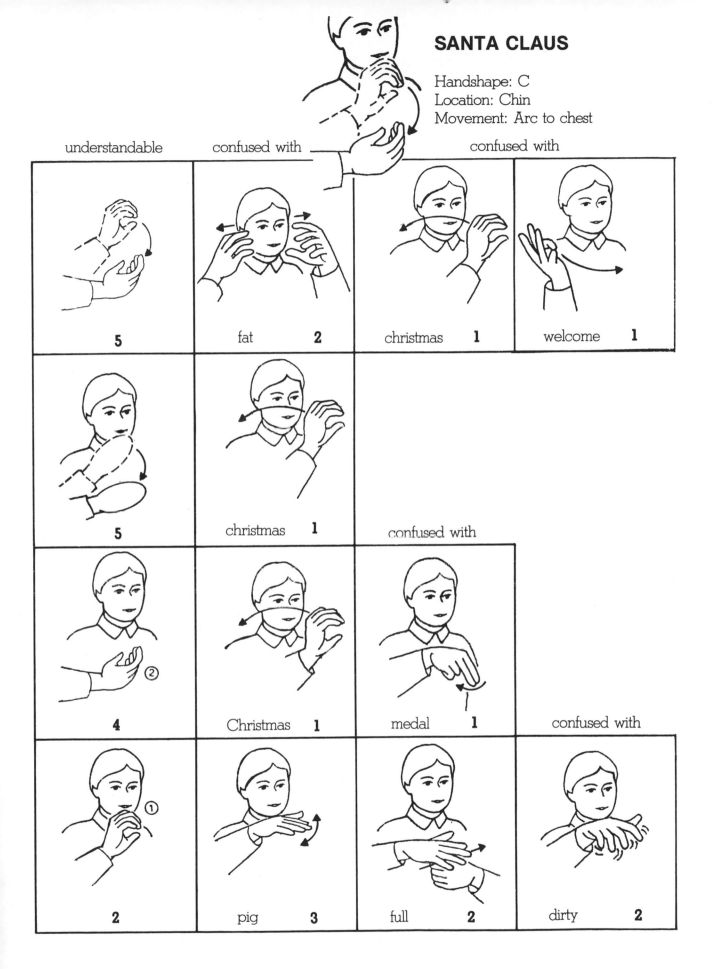

SANTA CLAUS

Handshape: C
Location: Chin
Movement: Arc to chest

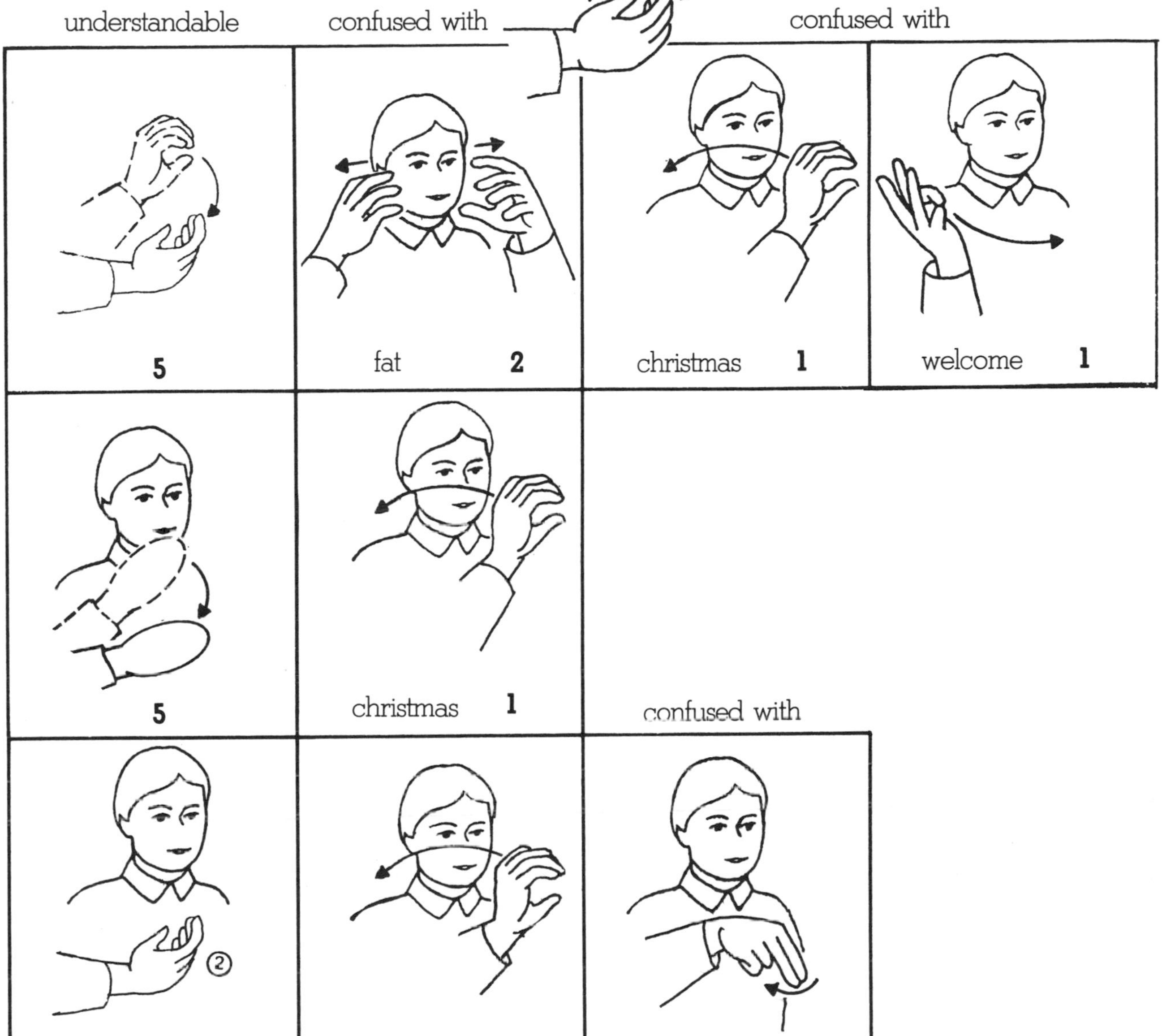

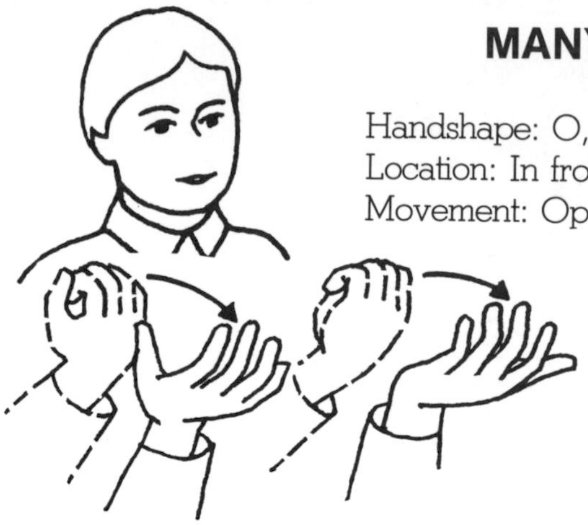

MANY

Handshape: O, both
Location: In front of body
Movement: Open fingers vigorously

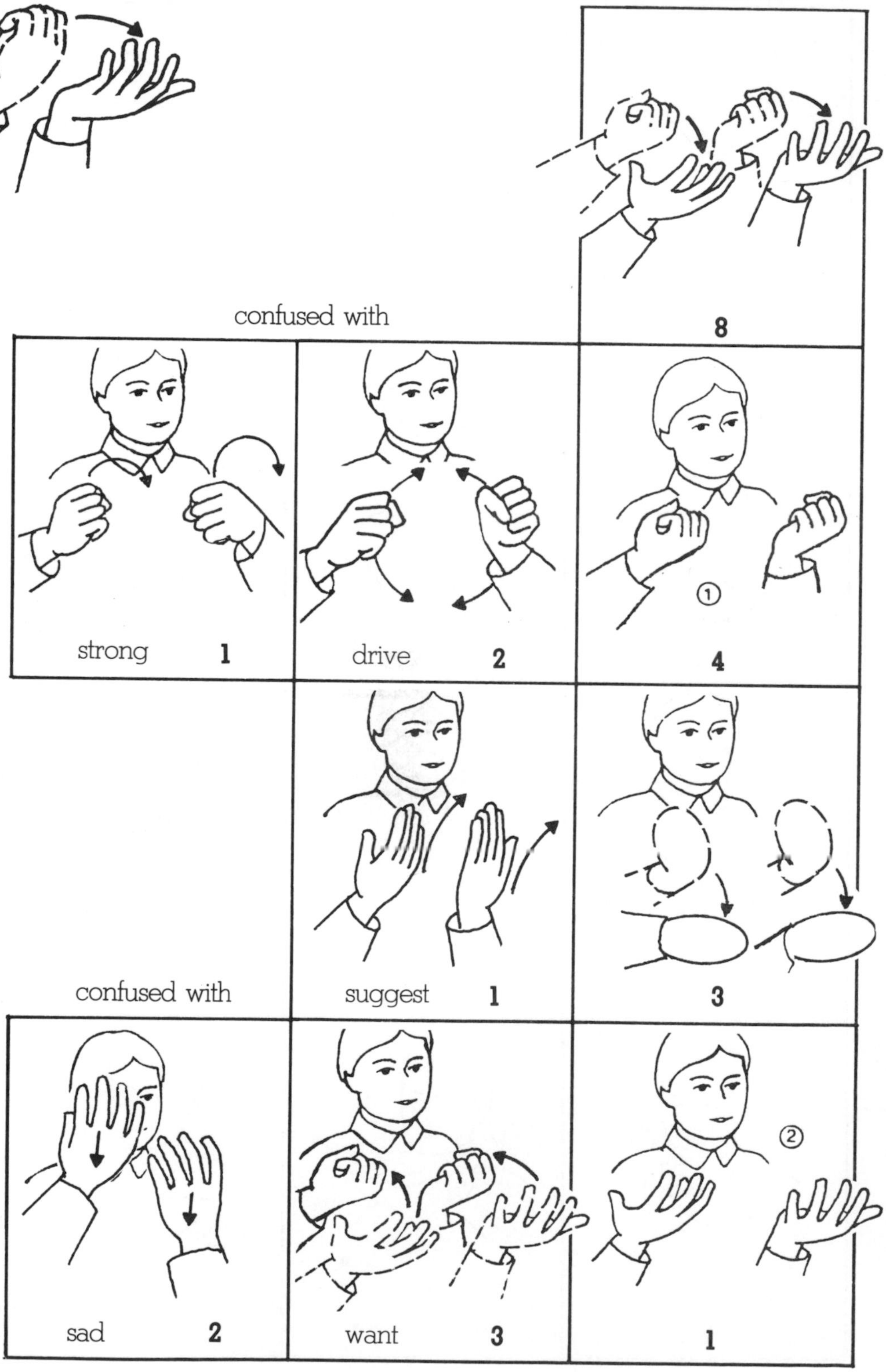

DOOR

Handshape: B, both
Location: In front of body
Movement: Turn hand toward self

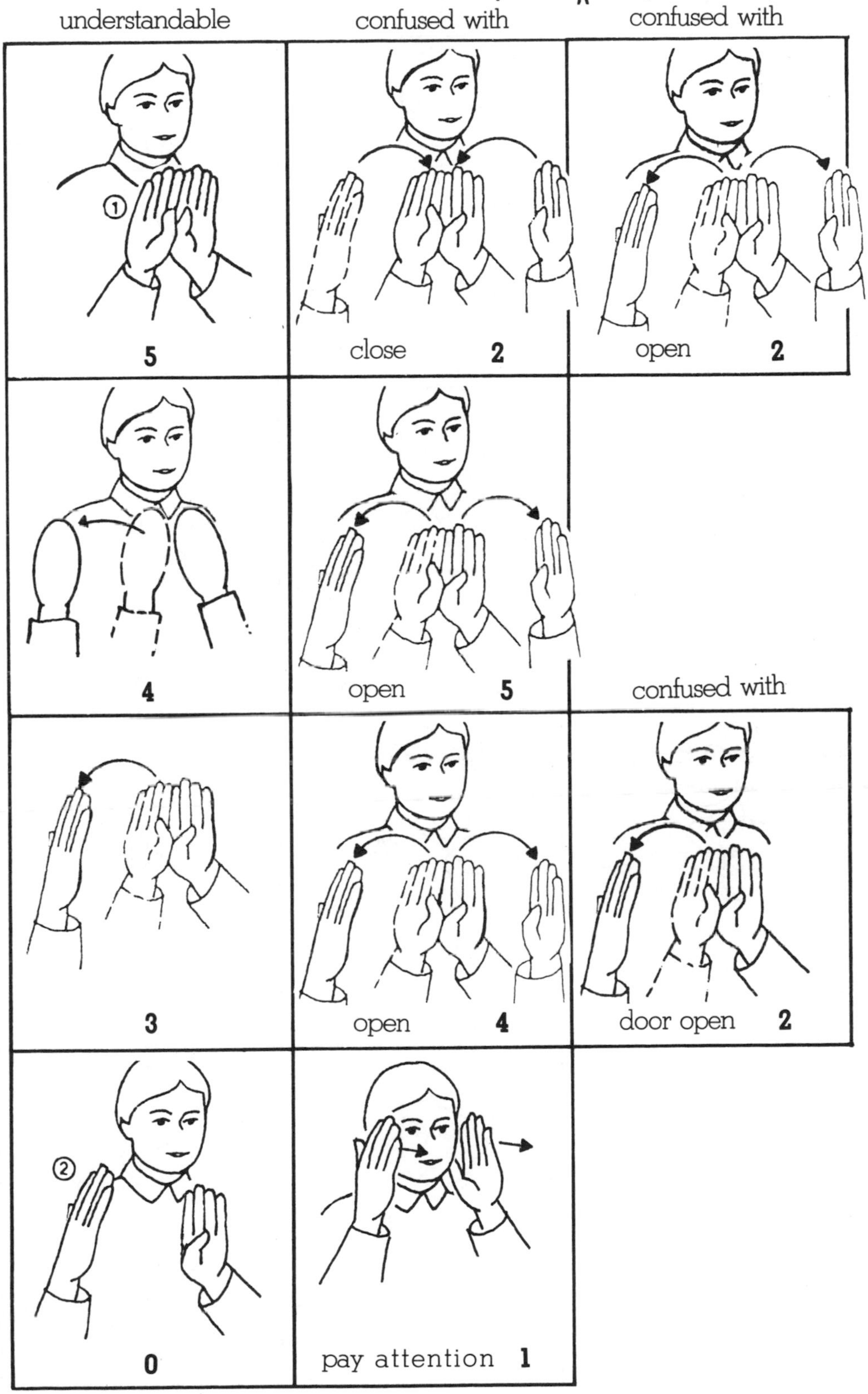

SLEEP

Handshape: 5
Location: Face
Movement: Slowly close fingers

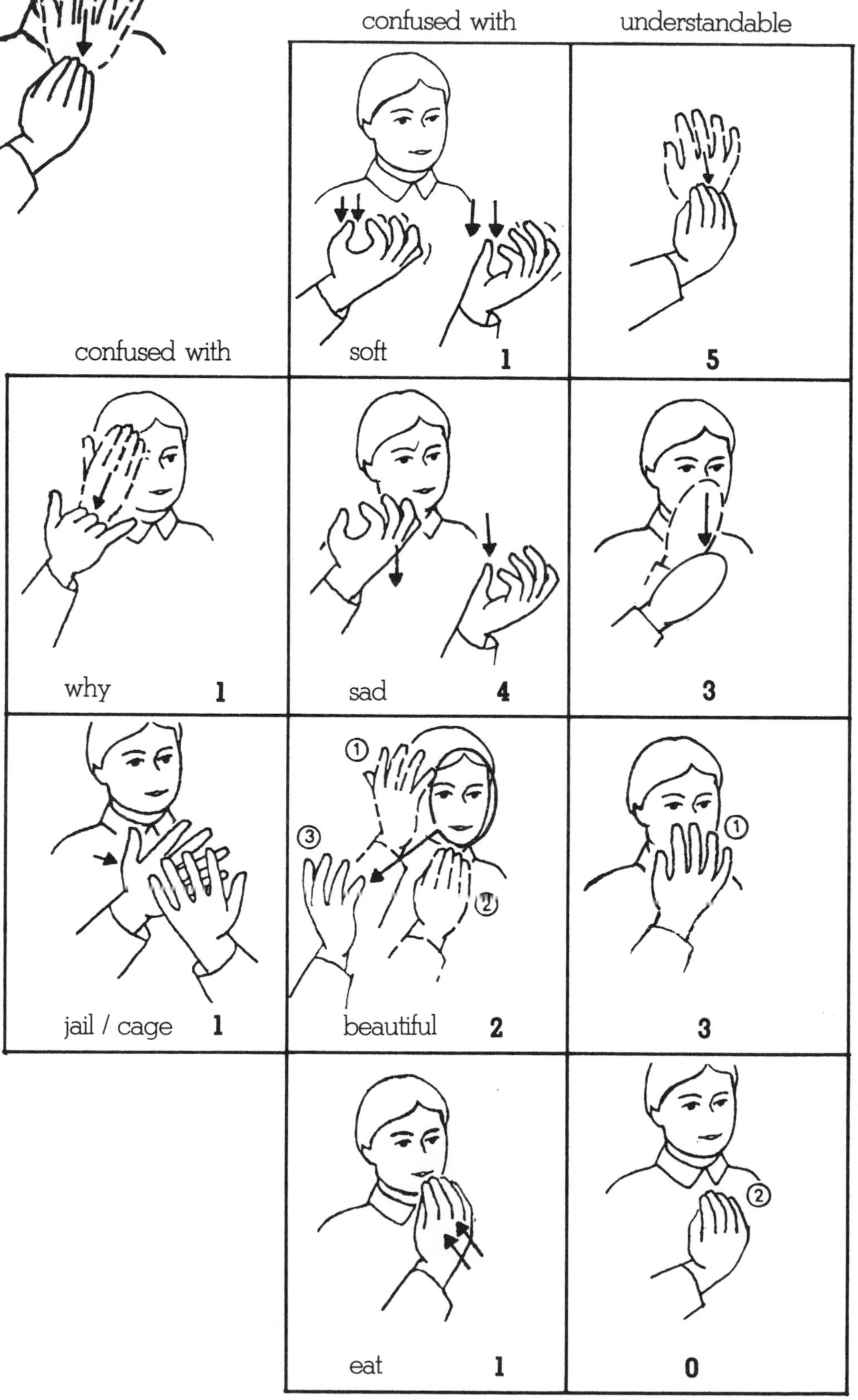

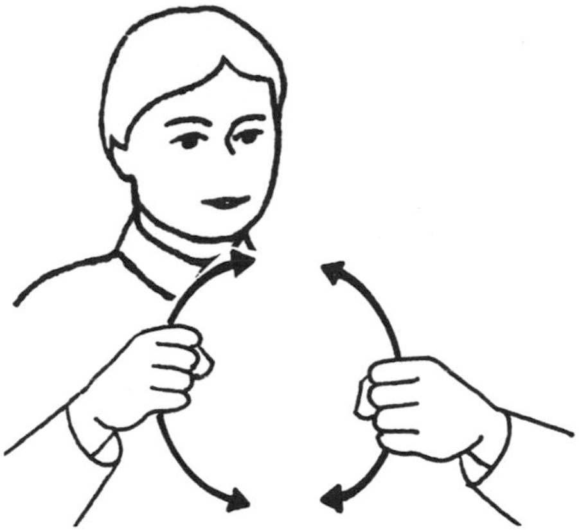

CAR

Handshape: S, both
Location: In front of body
Movement: Steer a car

understandable | confused with

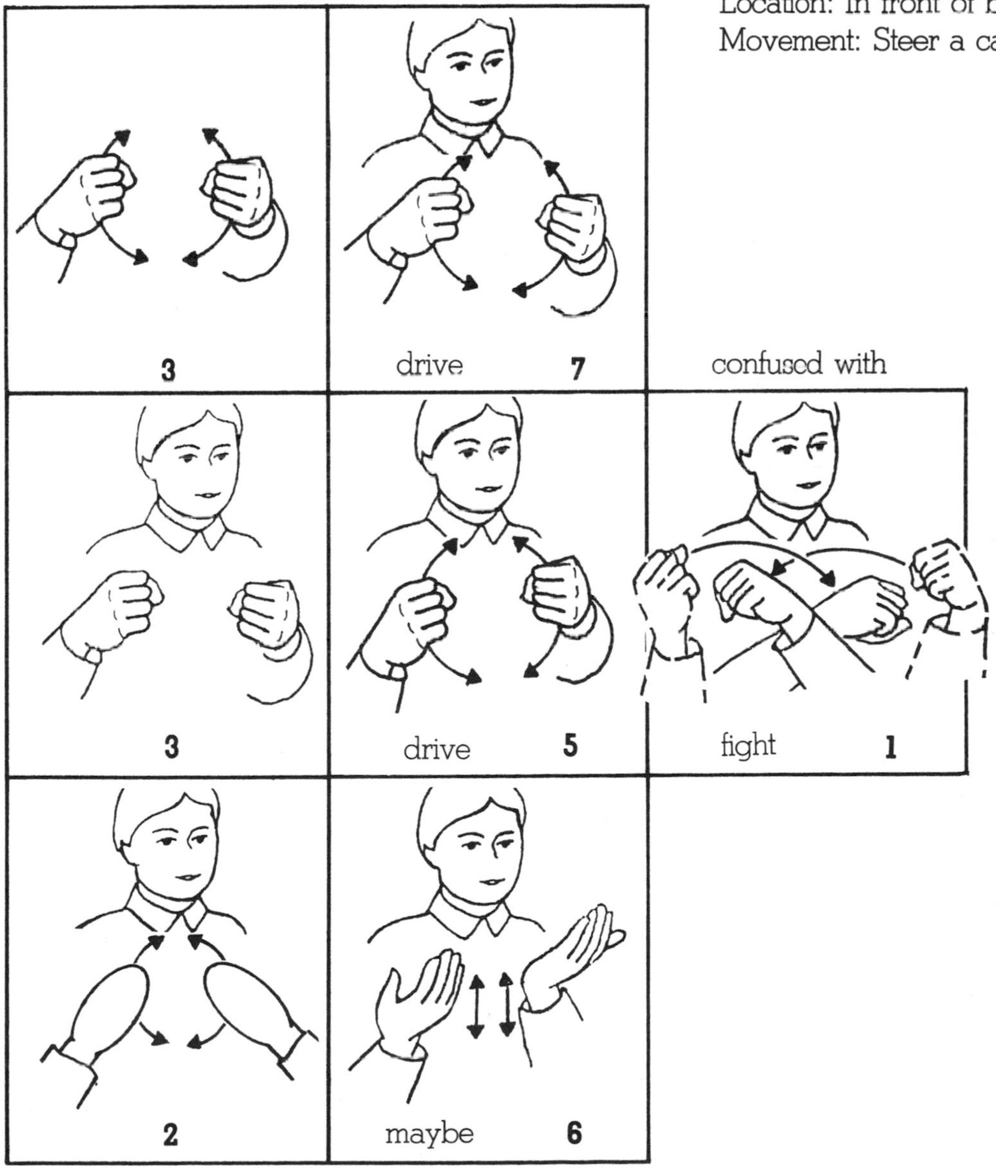

ZERO ROBUST

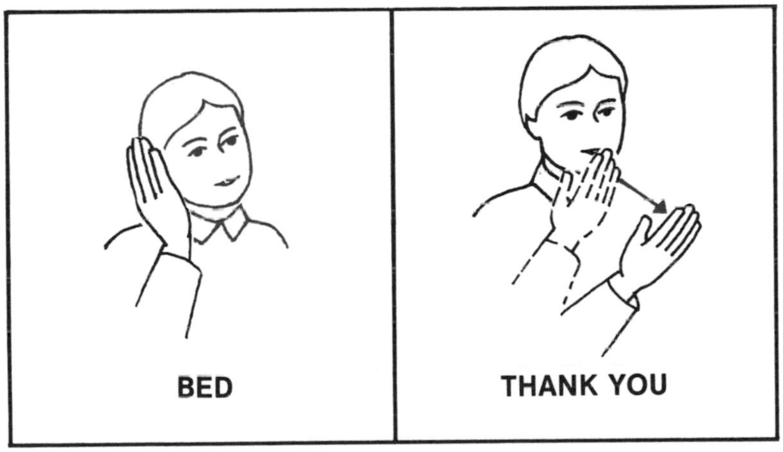

SINGLE FRAGILE

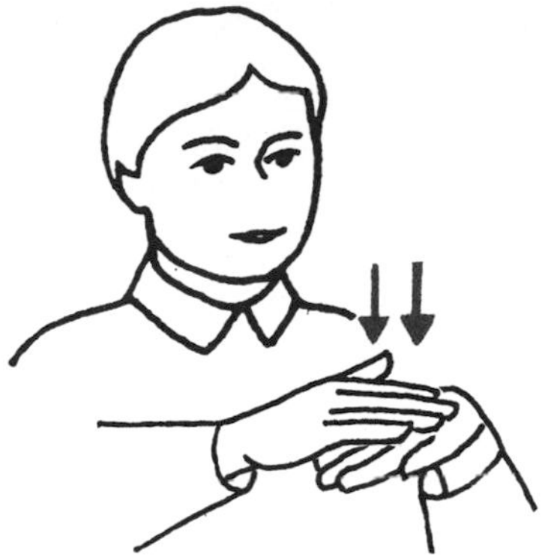

SCHOOL

Handshape: Open B, both
Location: In front of body
Movement: Tap several times

understandable

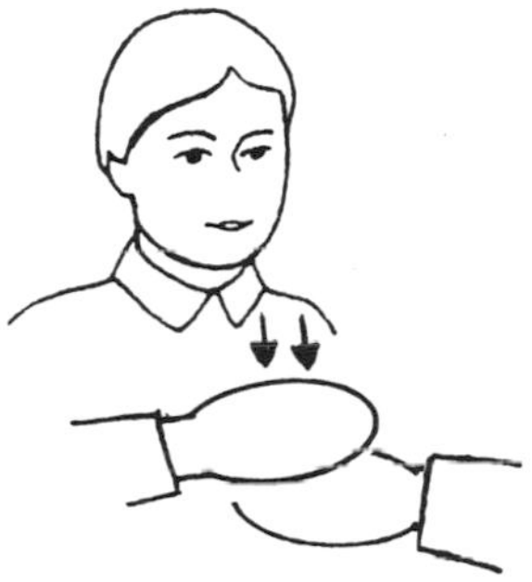

8

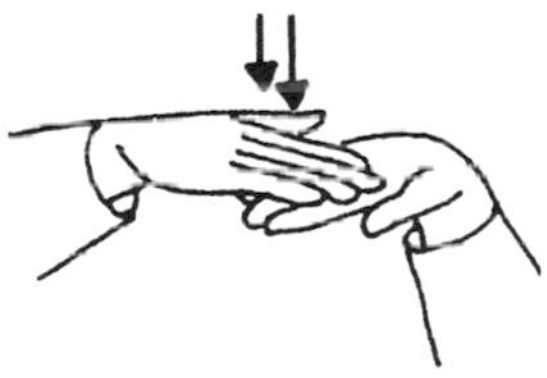

7

confused with

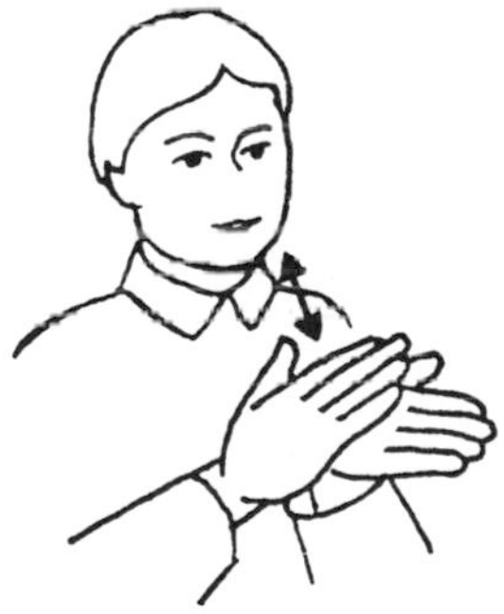

praise **2**

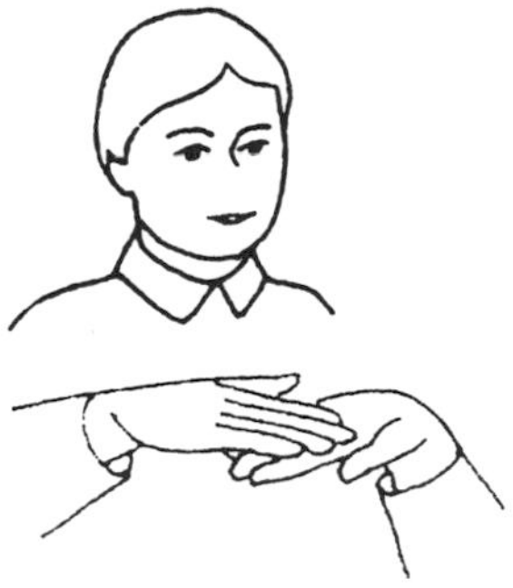

2

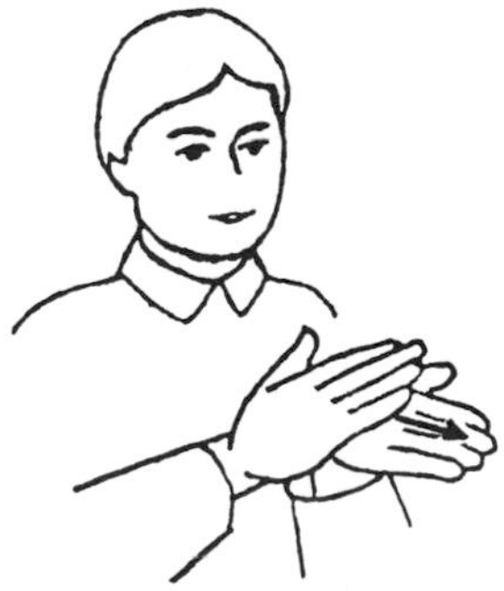

clean **2**

confused with

college **2**

HUNGRY

Handshape: C
Location: Chest
Movement: Down

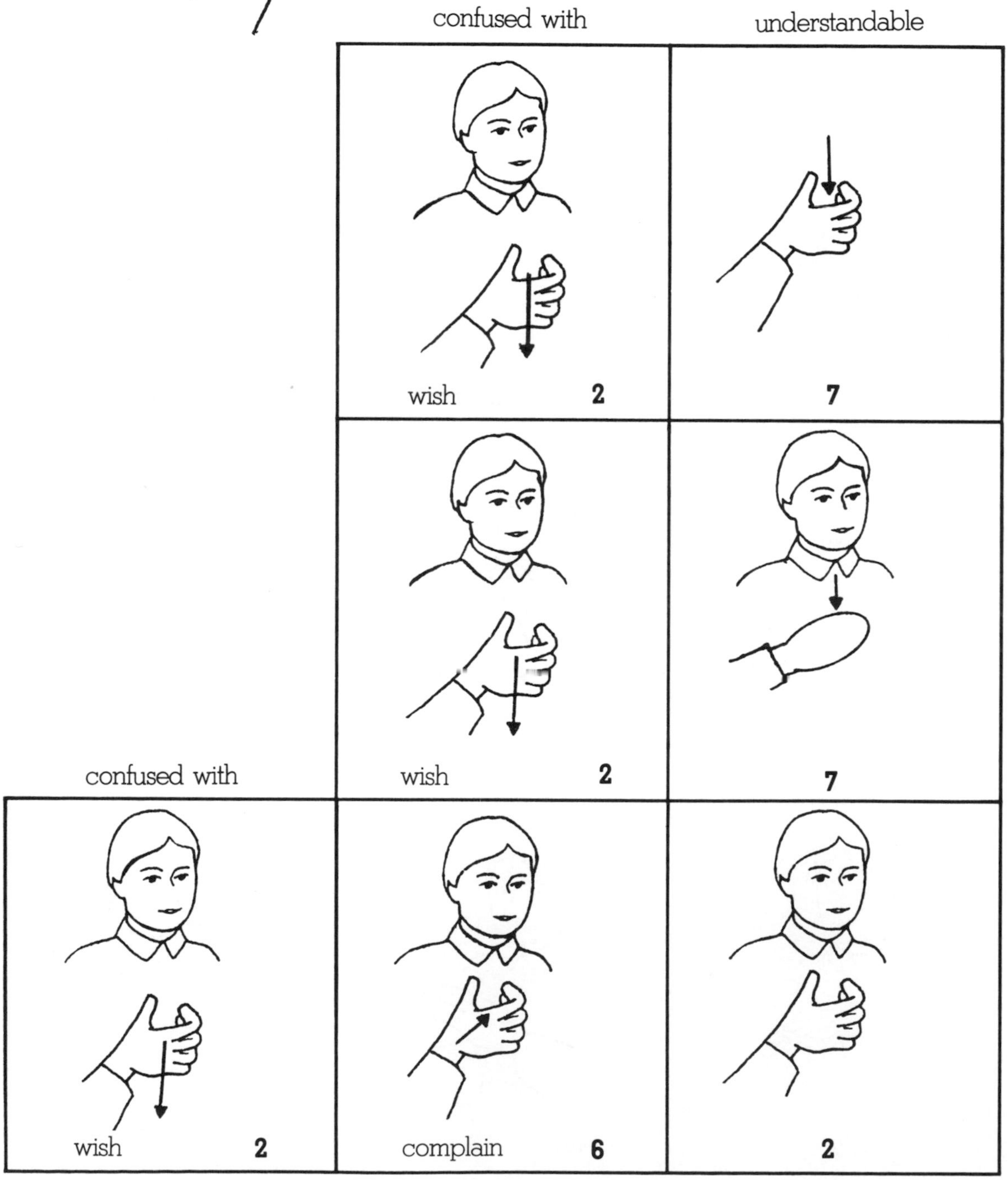

WITH

Handshape: A, both
Location: In front of body
Movement: Bring together

understandable

8

confused with

together 2

6

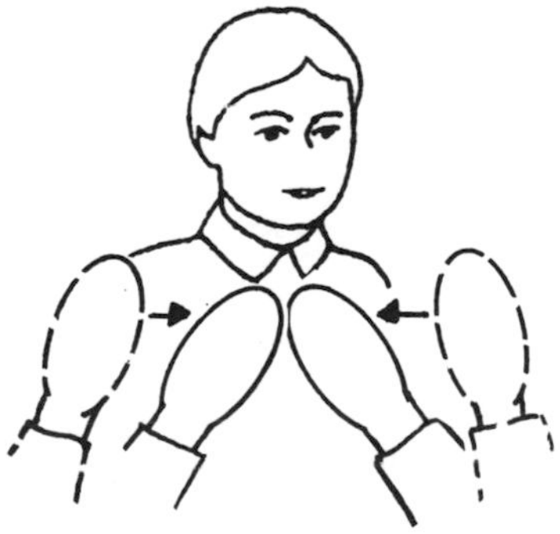

1

confused with

close 5

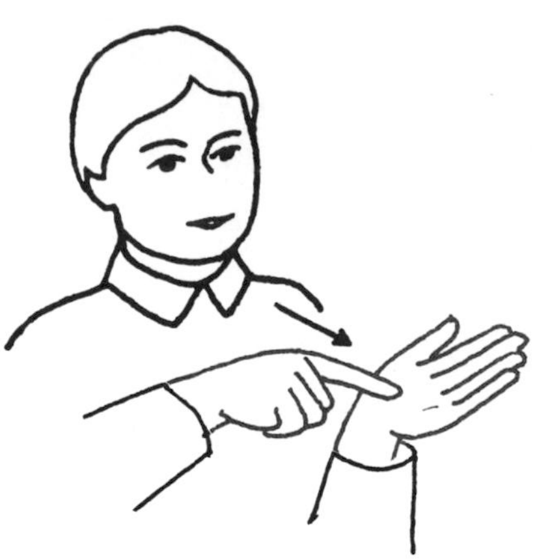

THIS

Handshape: 1; open B
Location: In front of body
Movement: Touch palm

understandable

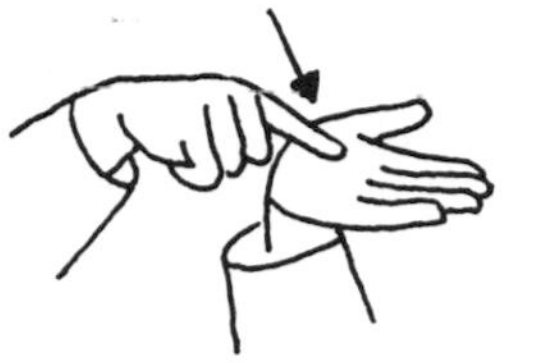

8

confused with

owe 2

6

center 8

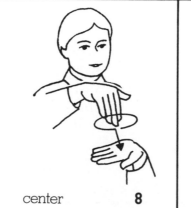

2

STAY

Handshape: A, thumb extended, both
Location: In front of body
Movement: Forward slightly

understandable | confused with

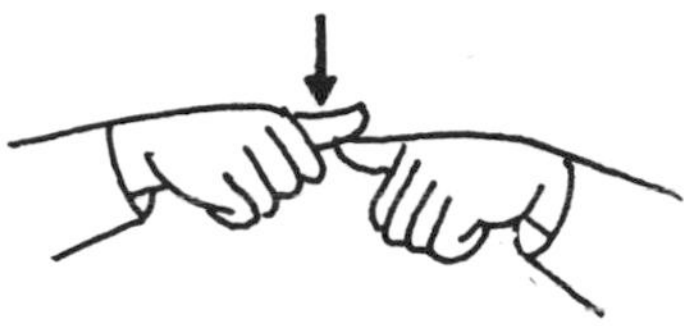

8

continue 1

5

continue 2

2

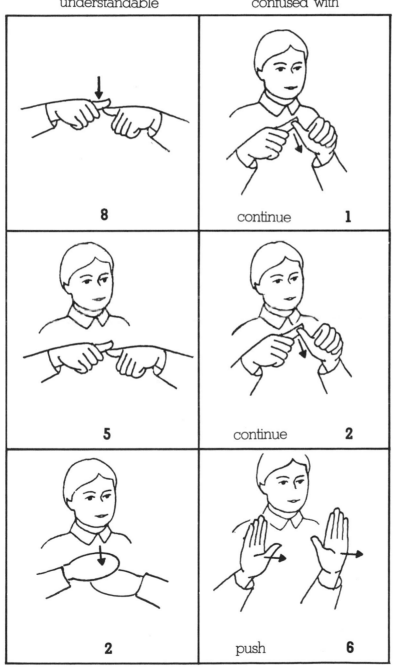

push 6

TEACH

Handshape: Flat O, both
Location: Temples
Movement: Outward twice

confused with	understandable
 none **2**	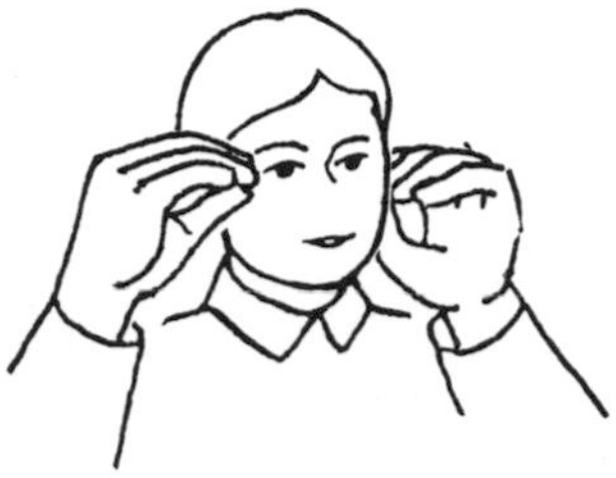 **7**
 none **2**	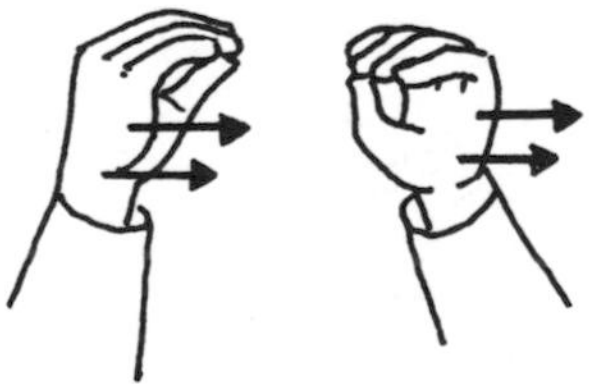 **6**
 attention **9**	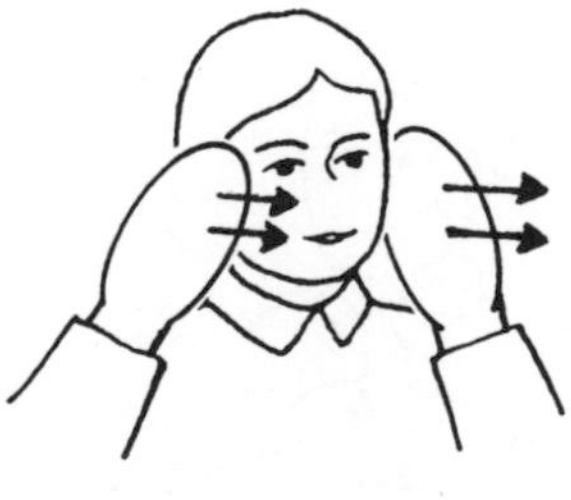 **1**

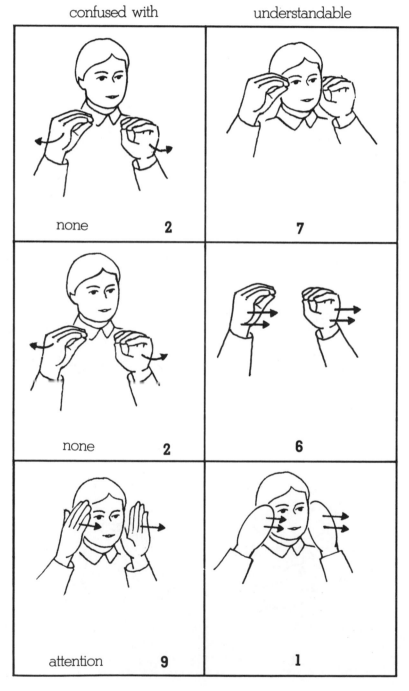

BOY

Handshape: Flat O
Location: Forehead
Movement: Open and close, twice

understandable	confused with
 8	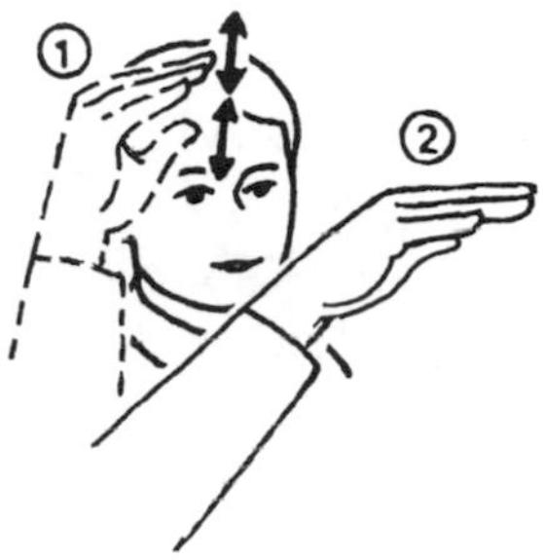 man 2
 6	 father 3
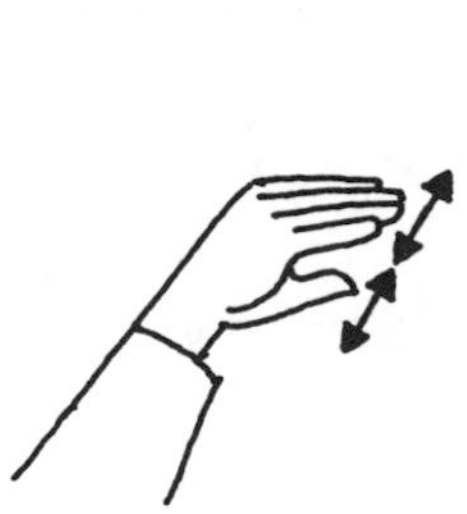 0	 talking 7

SUNDAY

Handshape: Open B, both
Location: In front of body
Movement: Circle away from each other

confused with | understandable

wonderful **2**

7

wonderful **2**

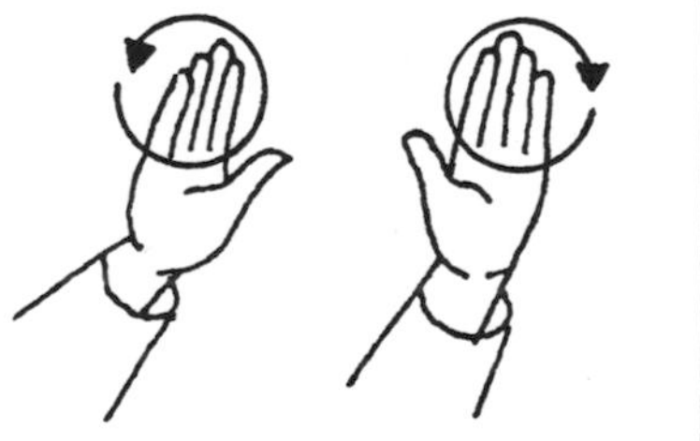

6

confused with

push **1**

wonderful **5**

1

LITTLE

Handshape: G
Location: In front of body
Movement: None

understandable

8

6

confused with

"G" **1**

0

good-bye **5**

ONE

Handshape: 1
Location: In front of body
Movement: None

confused with

understandable

wait **2**

7

wait **1**

6

confused with

stop **1**

yours **7**

0

TOOTHBRUSH

Handshape: 1
Location: Teeth
Movement: Back and forth

understandable	confused with
 8	 brush 2
 6	 brush 1
 0	 you 1

PENCIL

Handshape: Closed G
Location: Mouth
Movement: To palm, write on palm

understandable

confused with

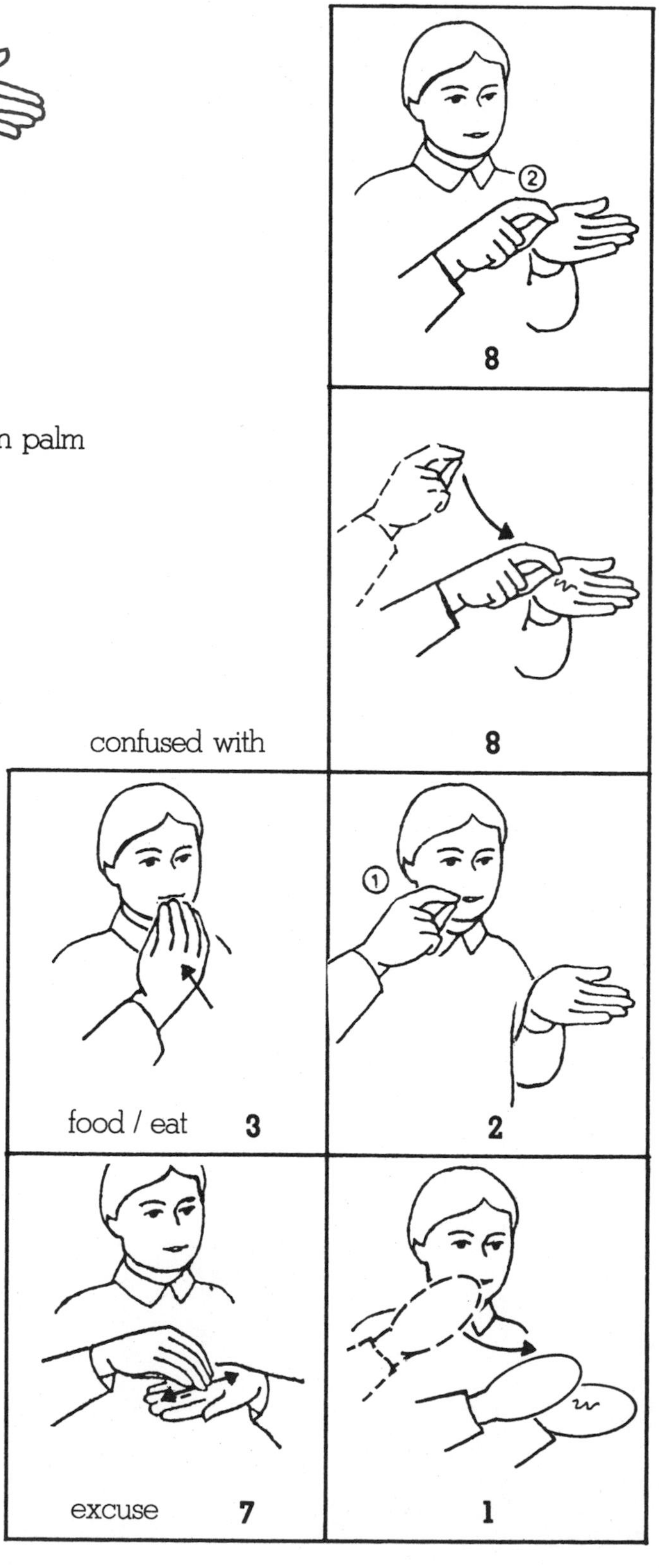

YELLOW

Handshape: Y
Location: In front of body
Movement: Shake

understandable / confused with

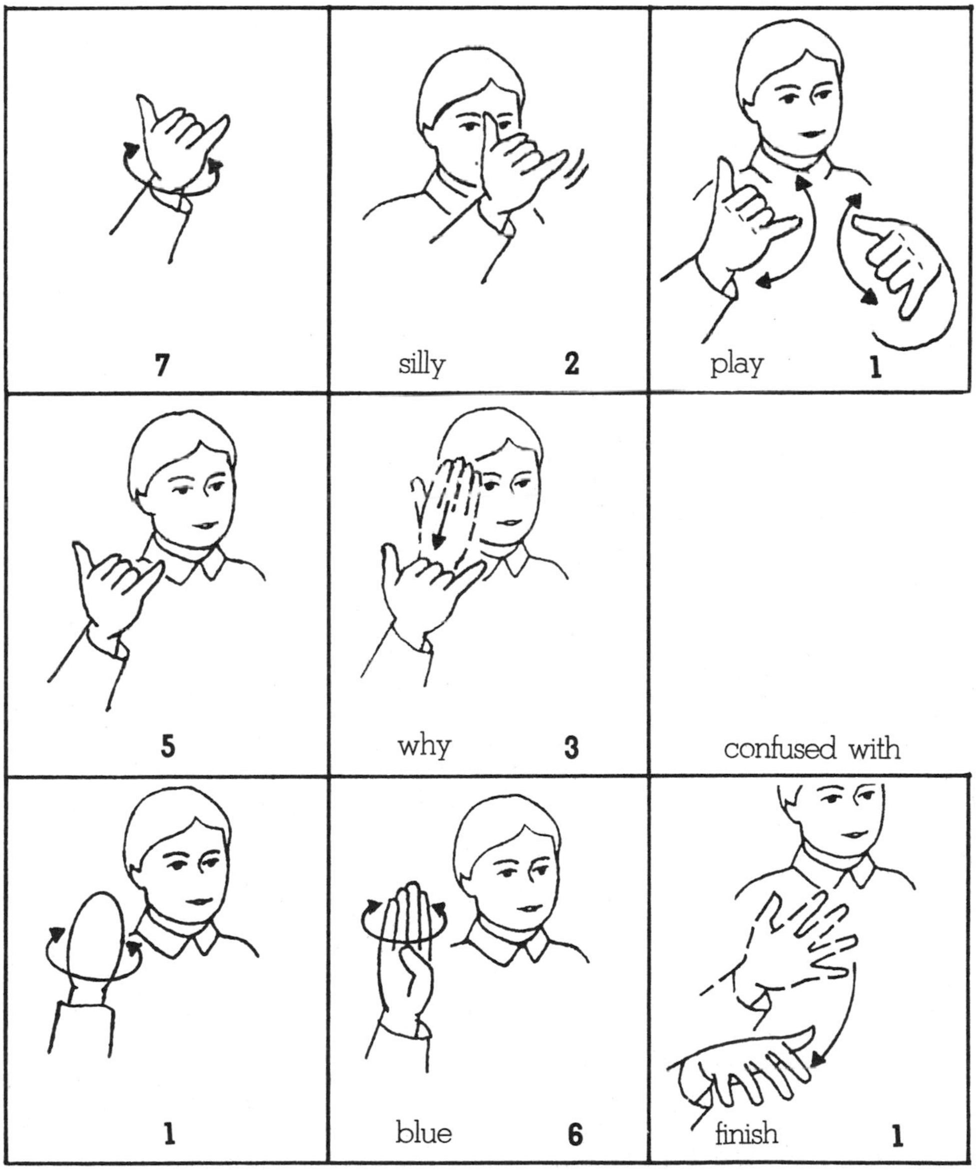

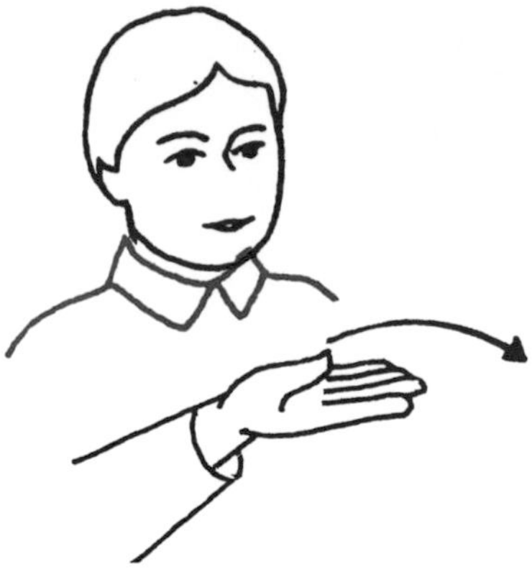

THERE

Handshape: B
Location: In front of body
Movement: Point out obliquely

confused with	understandable
 waiter 1	 7
	 5

confused with	
 give 1	 1

TEN

Handshape: A, thumb extended
Location: In front of body
Movement: Wiggle slightly

understandable

7

confused with

6

yourself **1**

0

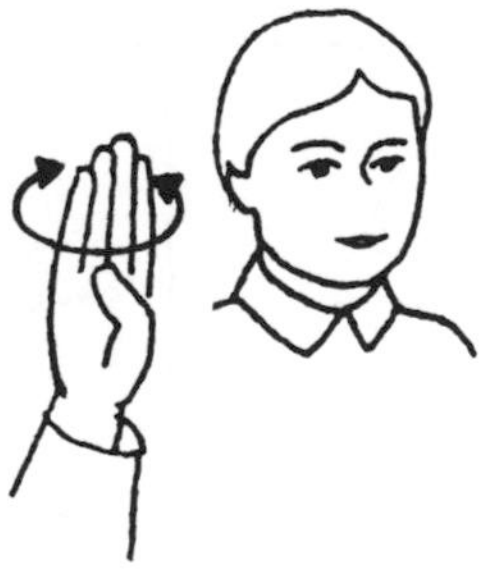

blue **2**

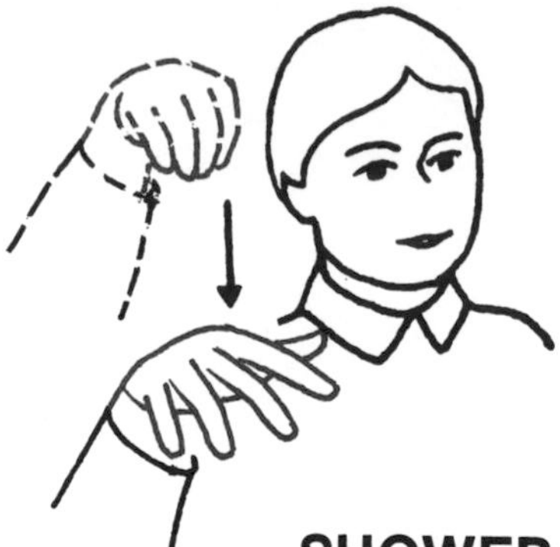

SHOWER

Handshape: Flat O
Location: Above head
Movement: Open twice

confused with

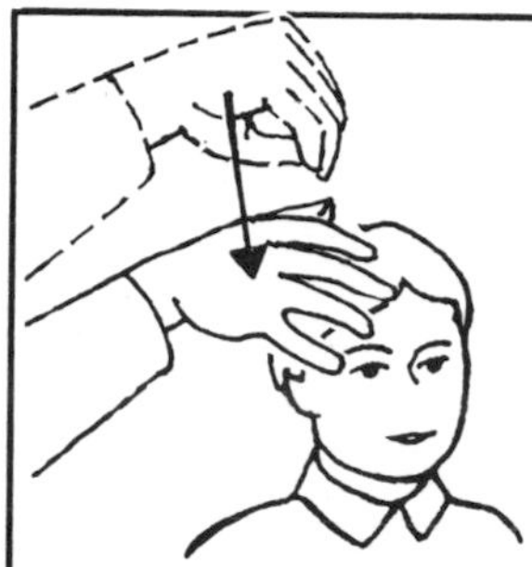

sun **2**

understandable

8

5

confused with

hope **1**

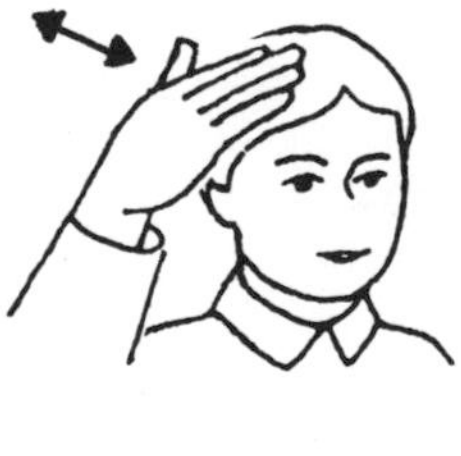

hat **3**

confused with

know **3**

1

THROW

Handshape: S
Location: In front of body
Movement: Forward vigoriously

understandable

confused with

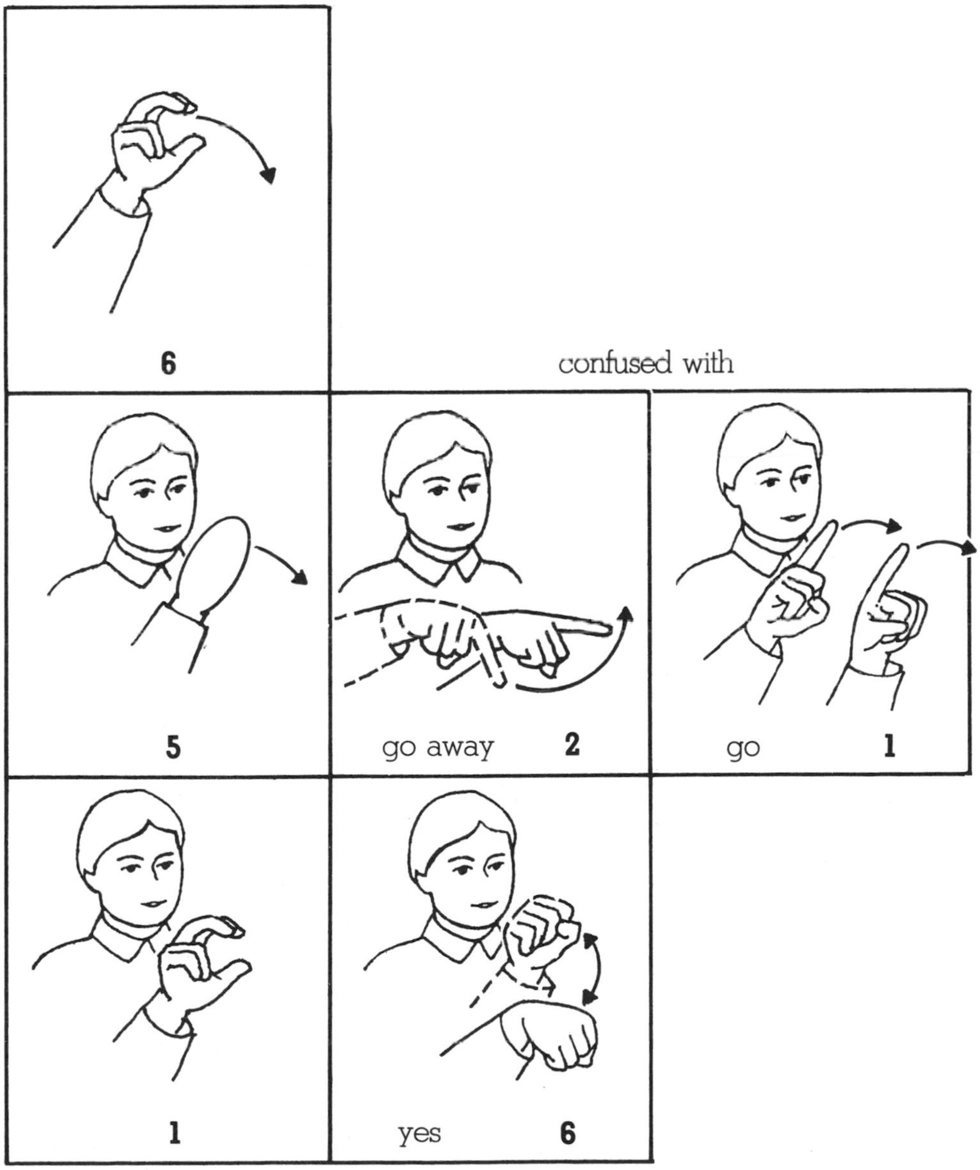

DOG

Handshape: B; G
Location: Side; in front of body
Movement: Touch pocket; snap finger twice

understandable

8

confused with

pay 1

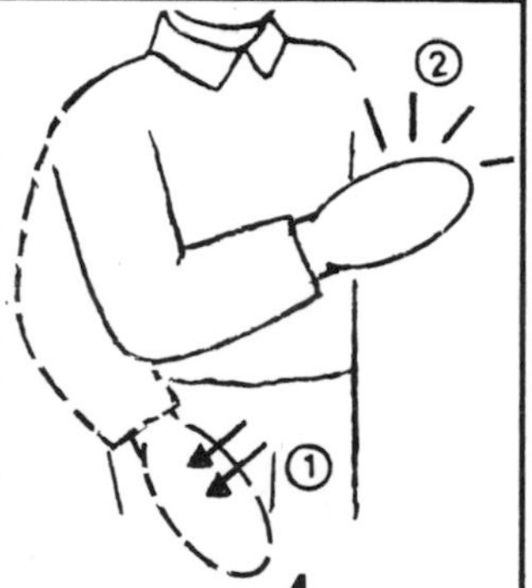

4

confused with

do, do 1

few 1

4

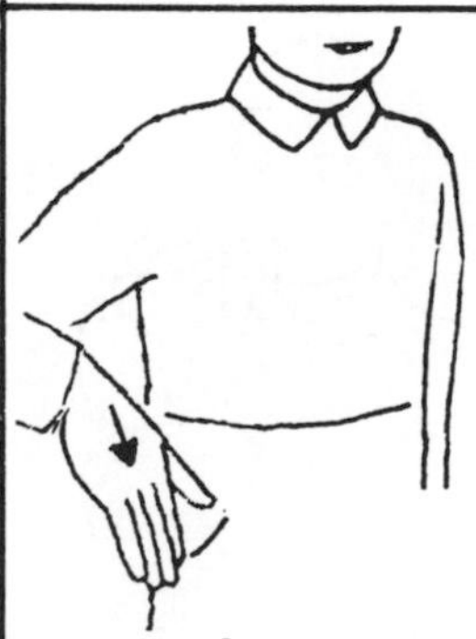

pocket 6

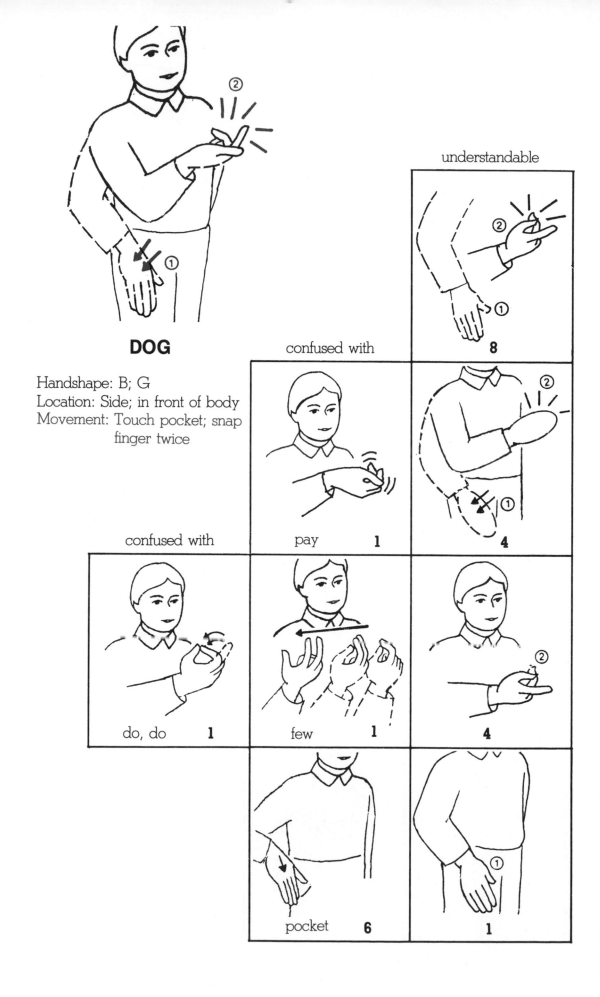

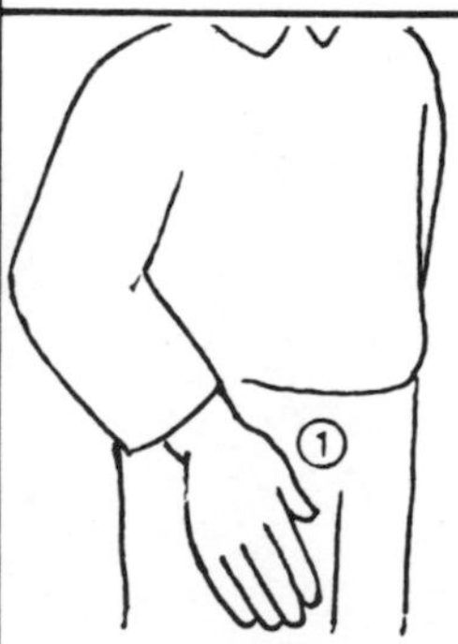

1

RED

Handshape: 1
Location: Lips
Movement: Brush down twice

understandable | confused with

6

quiet 2

6

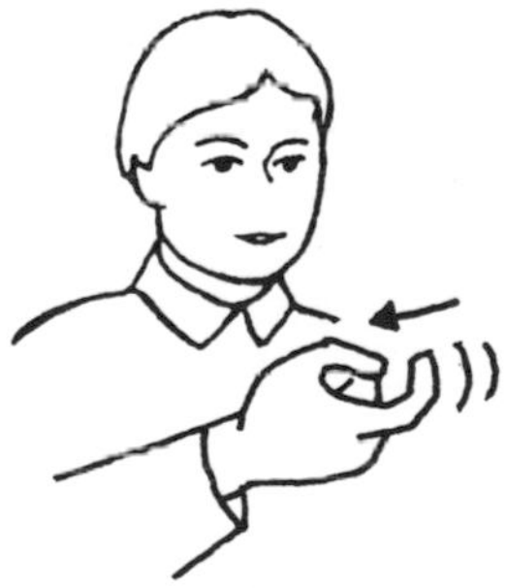

come 2

confused with

1

sweet 3

cute 2

candy 2

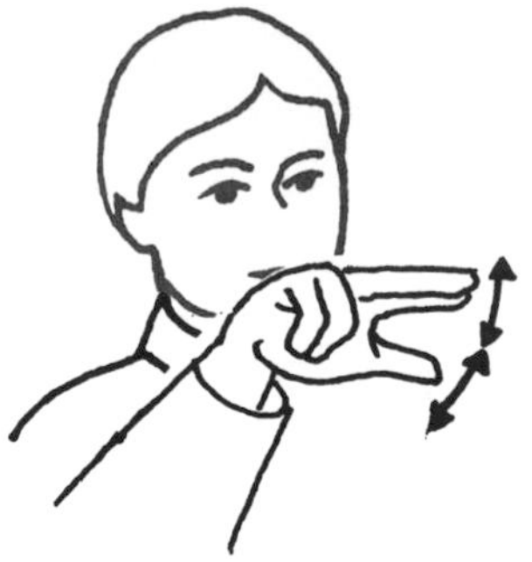

DUCK

Handshape: H, thumb extended
Location: Mouth
Movement: Snap together twice

confused with

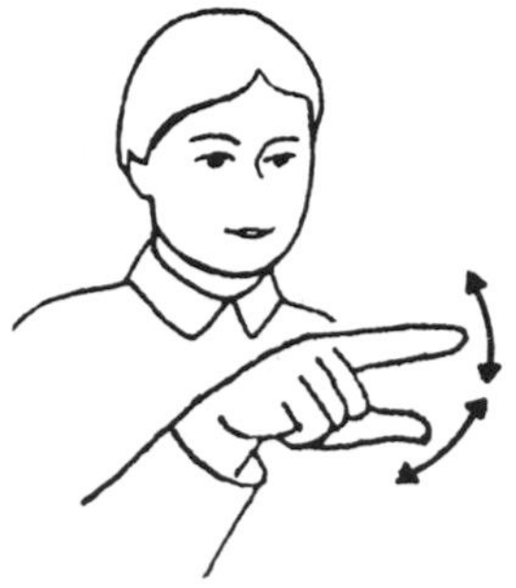

bird / chicken **1**

understandable

7

confused with

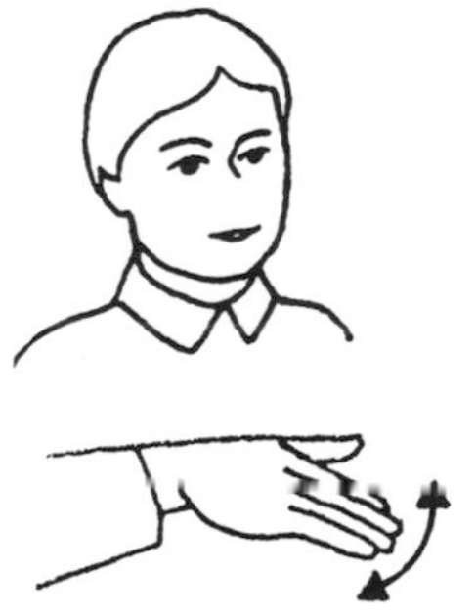

bye-bye **1**

bird / chicken **2**

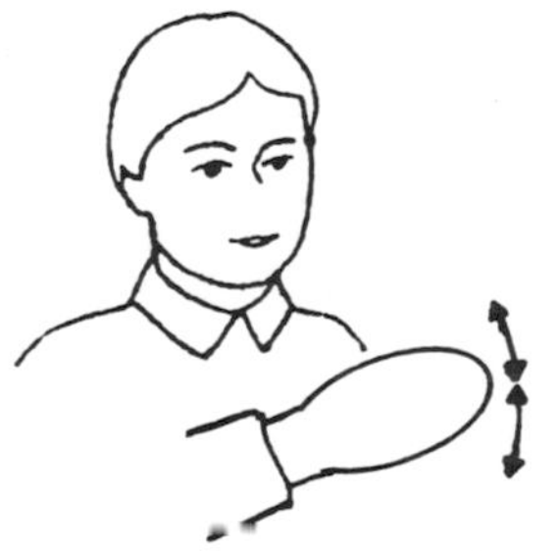

5

no **9**

0

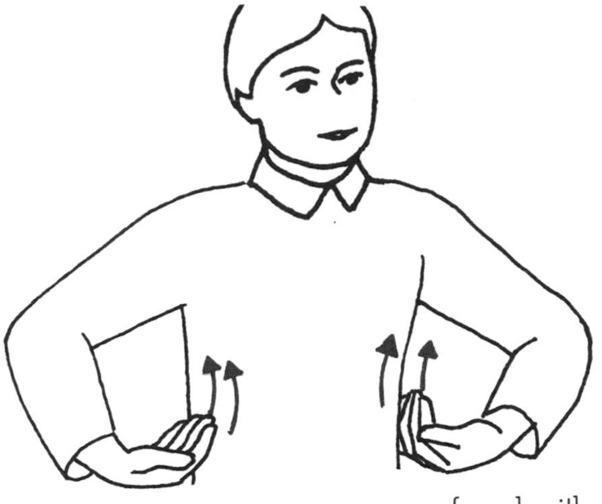

MONKEY

Handshape: C, both
Location: Sides
Movement: Scratch up twice

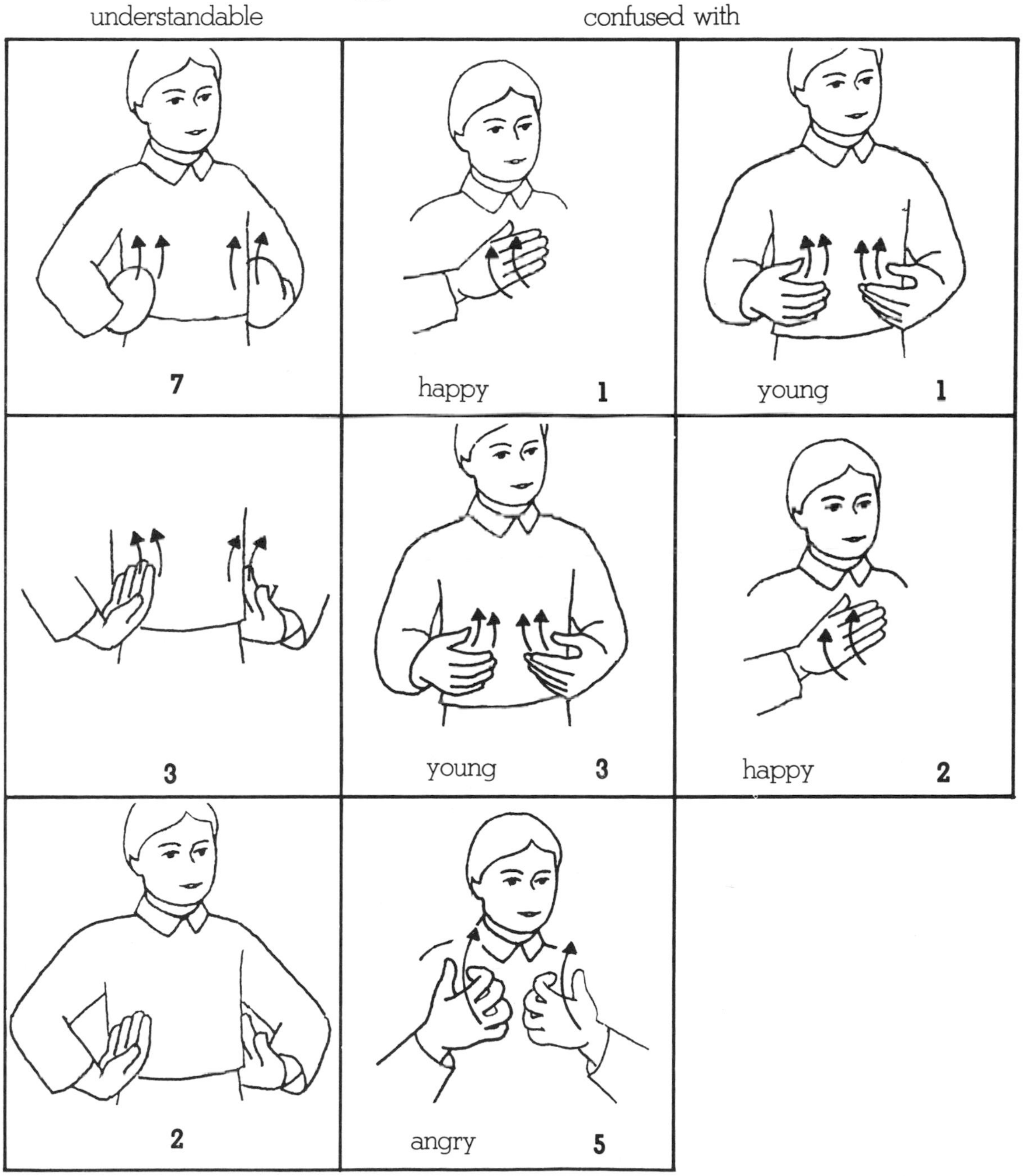

APPLE

Handshape: X
Location: Cheek
Movement: Twist forward and back

confused with understandable

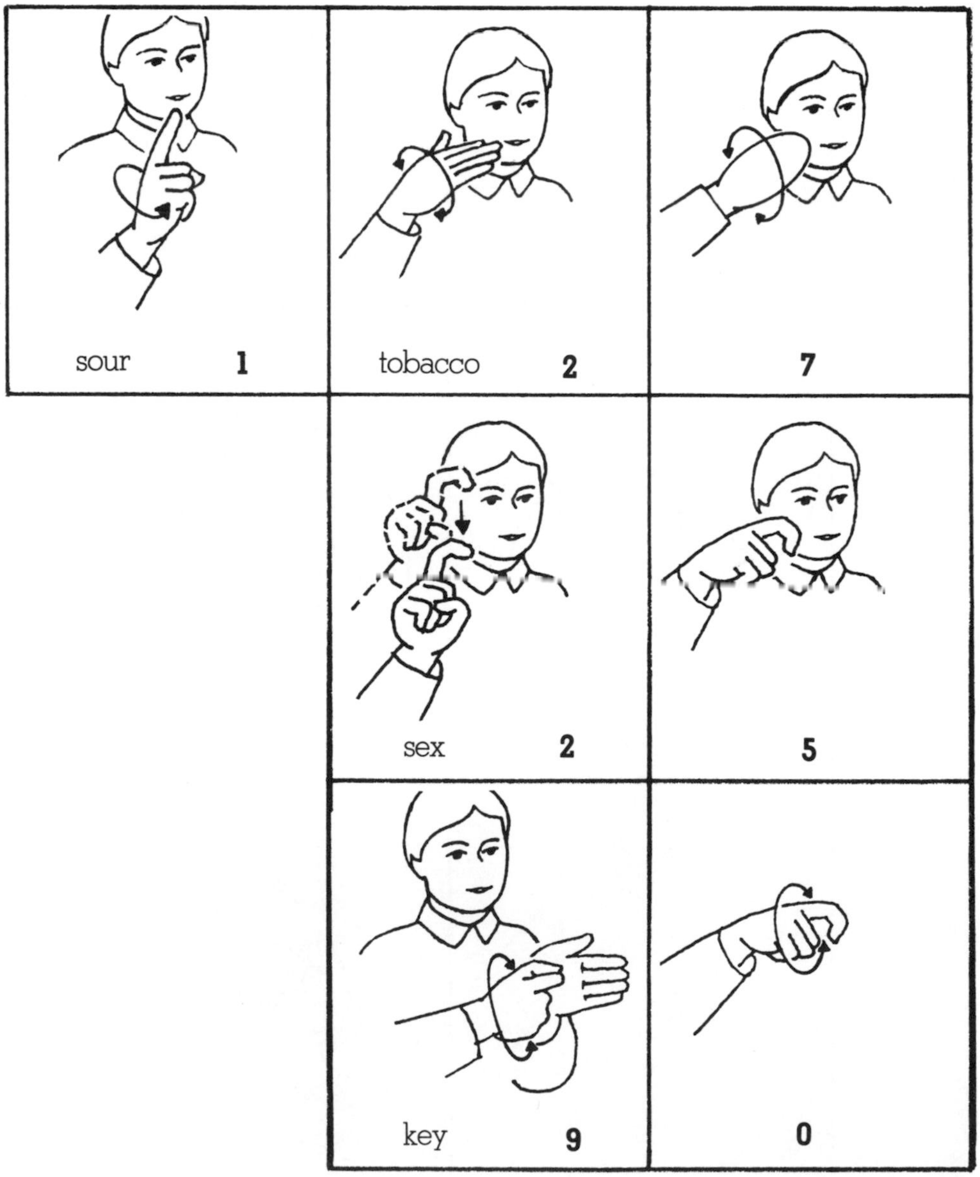

COAT

Handshape: A, both
Location: Chest
Movement: Trace Lapels

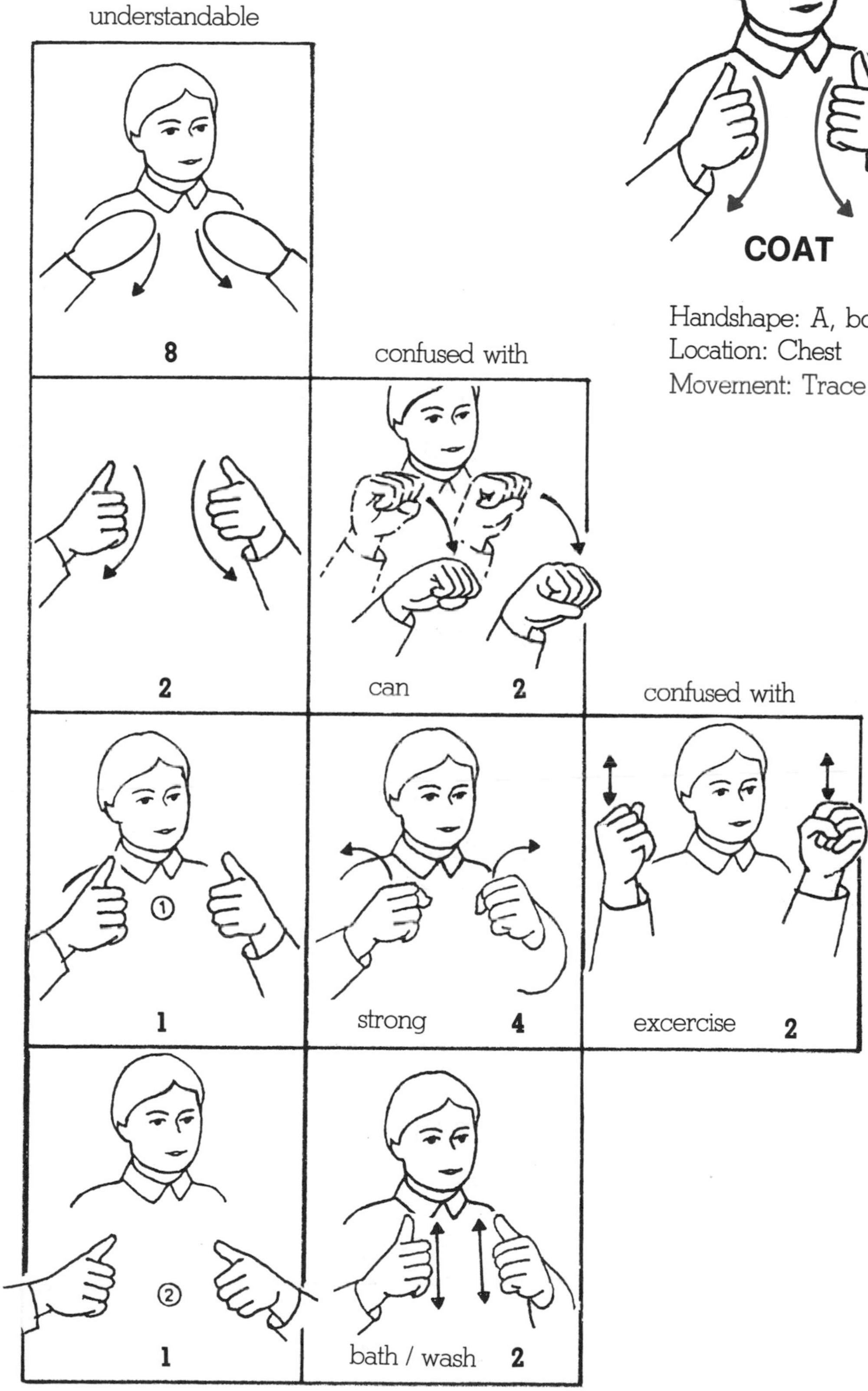

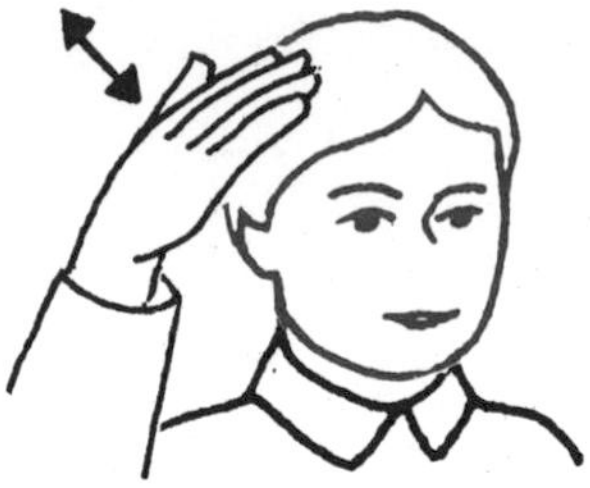

HAT

Handshape: Open B
Location: Top of head
Movement: Tap twice

confused with	understandable
 pat 1	 6
 head 2	 5
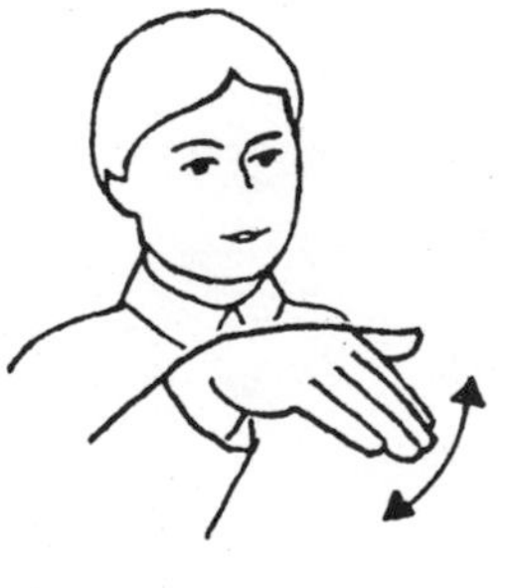 good-bye 7	 1

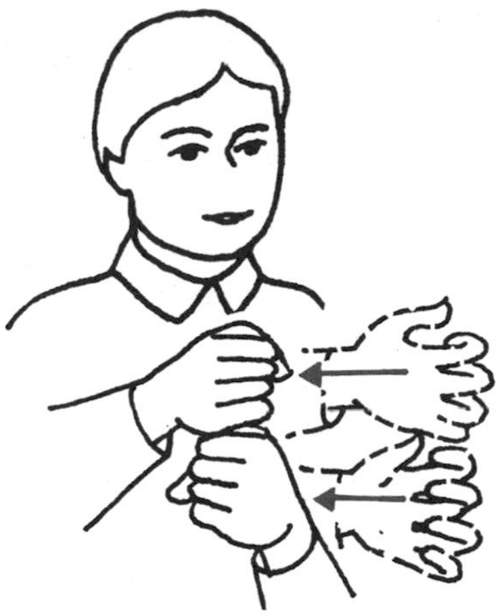

GET

Handshape: C, both
Location: In front of body
Movement: Close and toward self

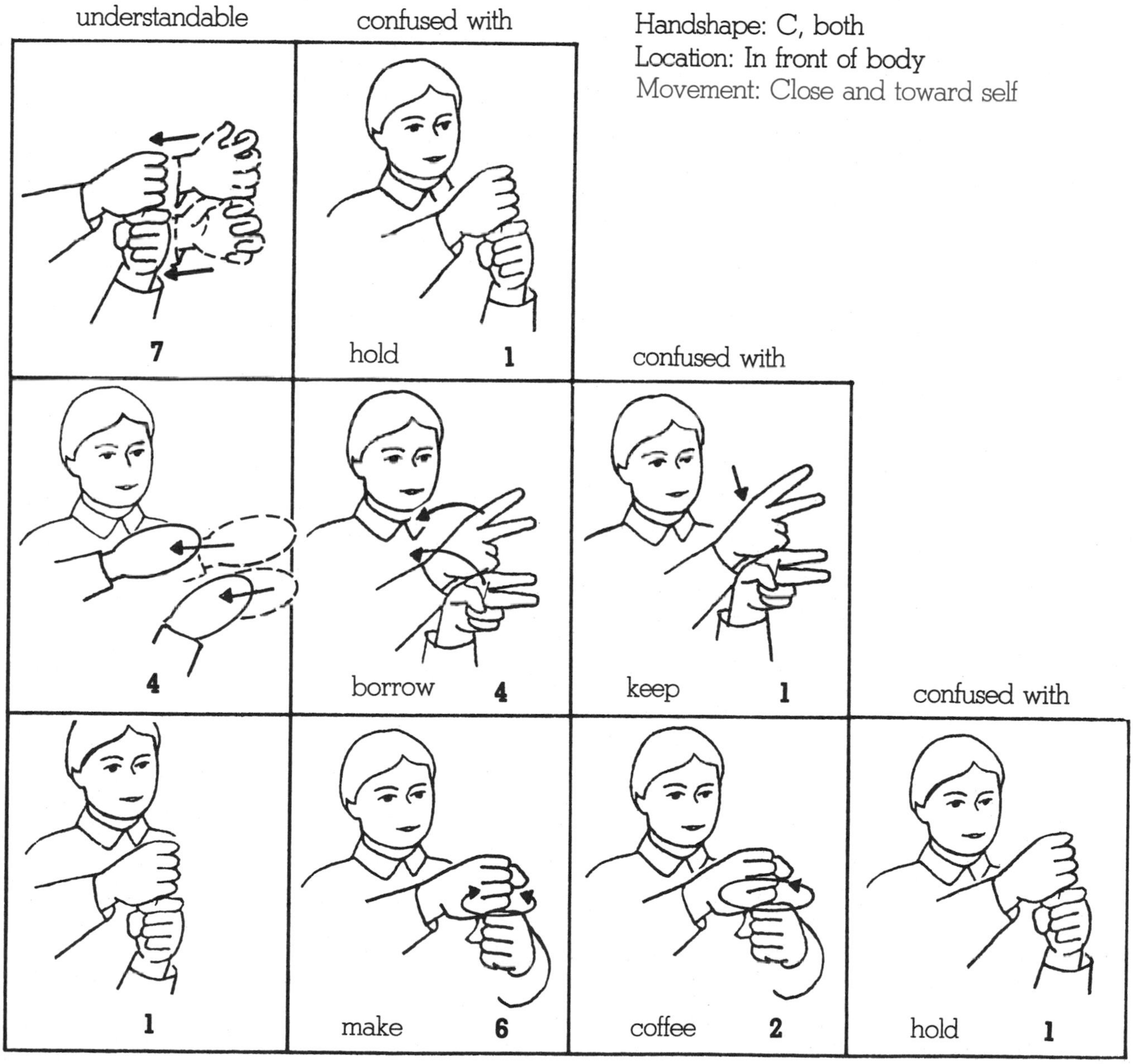

SNOW

Handshape: 5, both
Location: In front of body
Movement: Down, wiggling fingers

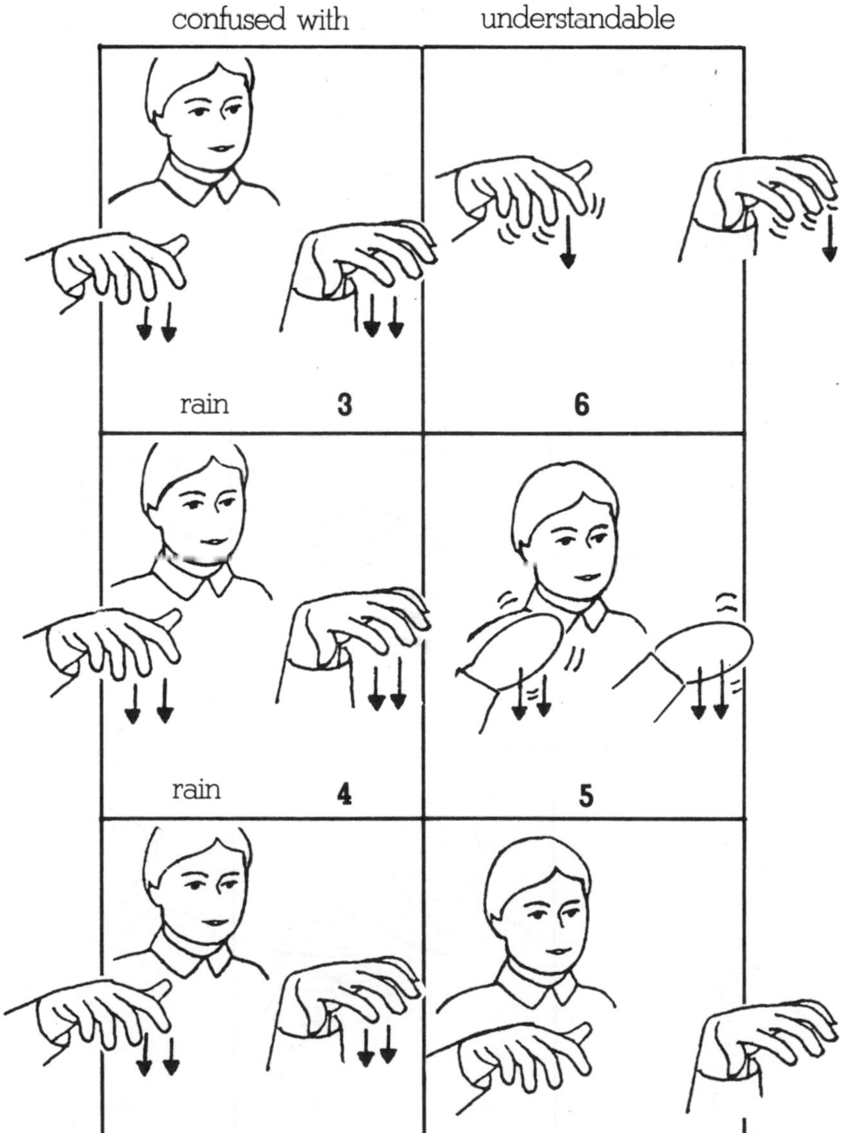

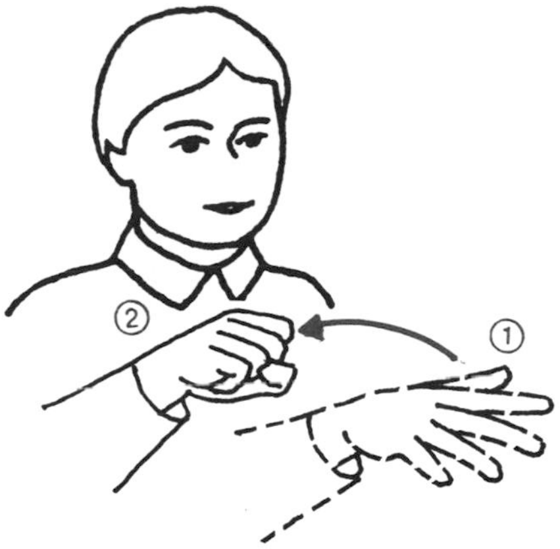

TAKE

Handshape: Open B, both
Location: In front of body
Movement: Close to fist, toward self

understandable

7

confused with

4

approve 1

confused with

2

catch 2

force 2

2

yes 2

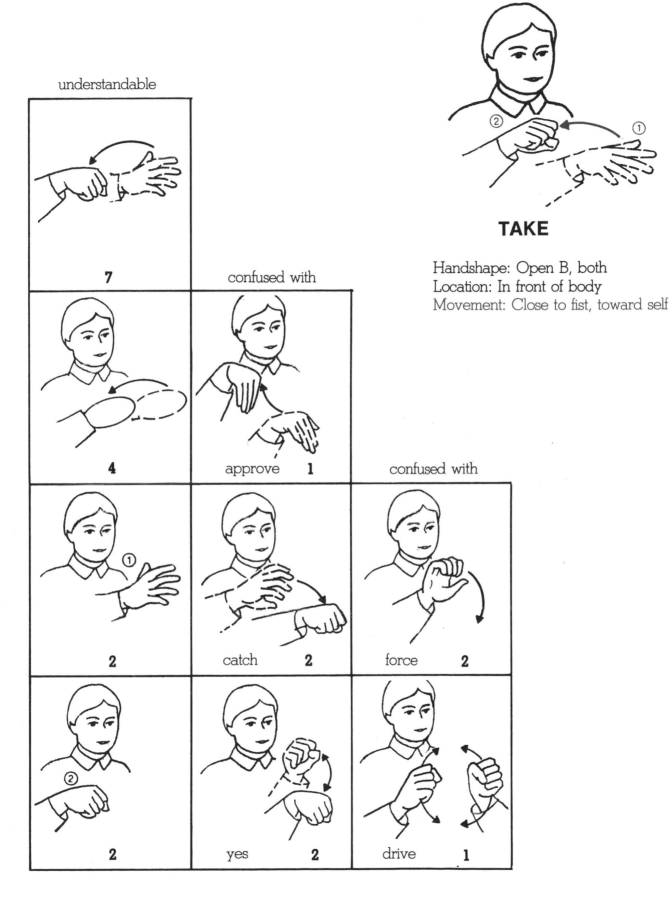

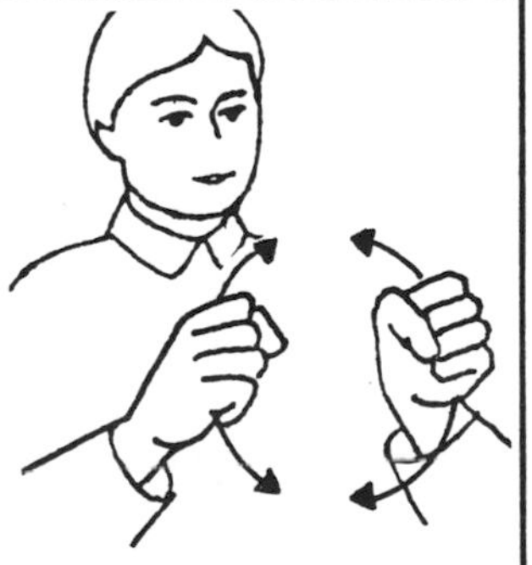

drive 1

DIRTY

Handshape: 5
Location: Chin
Movement: Wiggle fingers

	confused with	understandable
	 pig 1	 8
	 pig 6	 3
 hello 1	 fingerspell 3	 0

WOMAN

Handshape: A to 5
Location: Chin
Movement: Lower to chest

understandable

confused with

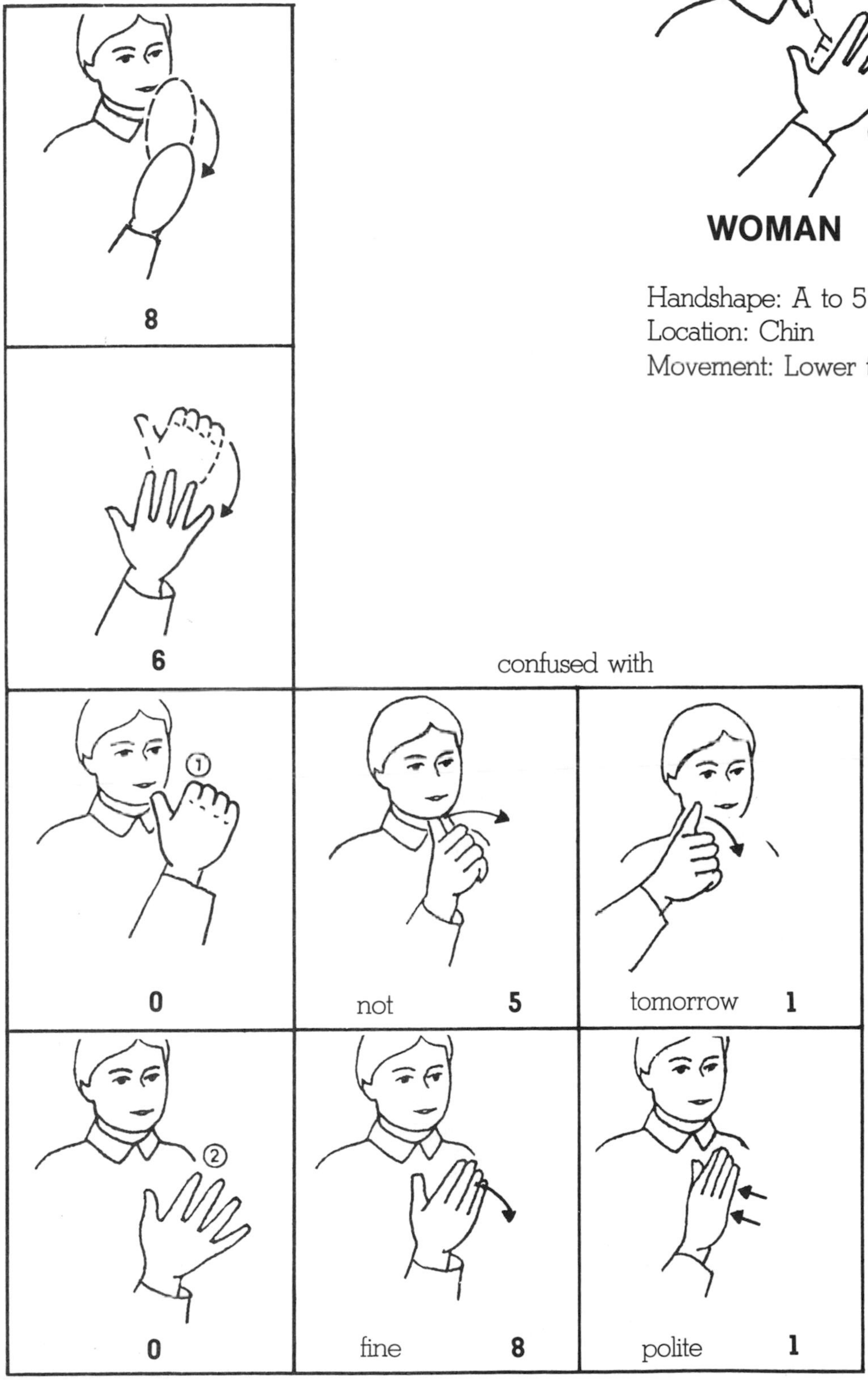

TUESDAY

Handshape: T
Location: In front of body
Movement: Small circle

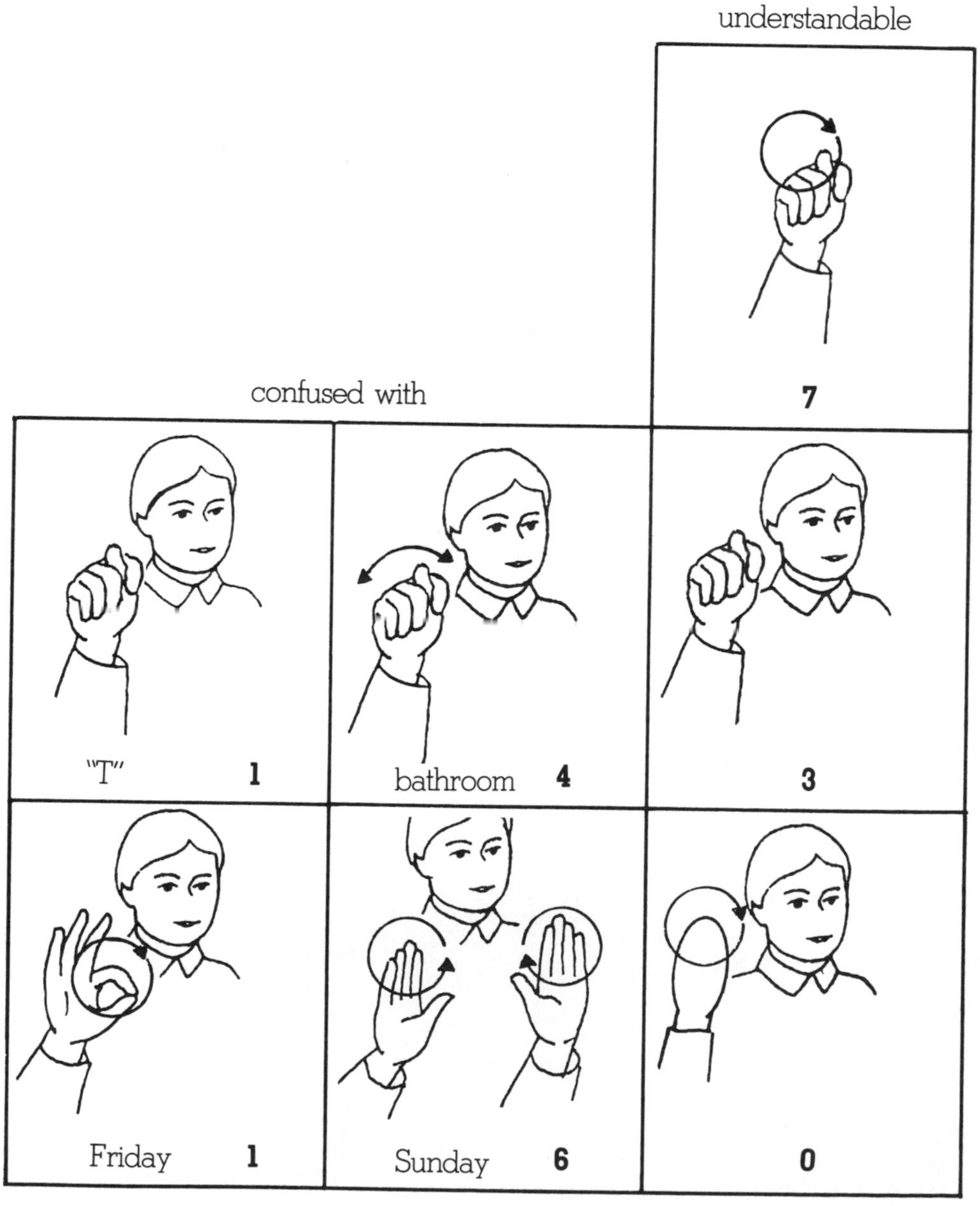

TURKEY

Handshape: G
Location: Chin
Movement: Down to chest and shake

understandable

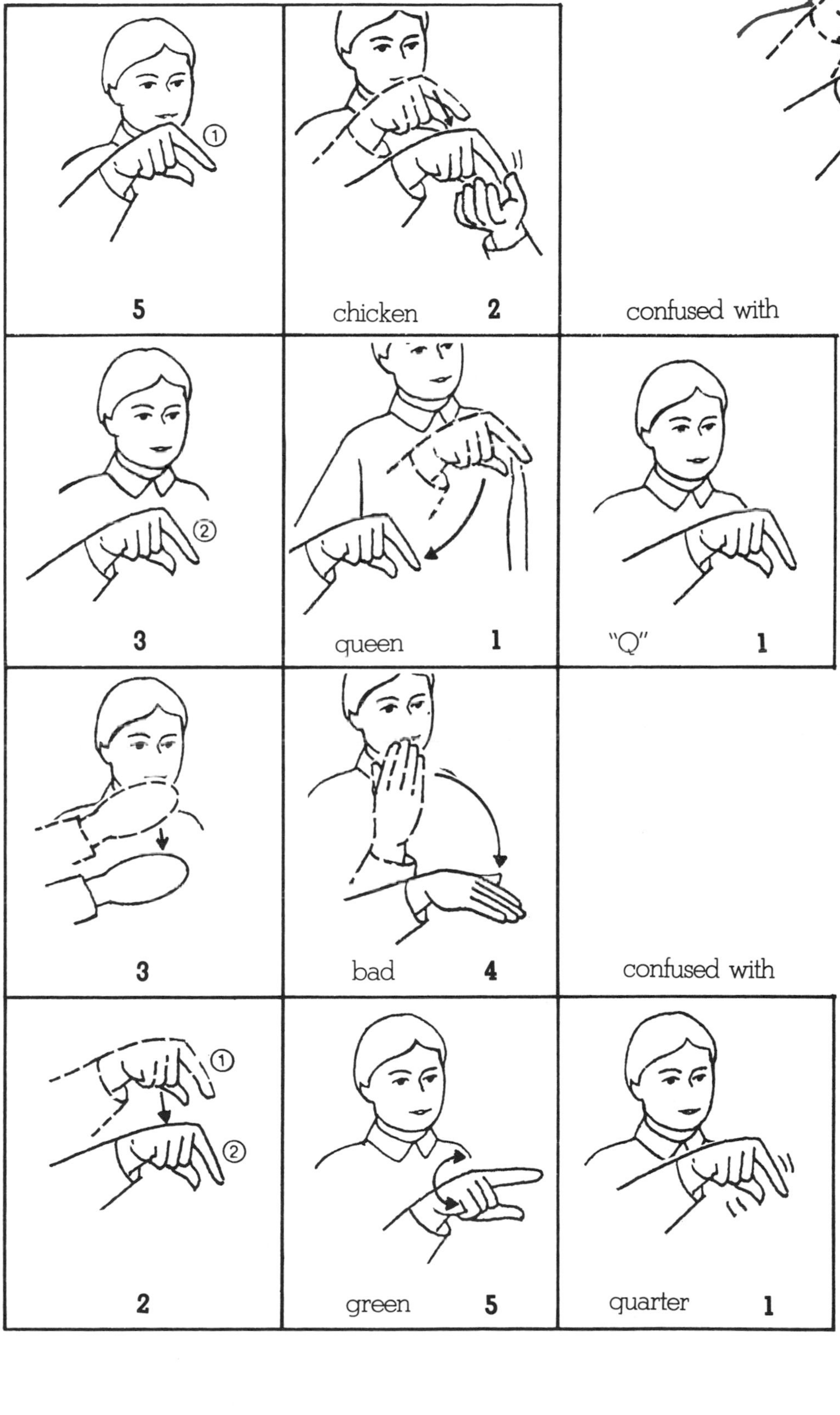

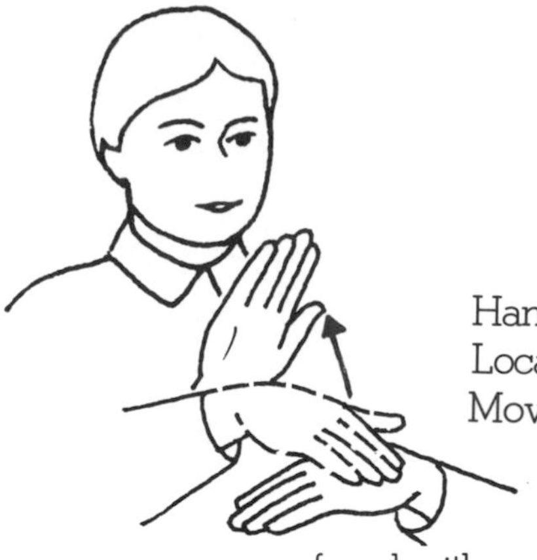

OFF

Handshape: Open B, both
Location: In front of body
Movement: Lift off back of hand

confused with | understandable

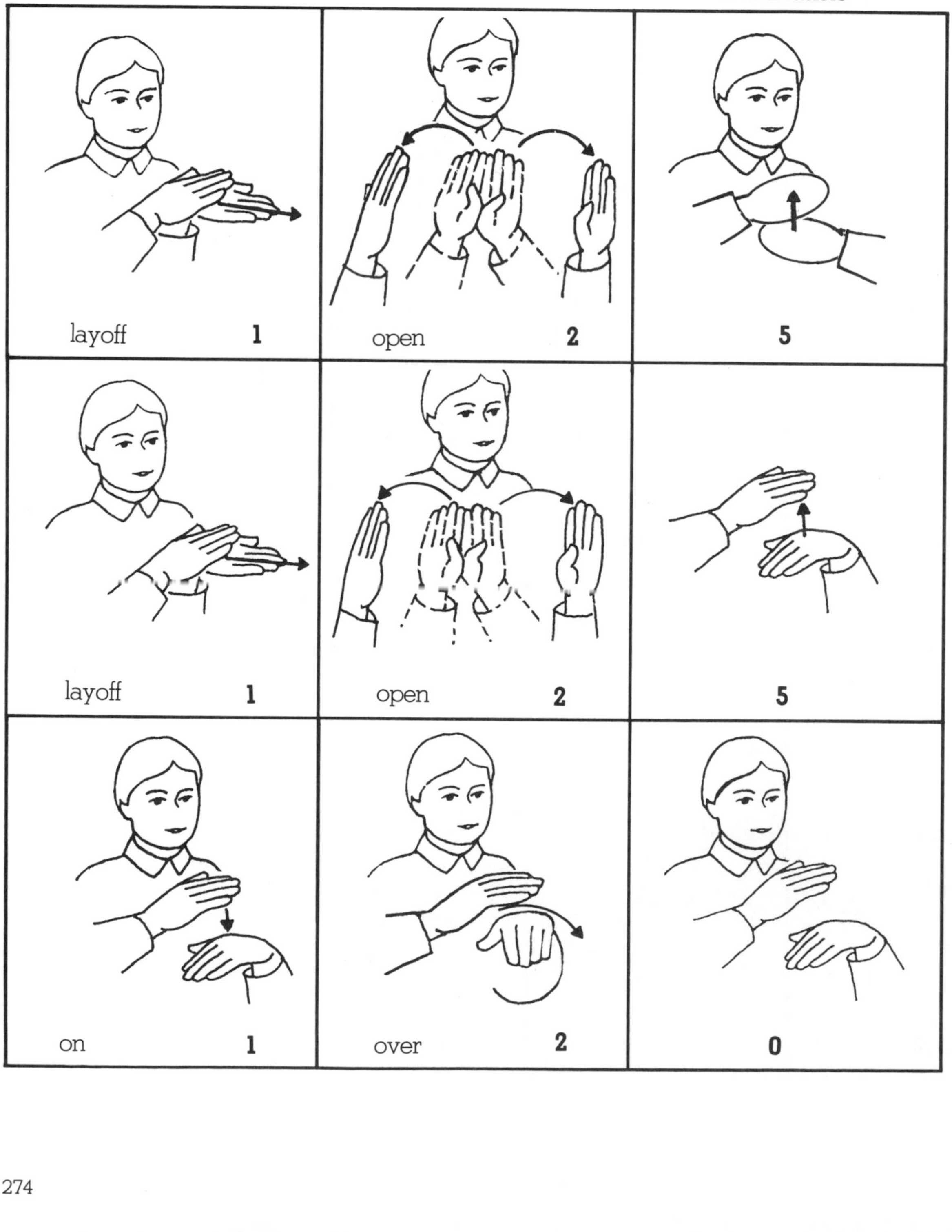

DOCTOR

Handshape: M; open B
Location: In front of body
Movement: Touch wrist

understandable	confused with	
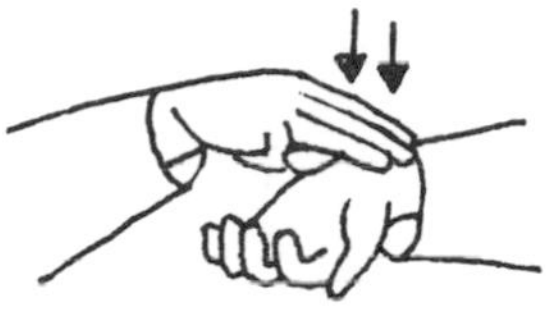 4	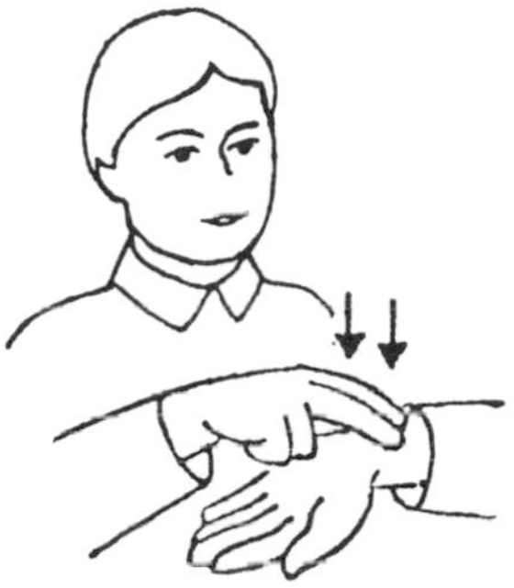 nurse 3	 medical 1
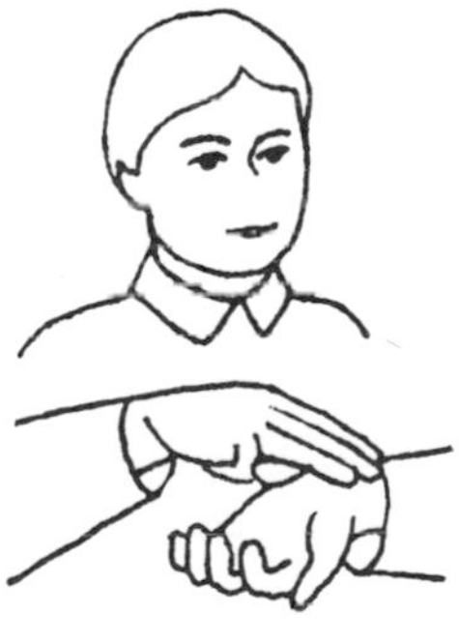 4	 nurse 4	 medical 2
 1	 school 4	 praise 2

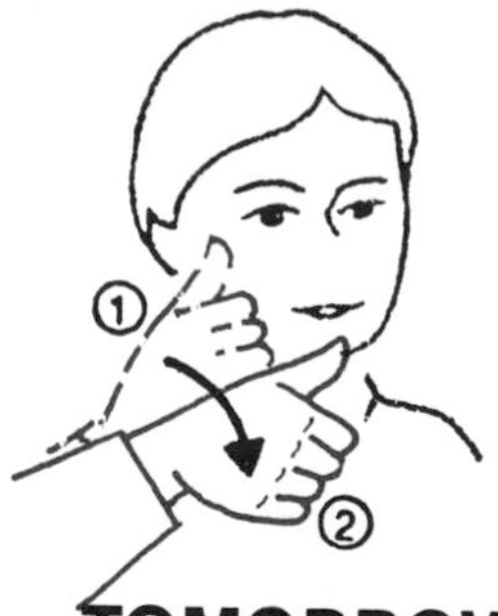

TOMORROW

Handshape: A, thumb extended
Location: Cheek
Movement: Turn forward

confused with | confused with | understandable

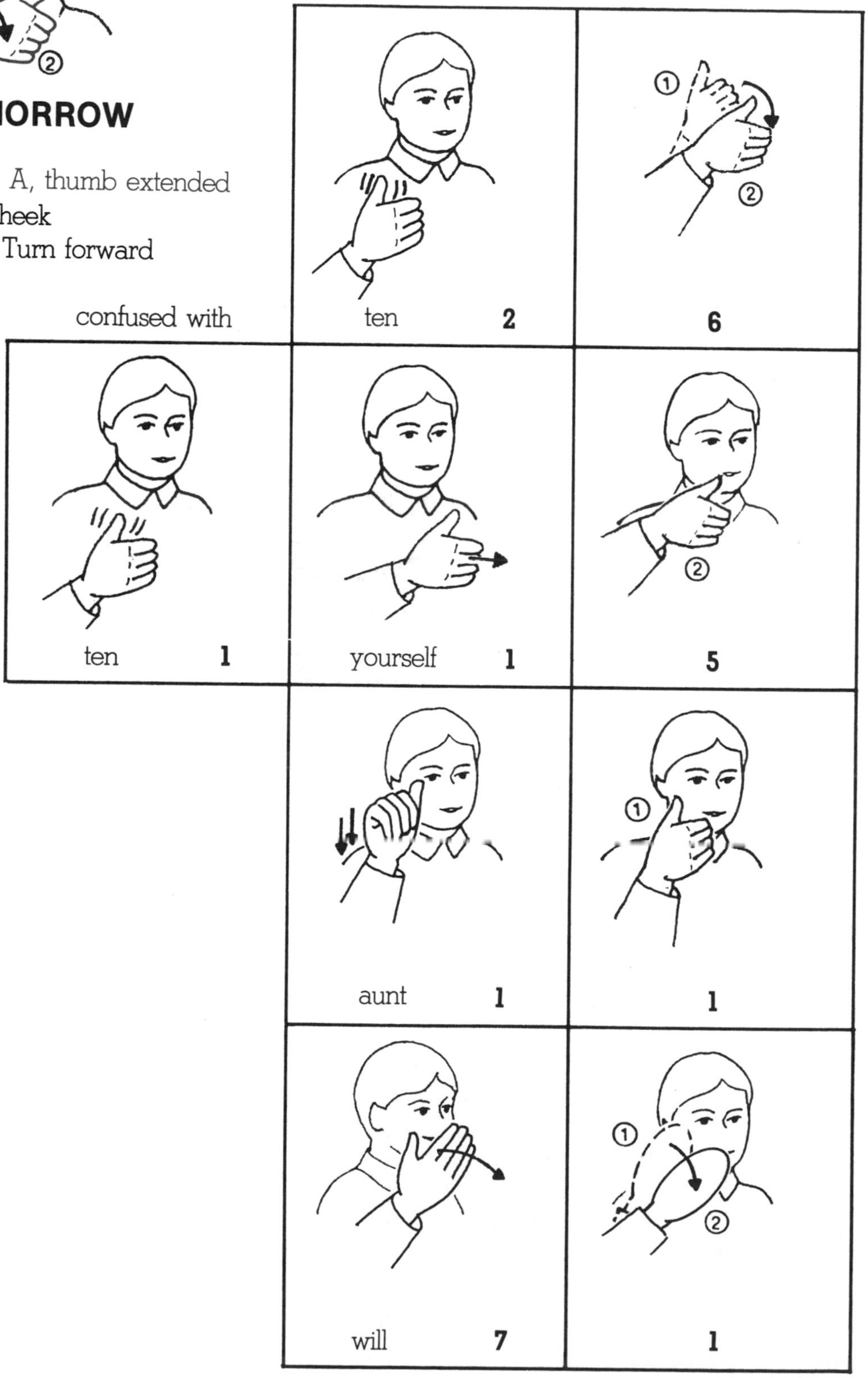

ARE

Handshape: R
Location: Mouth
Movement: Forward

understandable

6

confused with

red **1**

really **1**

4

respect **1**

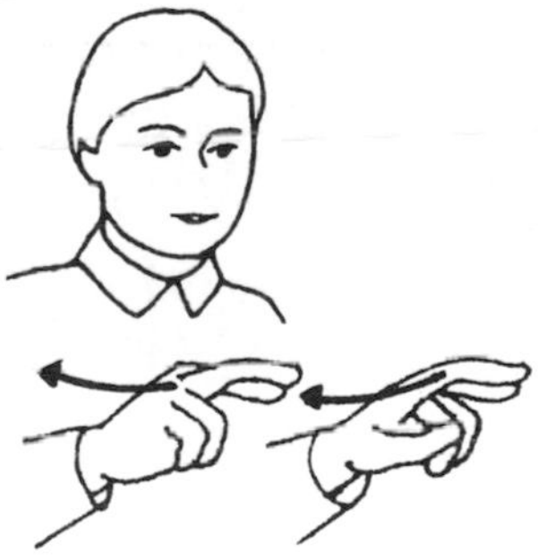

ready **1**

confused with

hope **1**

0

be **4**

HEAVY

Handshape: Open B, both
Location: In front of body
Movement: Lower slightly

understandable

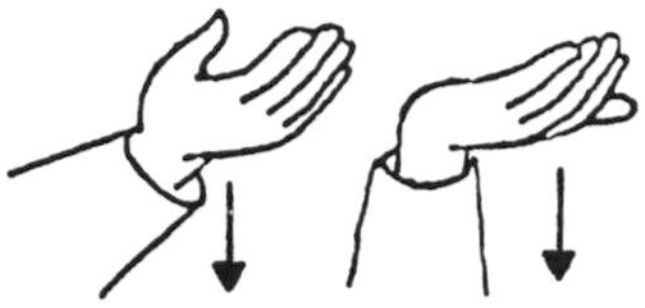

6

confused with

perhaps 1

now 4

3

now 3

here 5

1

TOILET

Handshape: T
Location: In front of body
Movement: Shake

understandable	confused with	confused with
 7	 bathroom 3	
 2	 bathroom 6	
 0	 hello / hi 7	 bye-bye 2

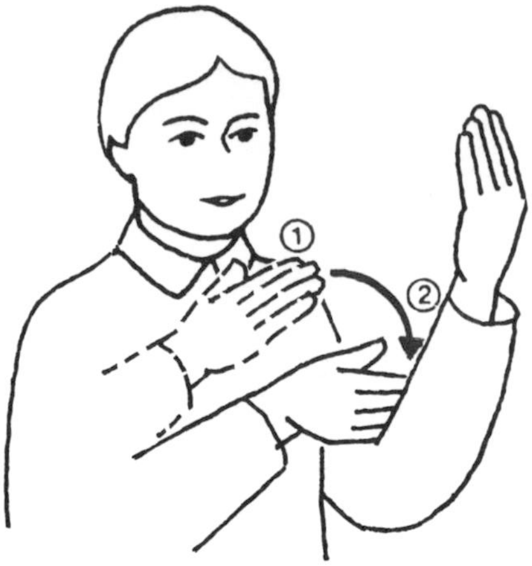

BIRTHDAY

Handshape: Open B, both; 1, both
Location: In front of body
Movement: Stomach to palm, lower from elbow

THESE

Handshape: 1; open B
Location: In front of body
Movement: Tap several times

understandable

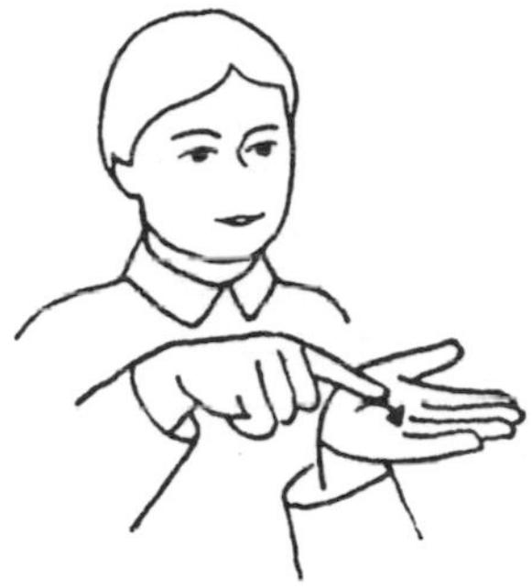

6

3

confused with

calculator **4**

0

study **3**

confused with

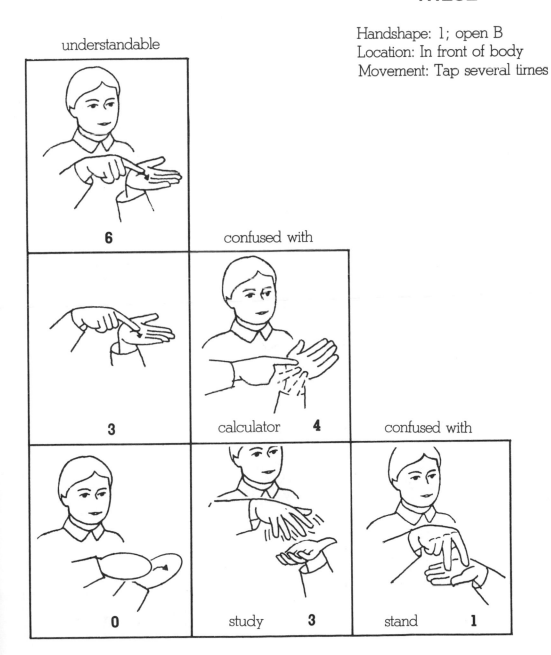

stand **1**

THINK

Handshape: 1
Location: Forehead
Movement: Small circle

understandable

confused with

confused with

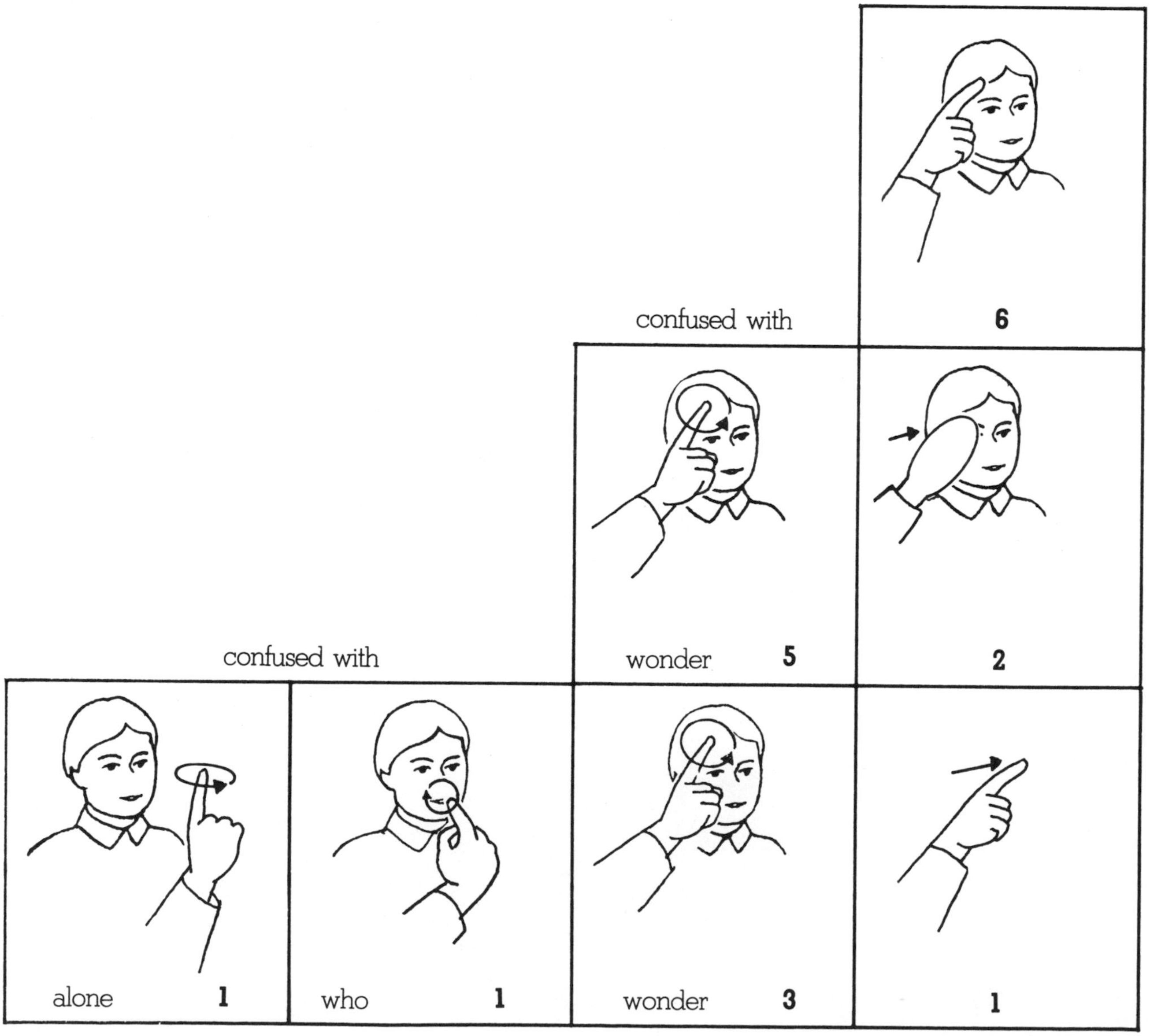

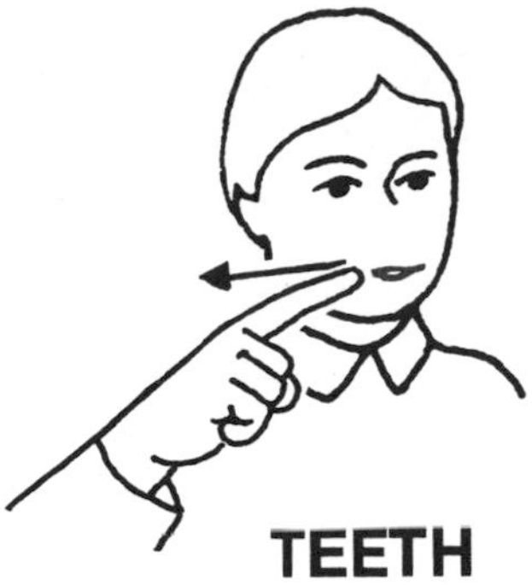

TEETH

Handshape: 1
Location: Teeth
Movement: Trace teeth

understandable

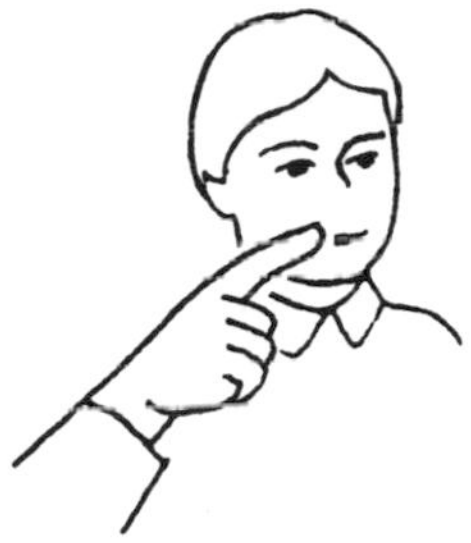

7

confused with

2

toothbrush **5**

0

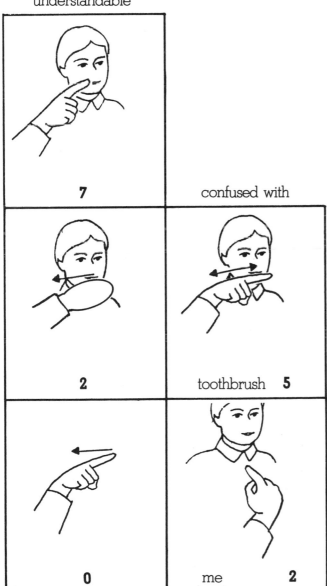

me **2**

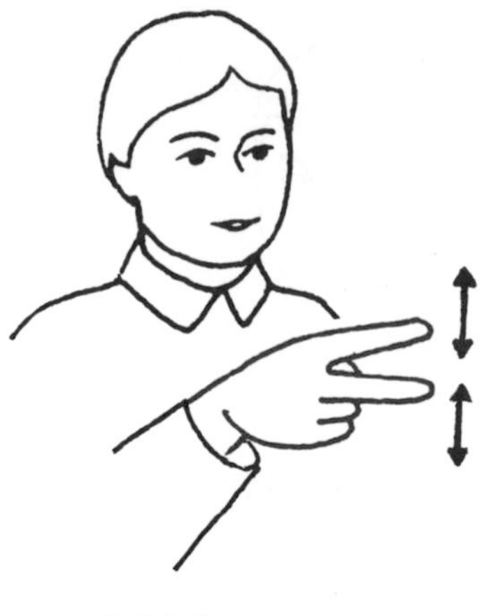

SCISSORS

Handshape: V
Location: In front of body
Movement: Open and close twice

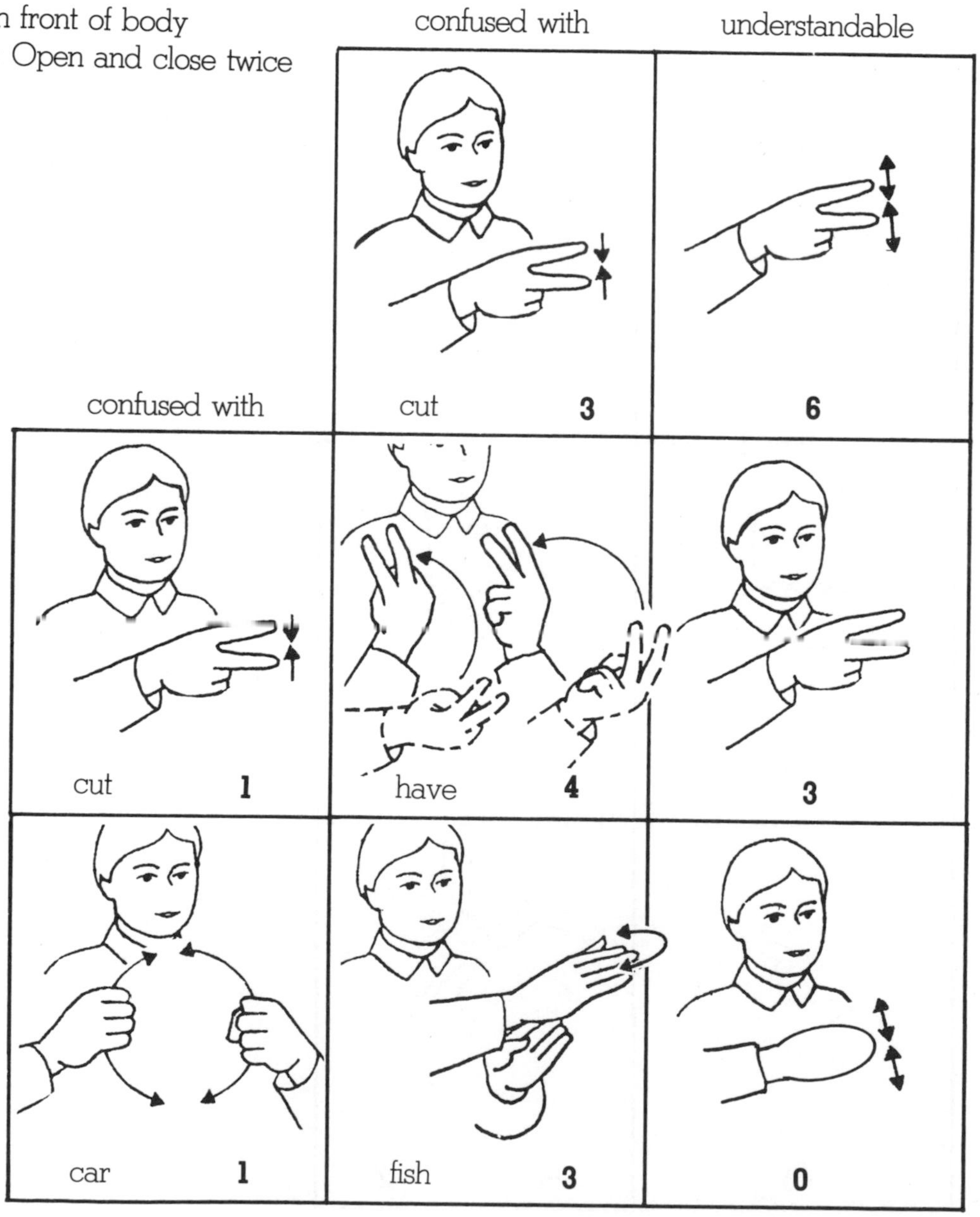

BED

Handshape: Open B, both
Location: Cheek
Movement: Rest head on hands

understandable | confused with

4

sleep **5**

3

sleep **4**

confused with

1

praise **2**

school **1**

sleep **1**

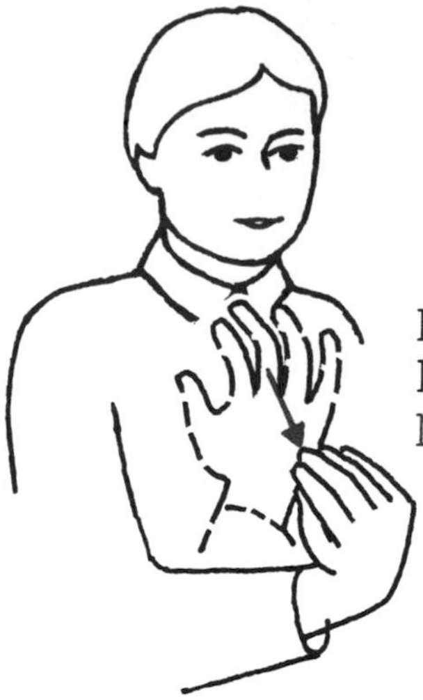

WHITE

Handshape: 5
Location: Chest
Movement: Out and close fingers

confused with

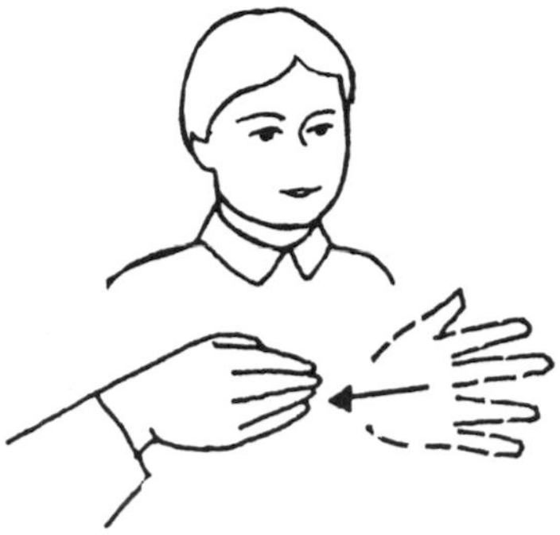

and 3

understandable

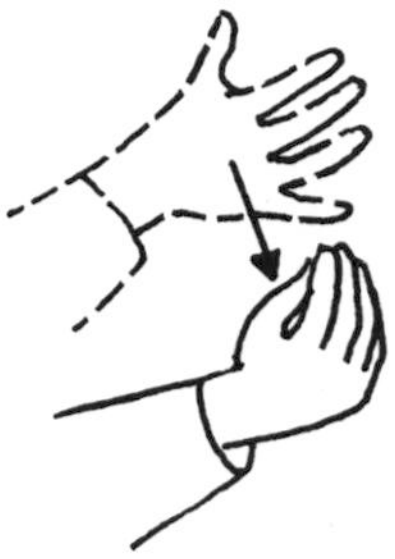

5

breath 1

3

confused with

my 3

sorry 4

0

WHO

Handshape: L
Location: Mouth
Movement: Circle mouth

understandable | confused with

4

"21" 2

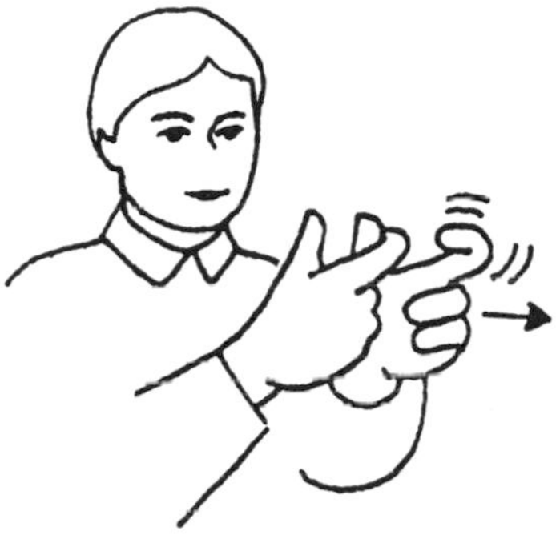

run 1

3

confused with

1

mother 4

DIFFERENT

Handshape: 1, both
Location: In front of body
Movement: Uncross and move fingers apart

confused with   understandable

confused with

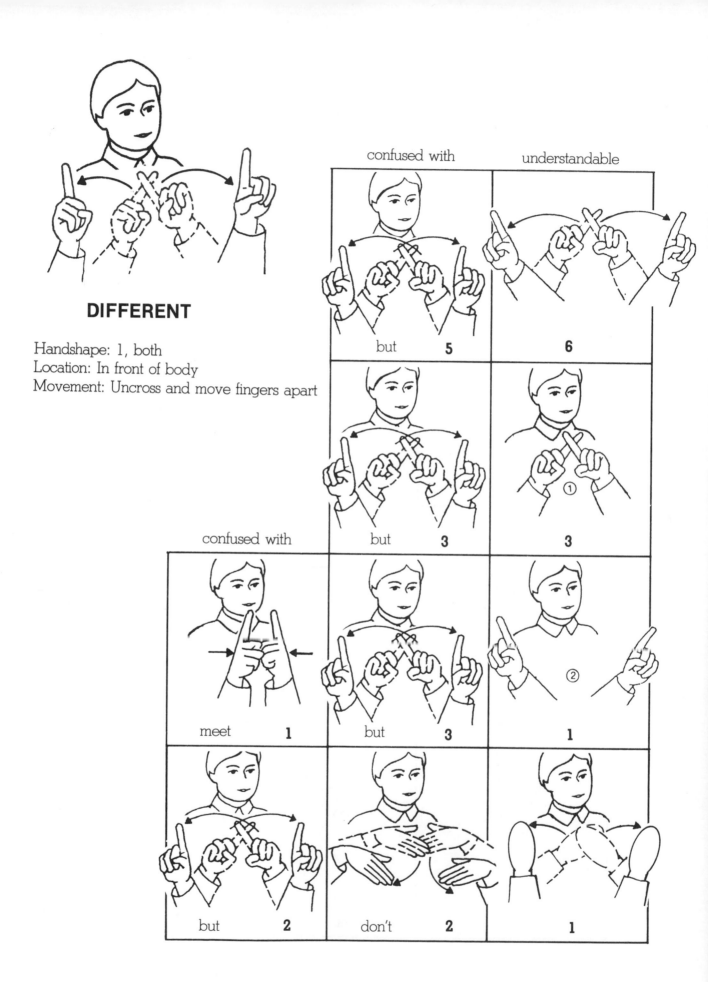

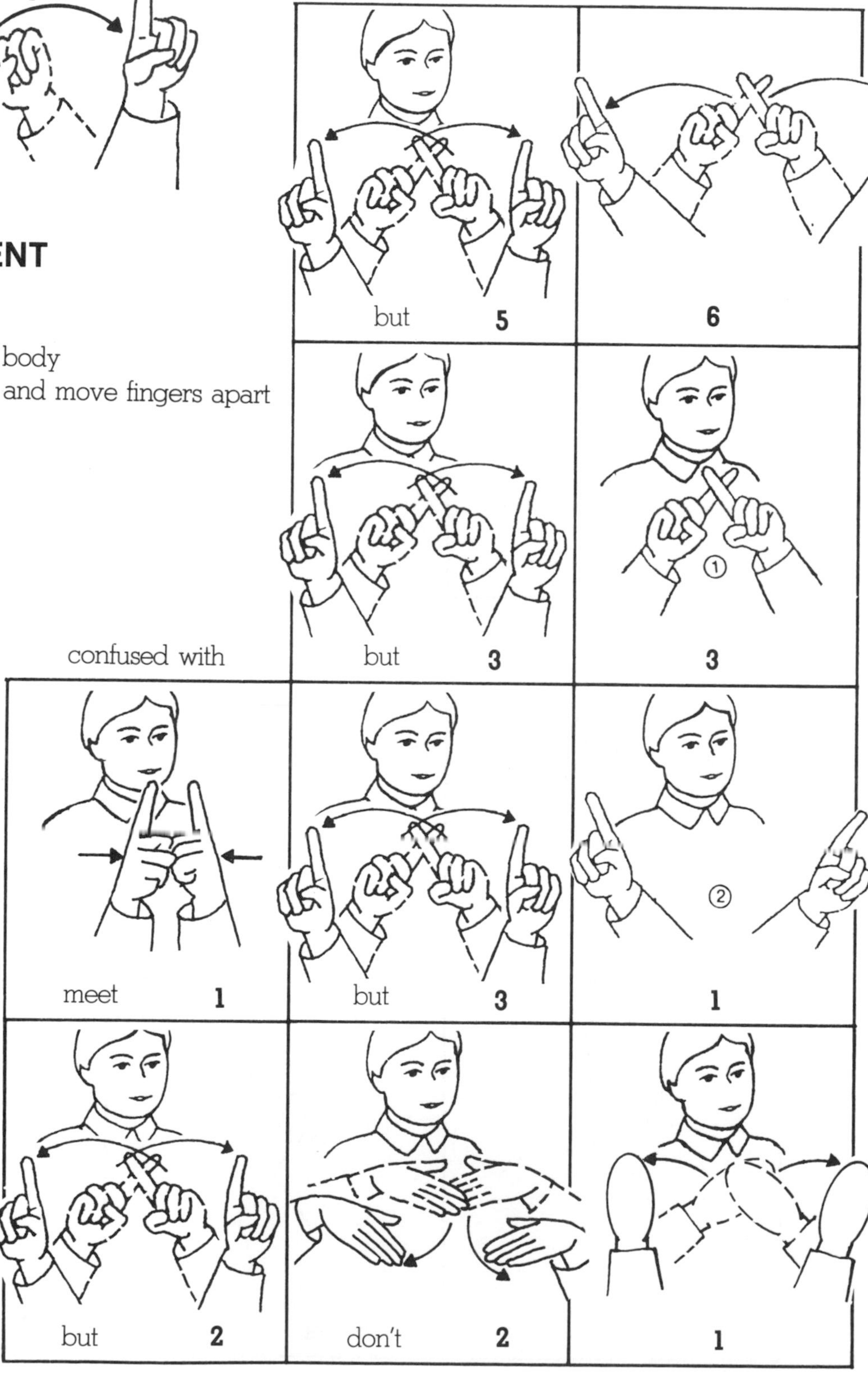

BREAKFAST

Handshape: B
Location: Chin
Movement: Tap twice

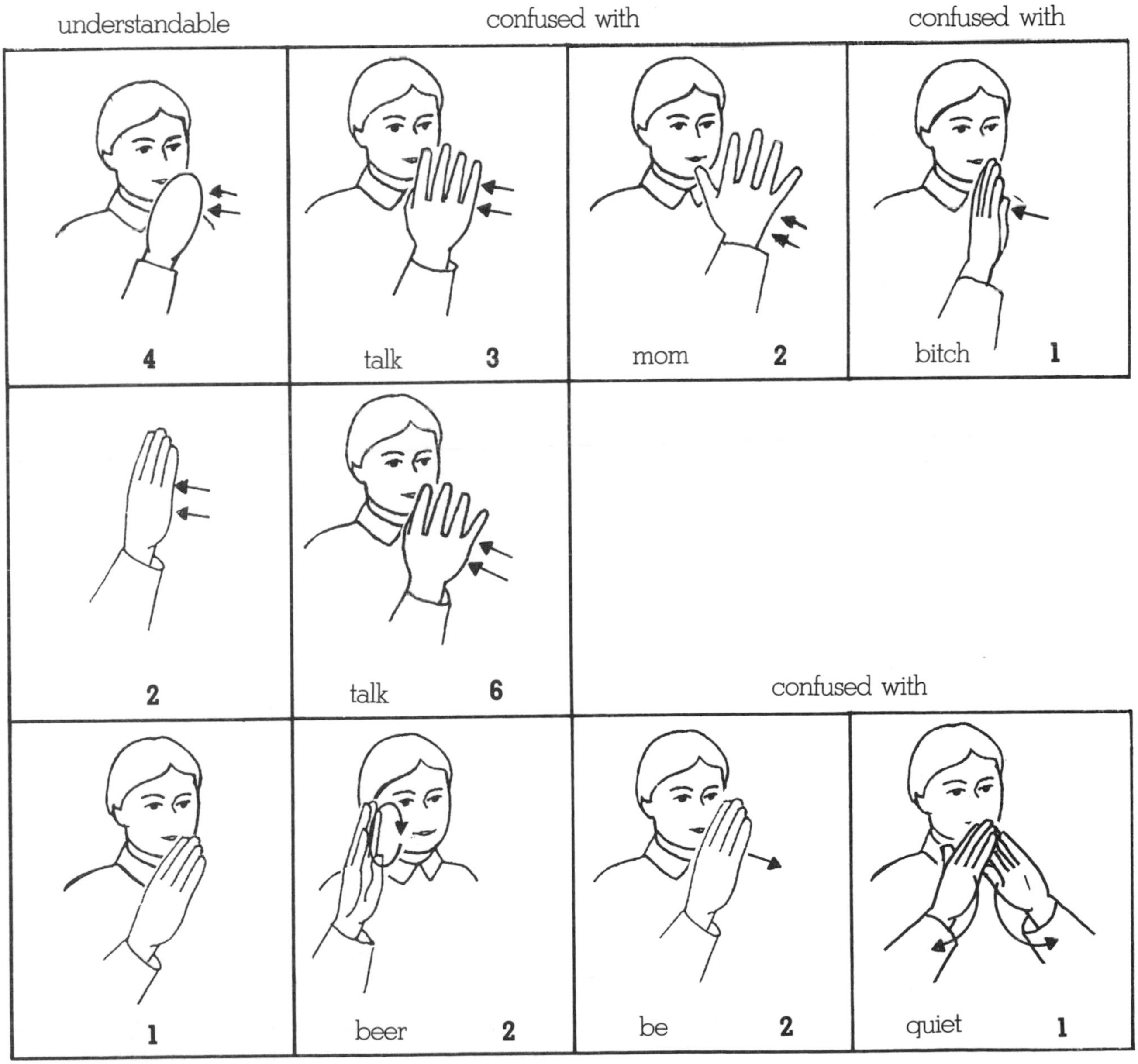

BROTHER

Handshape: 1, both
Location: Forehead
Movement: To front of body, tap fingers together

confused with | understandable

son 3

6

confused with

right 2

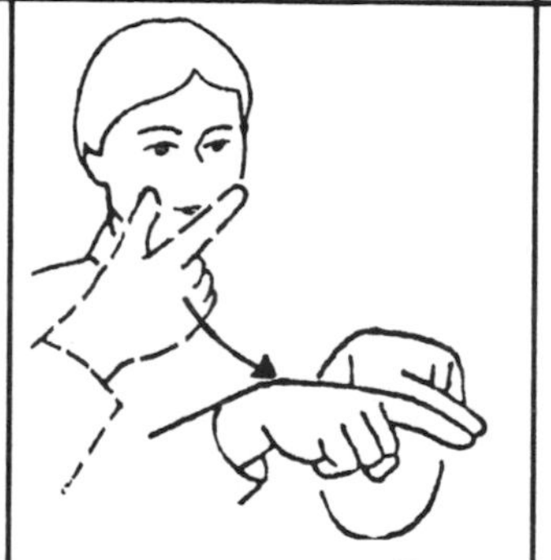

sister 5

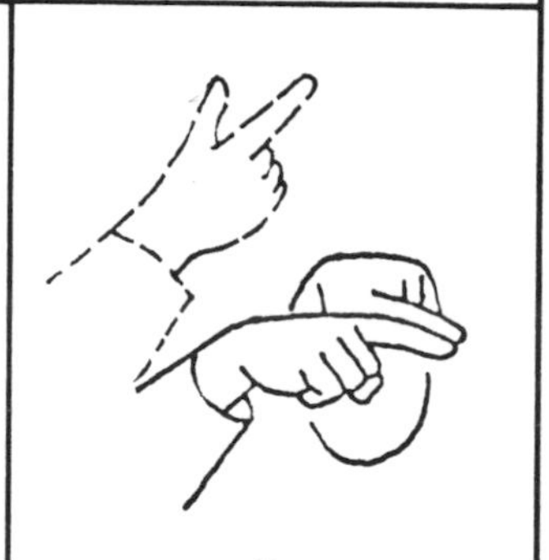

3

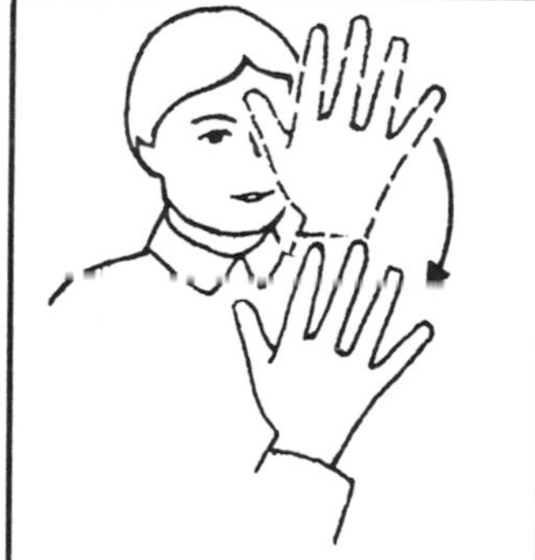

man 1

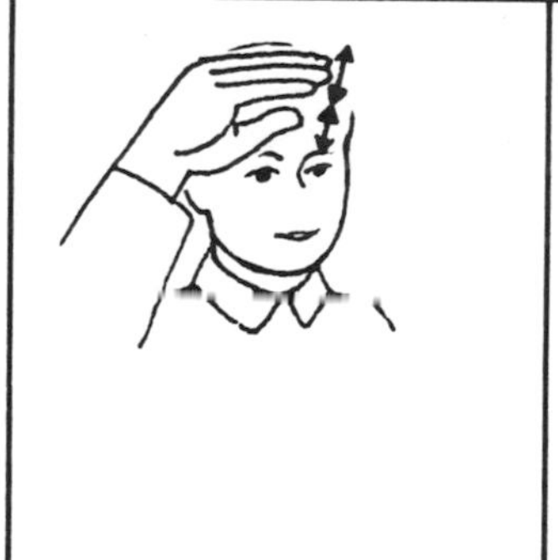

boy 6

0

same 9

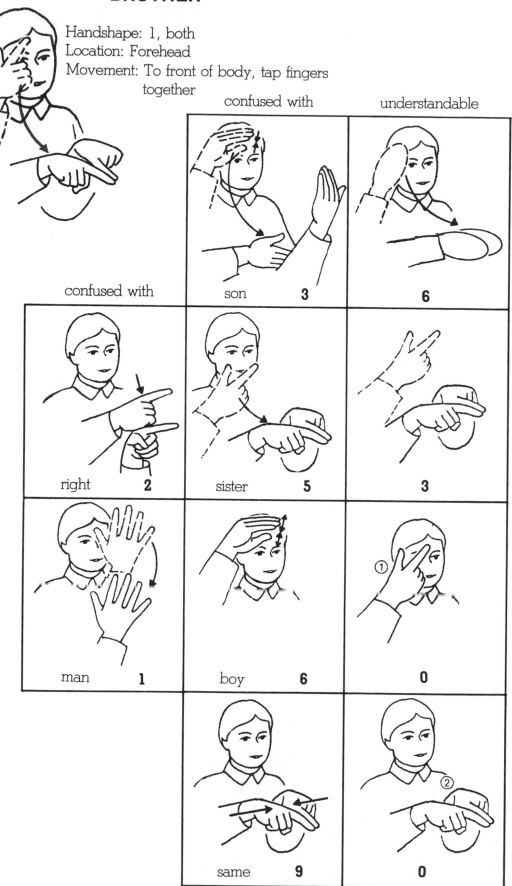

0

CHILD

Handshape: Open B
Location: Waist
Movement: Down slightly, twice

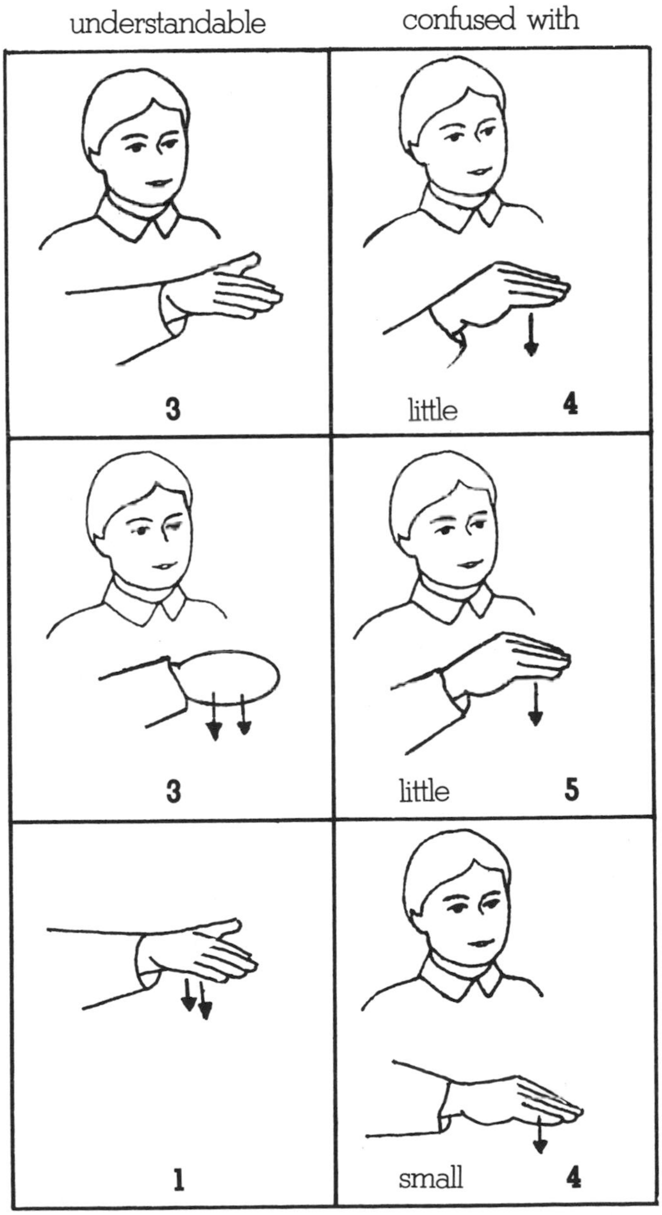

PILLOW

Handshape: Open B, both
Location: Cheek
Movement: Tilt head

confused with		understandable
 bed **3**	 sleep **3**	 **3**
 bed **2**	 sleep **4**	 **3**
 decrease **1**	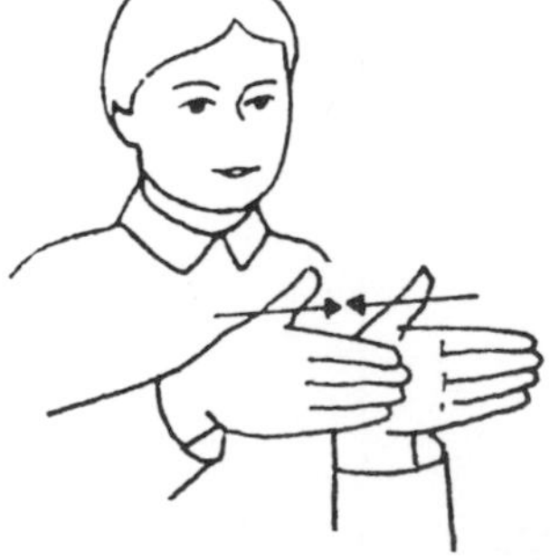 small **5**	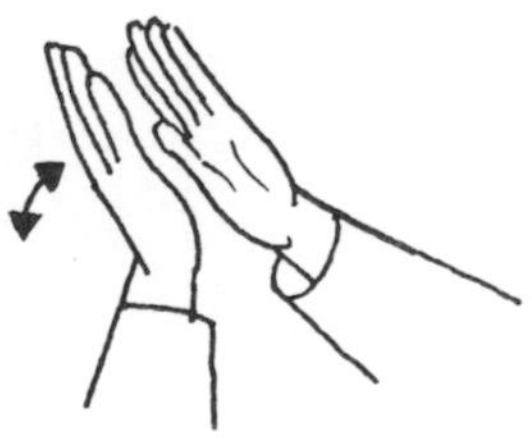 **0**

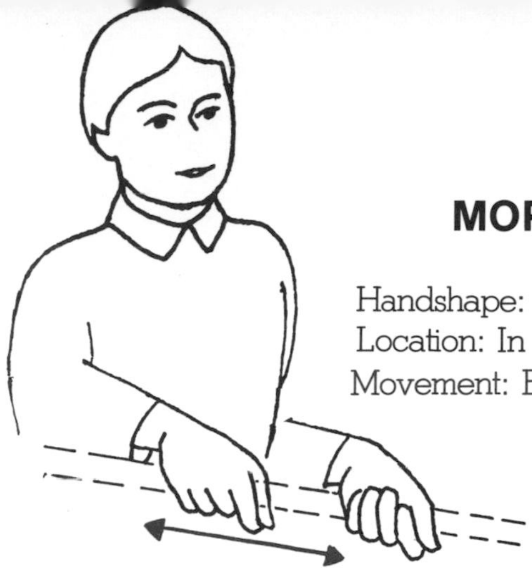

MOP

Handshape: A, both
Location: In front of body
Movement: Back and forth

understandable confused with

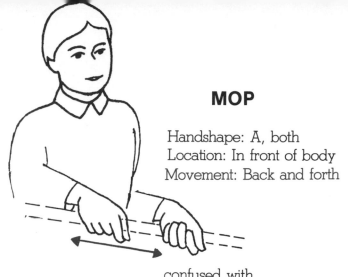

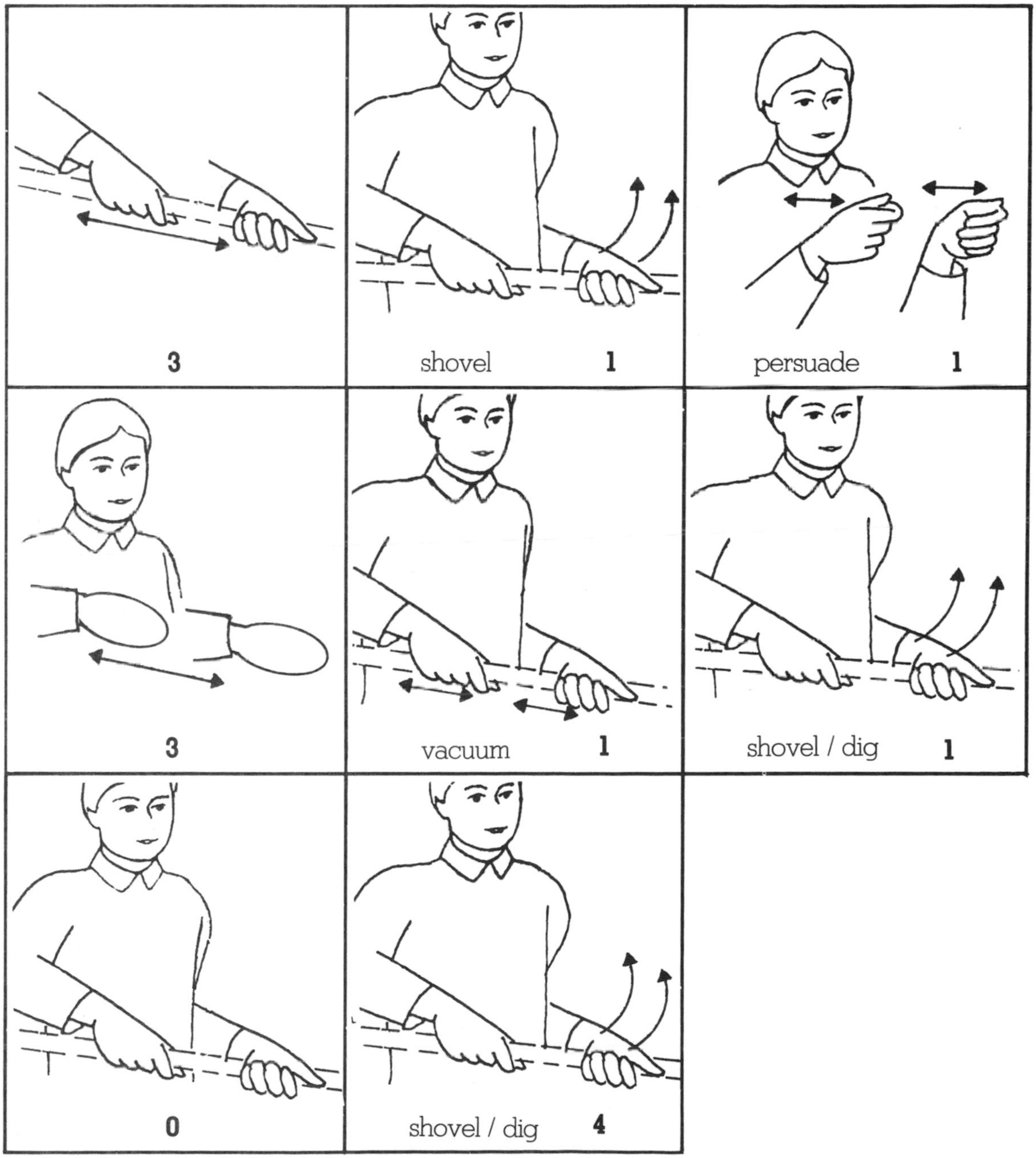

SPOON

Handshape: H; open B
Location: In front of body
Movement: Palm to mouth

confused with — understandable

confused with

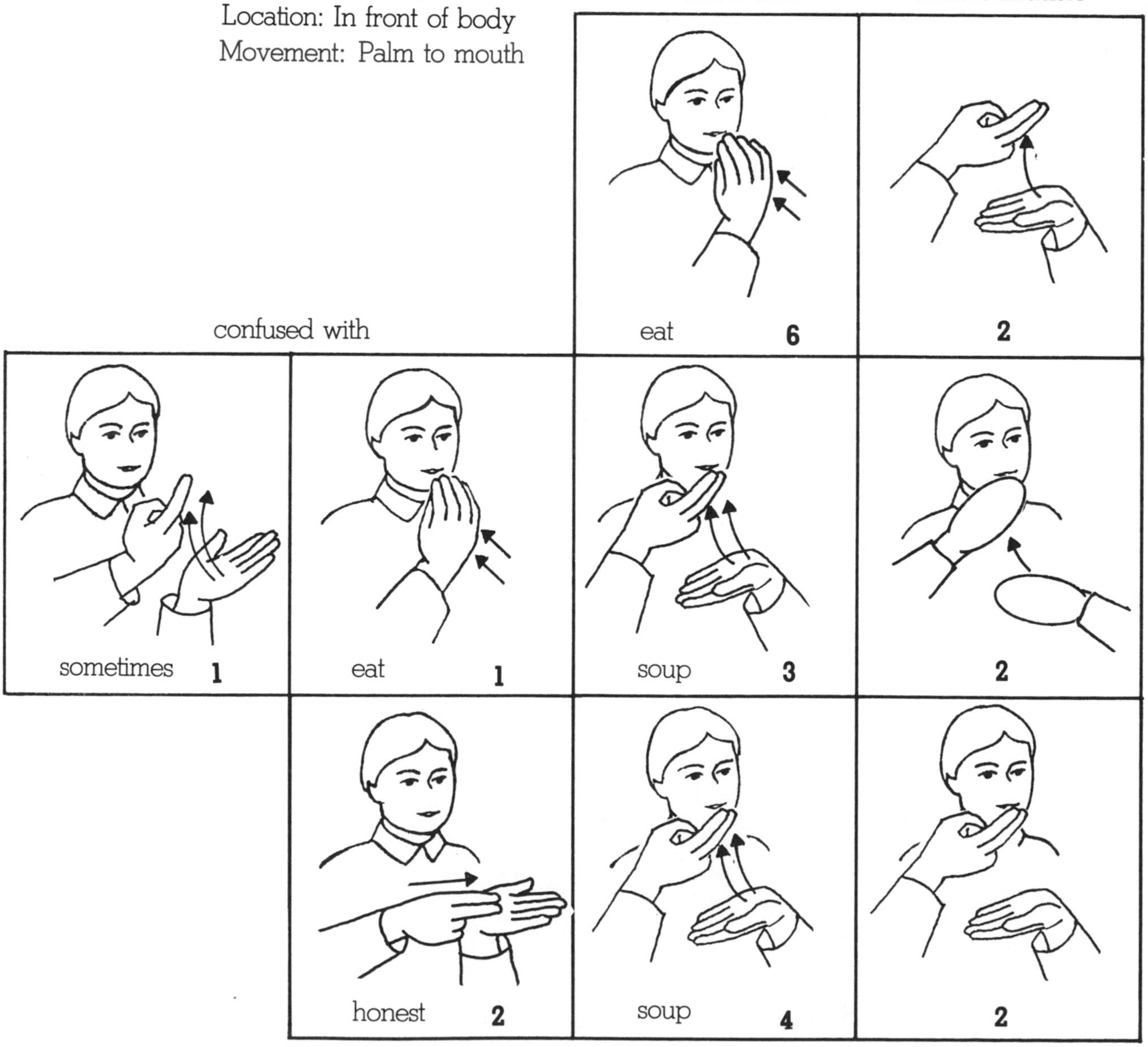

PINK

Handshape: P
Location: Lips
Movement: Rub lips

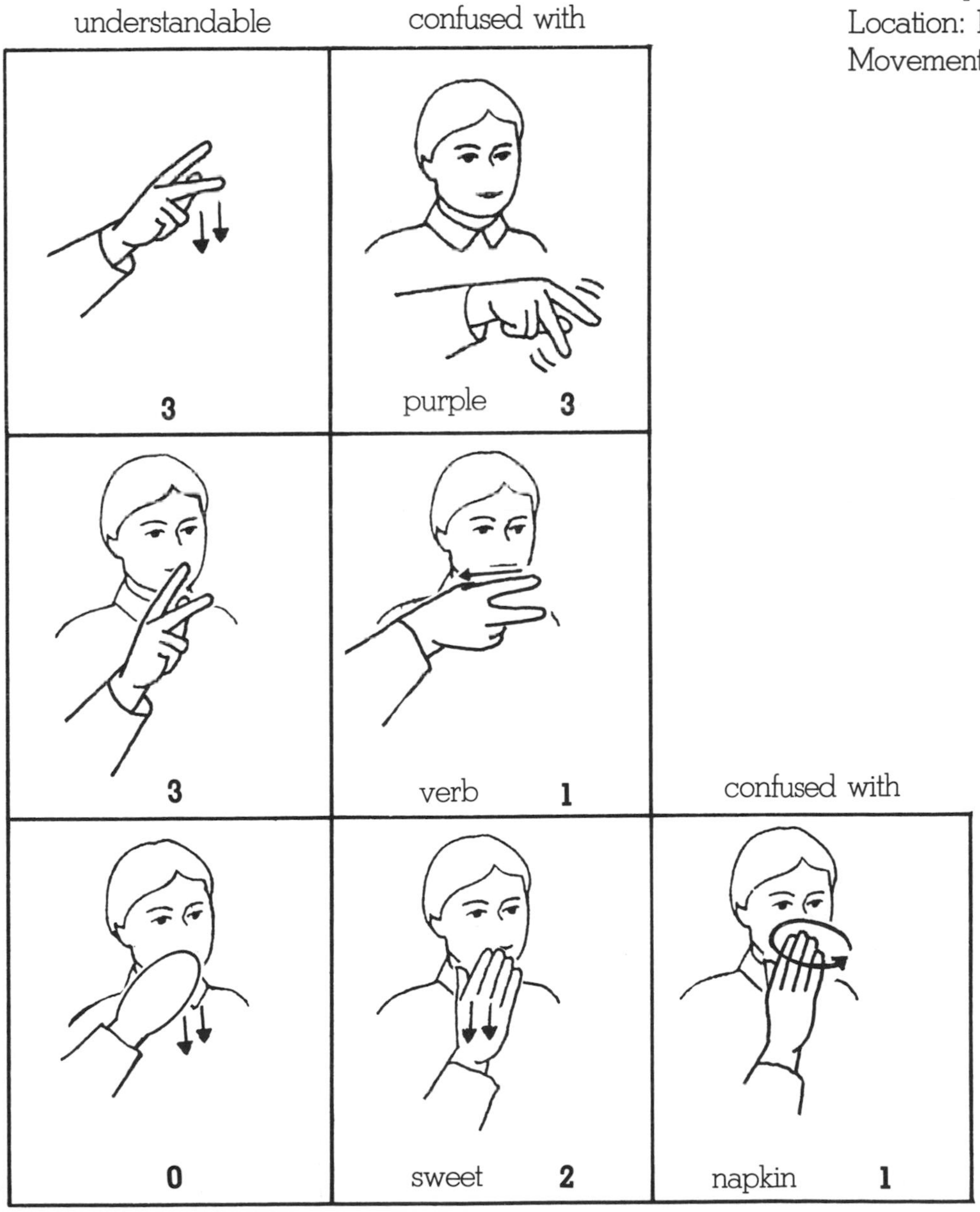

THANK YOU

Handshape: Open B
Location: Mouth
Movement: Outward

confused with

admit **2**

understandable

3

confused with

good afternoon **1**

good night **1**

bad **3**

2

bad **1**

oops **3**

good **3**

1

here **1**

give them to me **1**

beg **3**

0

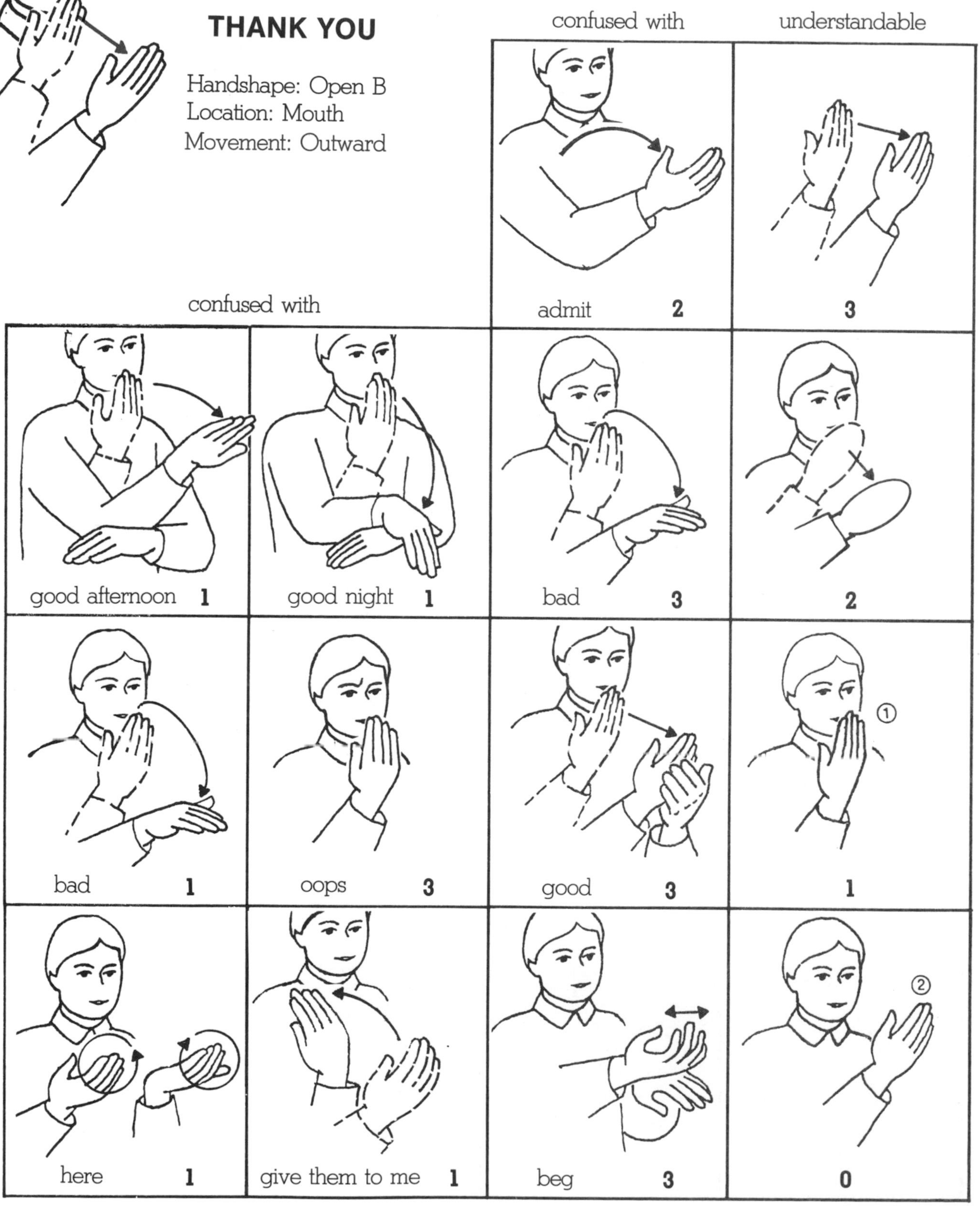

understandable | confused with

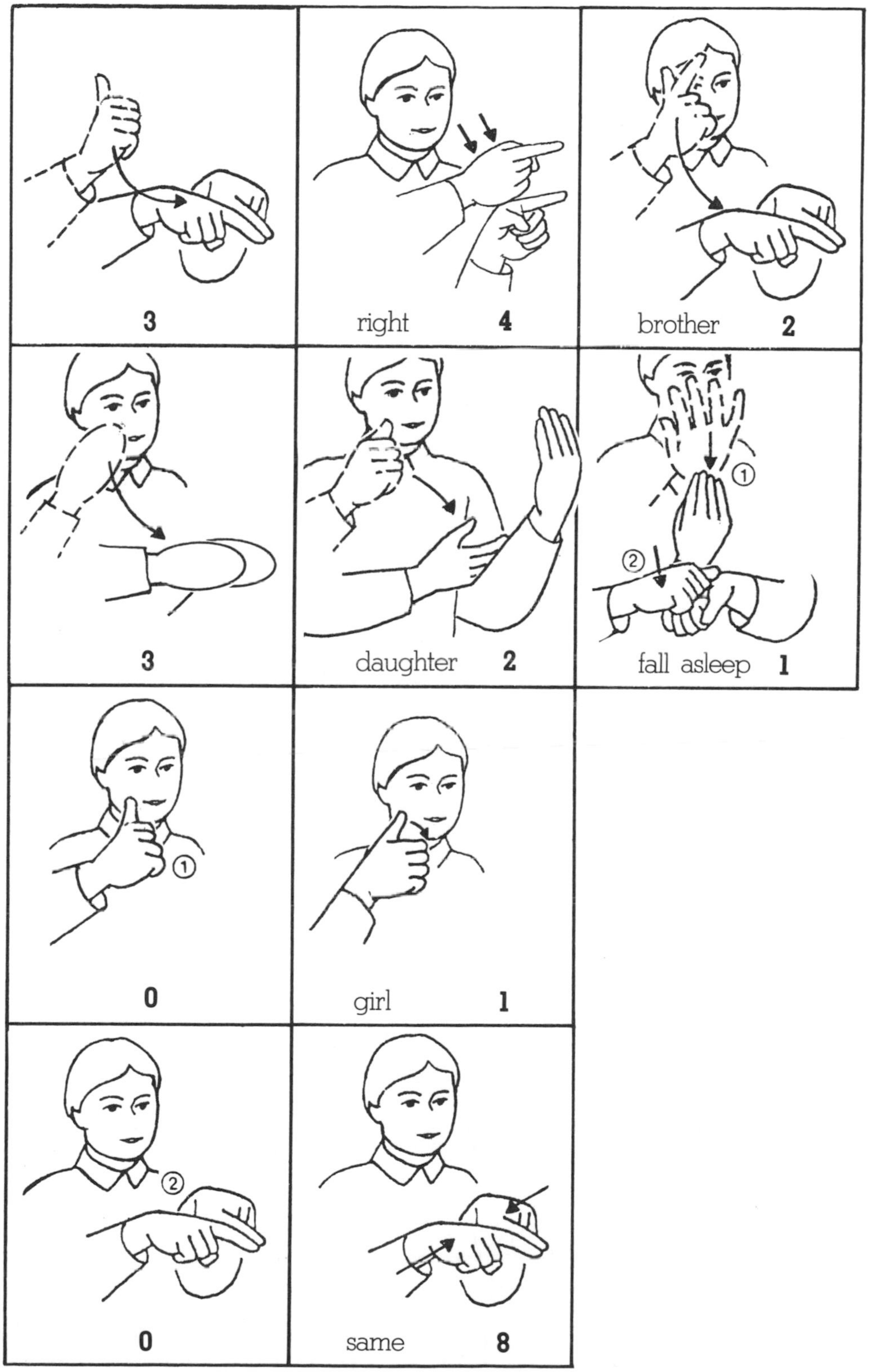

SISTER

Handshape: 1, both
Location: Cheek
Movement: To front of body, tap fingers together

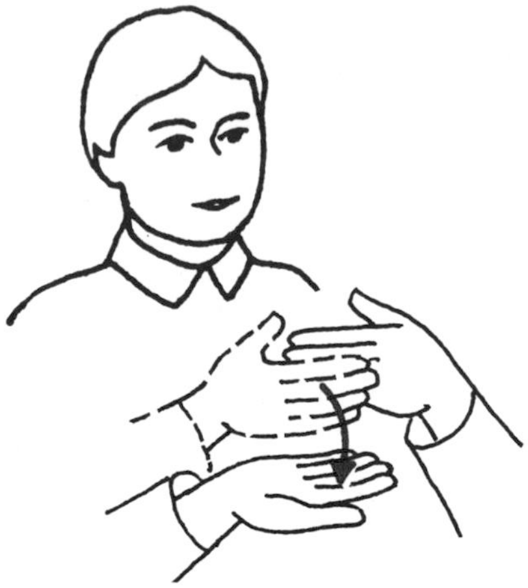

AFTER
Handshape: Open B, both
Location: In front of body
Movement: Far hand outward
confused with
understandable
next 5
2
from now on 3
2
confused with
against 1
from now on 1
close / near 4
1
close / near 3
against 4
0
①
②

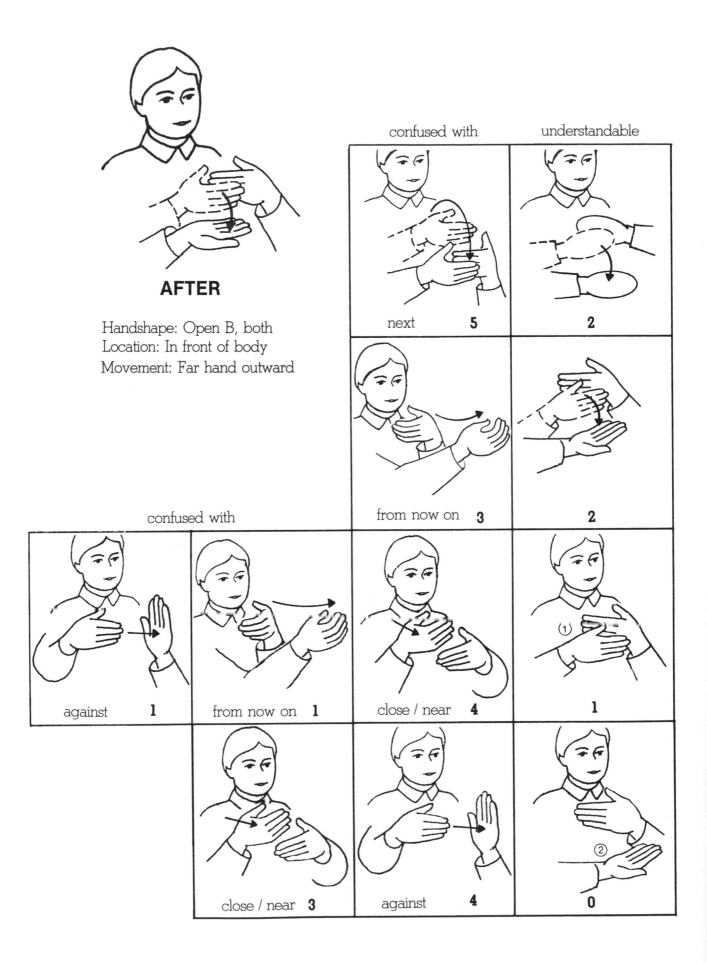

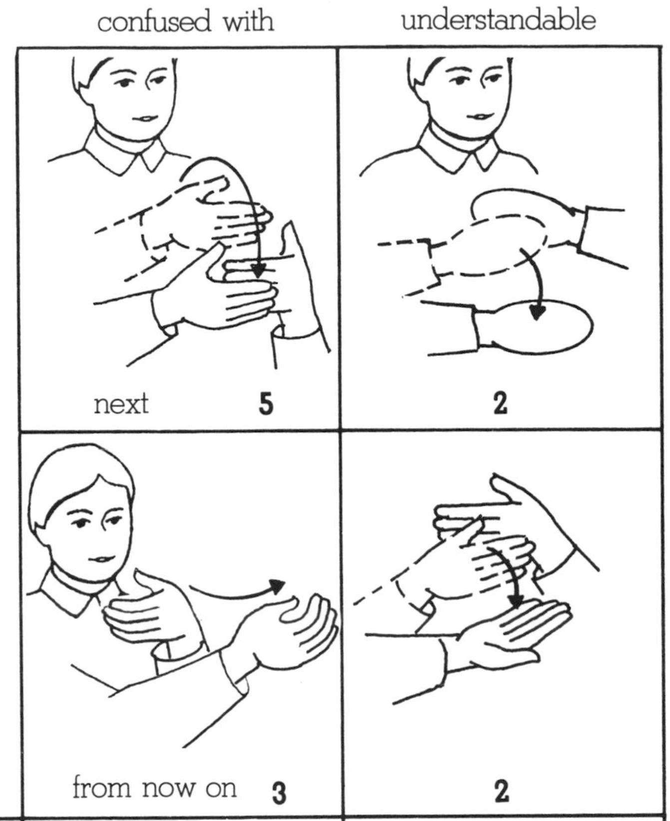

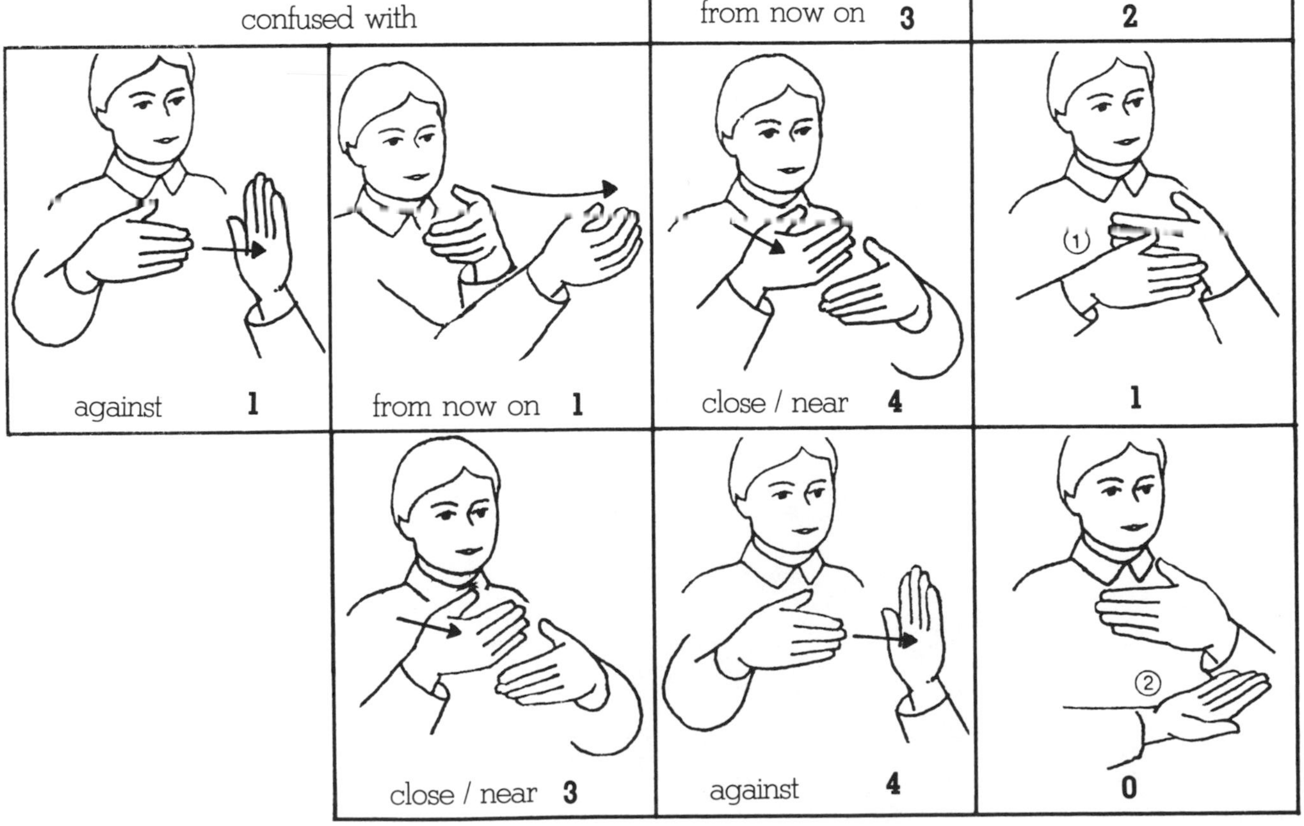

ZERO ROBUST

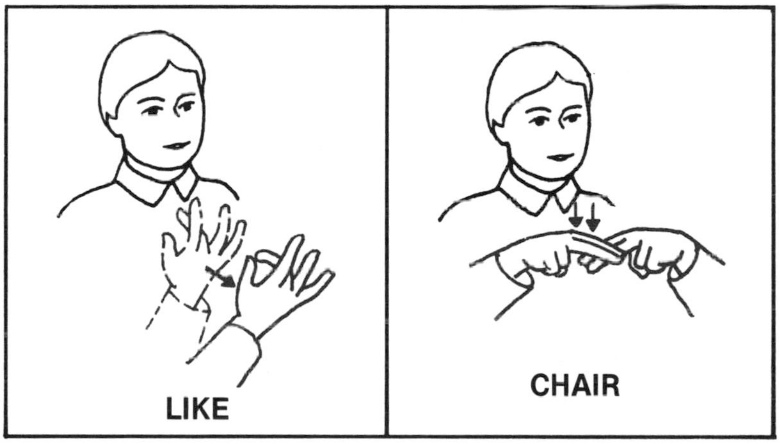

DOUBLE FRAGILE

GRANDMOTHER

Handshape: 5
Location: Chin
Movement: Two short arcs forward

understandable

8

confused with

speak 1

2

creative 2

confused with

Christmas 1

1

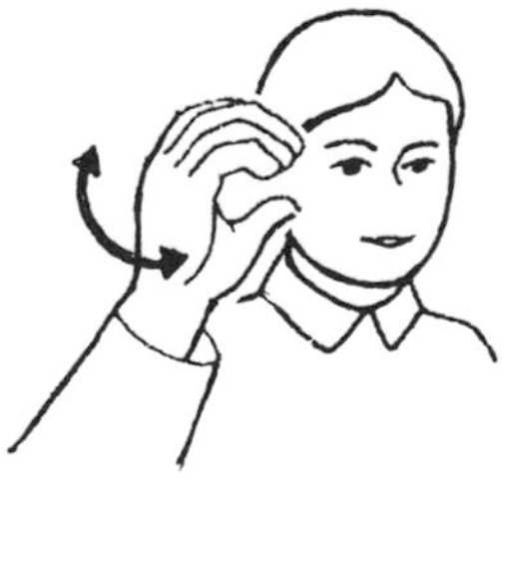

cousin 1

drink 1

HE

Handshape: E
Location: Temple
Movement: Forward

confused with

Europe **1**

understandable

4

confused with

moon **1**

opinion **1**

3

east **3**

2

hello **7**

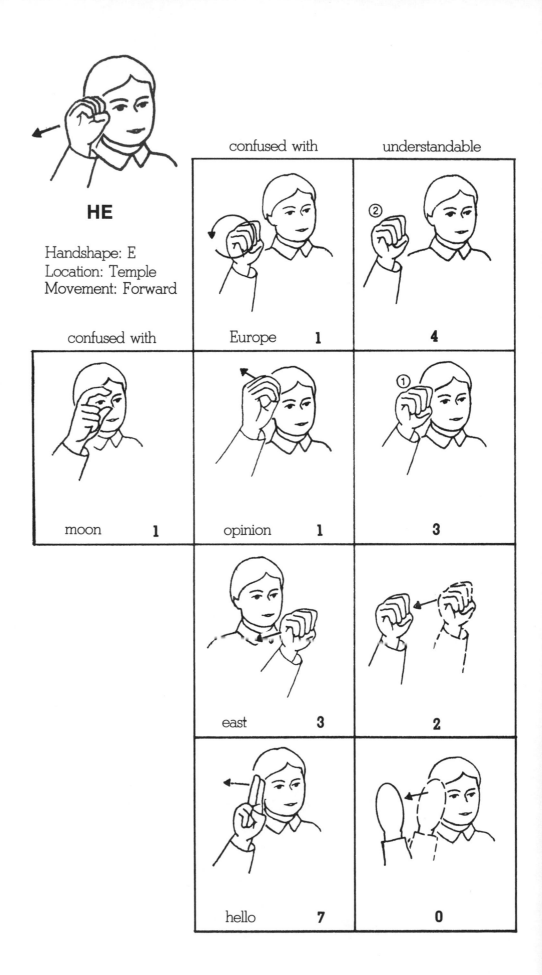

0

FOR

Handshape: 1
Location: Forehead
Movement: Twist forward

understandable | confused with

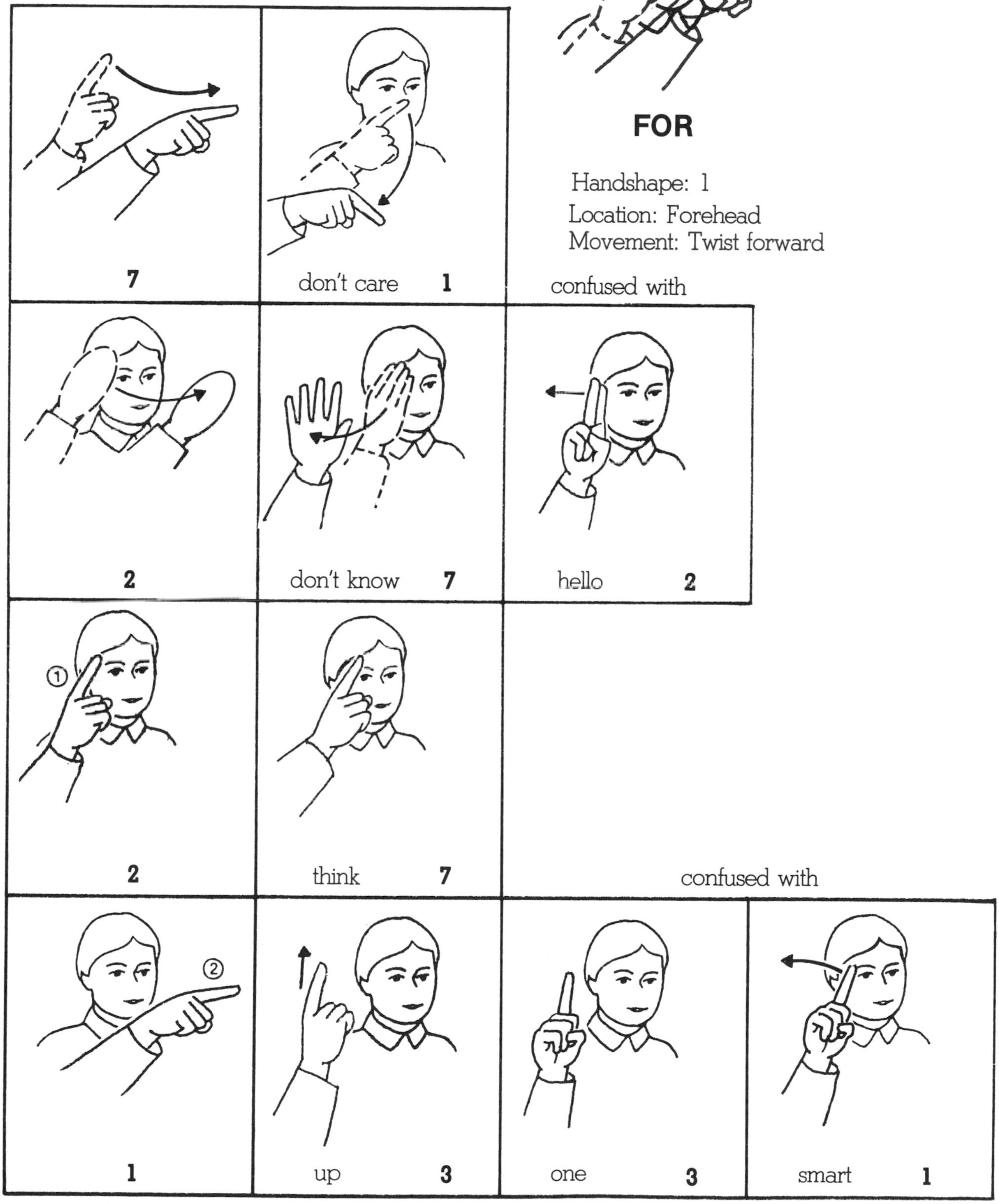

TOWEL

Handshape: S, both
Location: In front of body
Movement: Alternately in and out

understandable

5

confused with

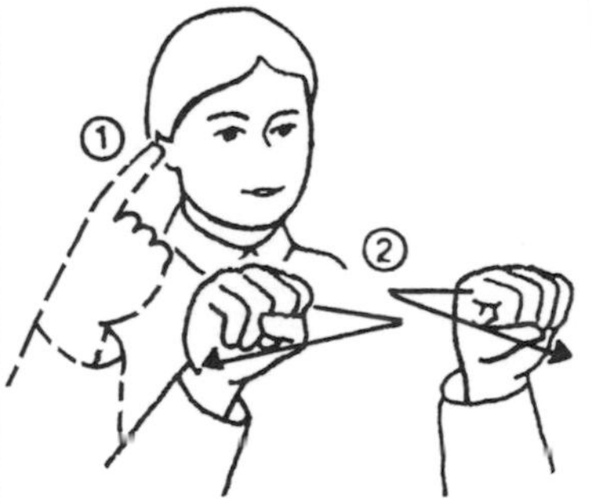

loud 6

2

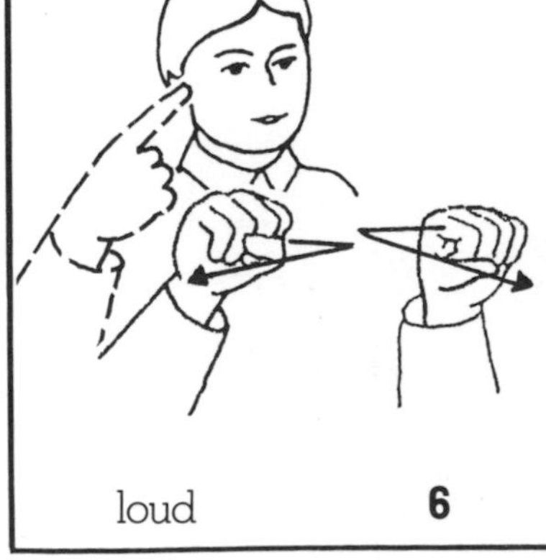

loud 6

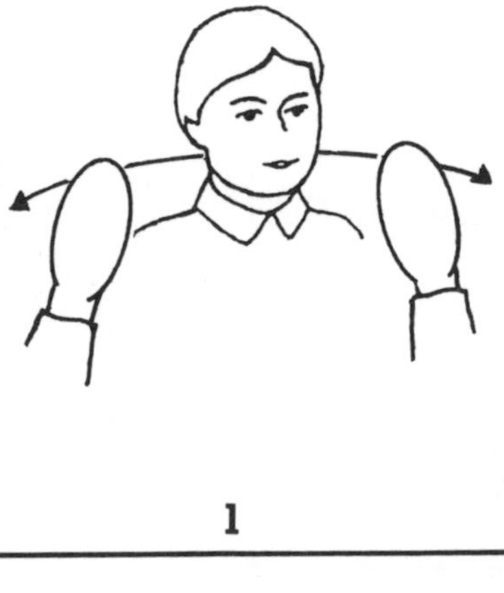

1

ALL GONE

Handshape: C; open B
Location: In front of body
Movement: Cross palm, closing to fist

understandable

7

1

confused with

reject 3

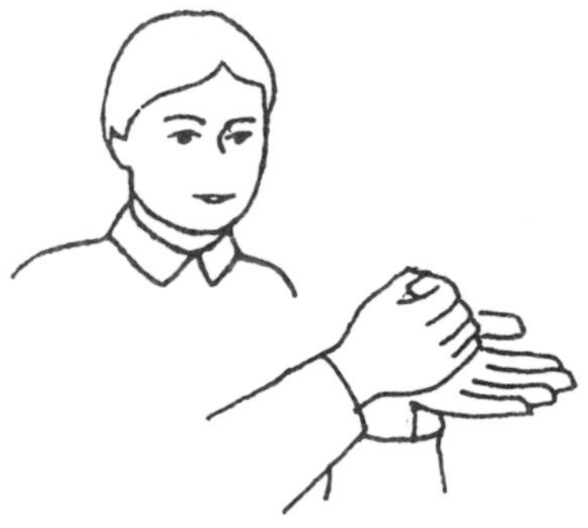

0

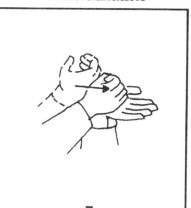

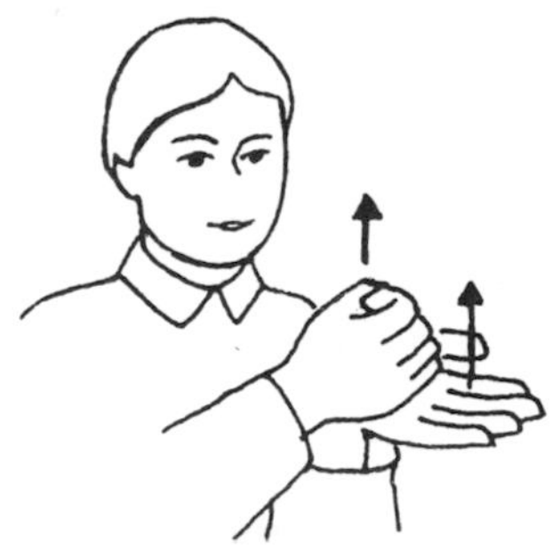

help 4

WEDNESDAY

Handshape: W
Location: In front of body
Movement: Small circle

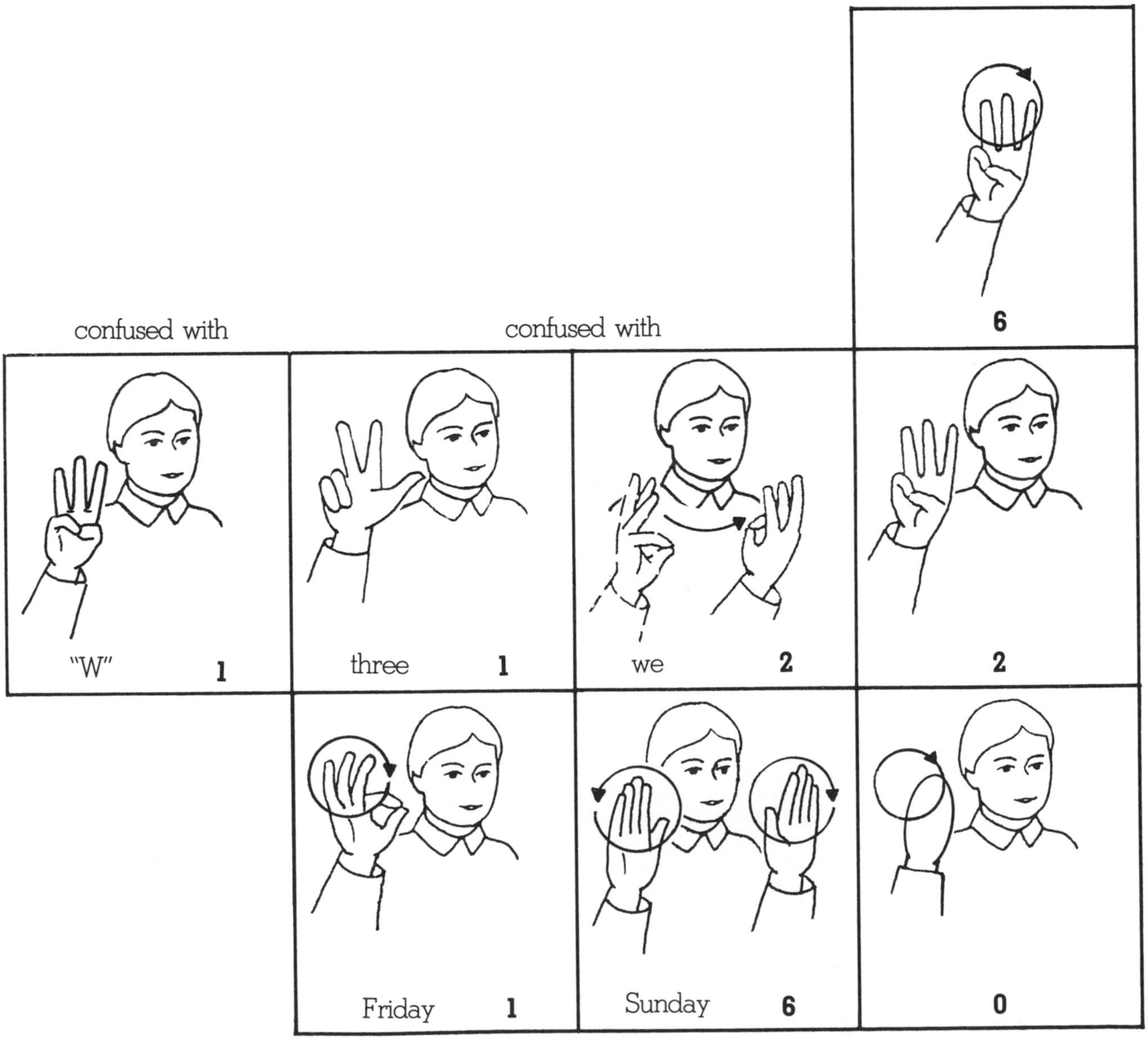

GLASSES

Handshape: G, both
Location: Eyes
Movement: Twist back

OLD

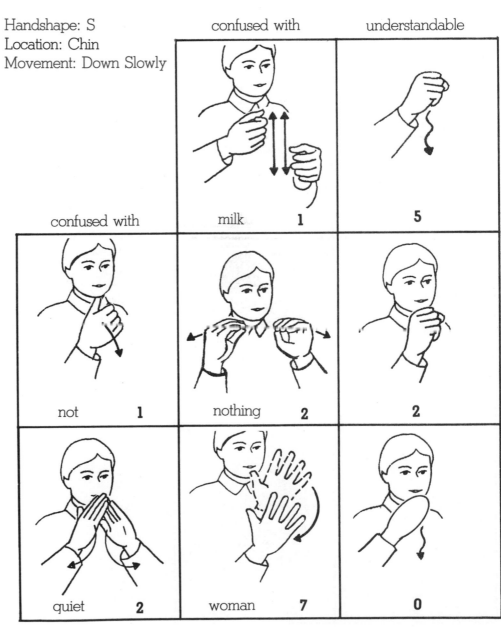

Handshape: S
Location: Chin
Movement: Down Slowly

confused with | understandable

confused with

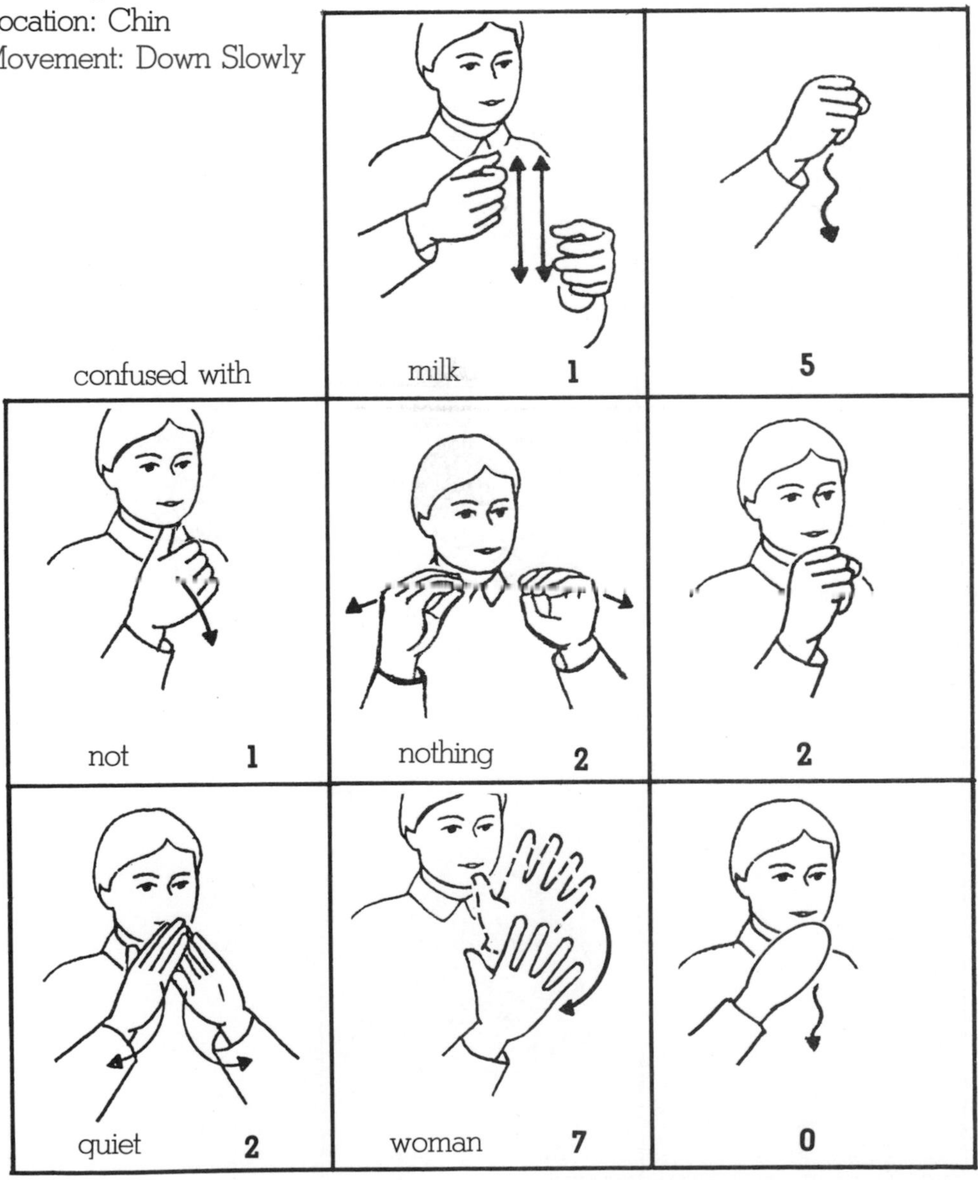

COW

Handshape: Y
Location: Temple
Movement: Twist forward

understandable | confused with

5

idiot 2

1

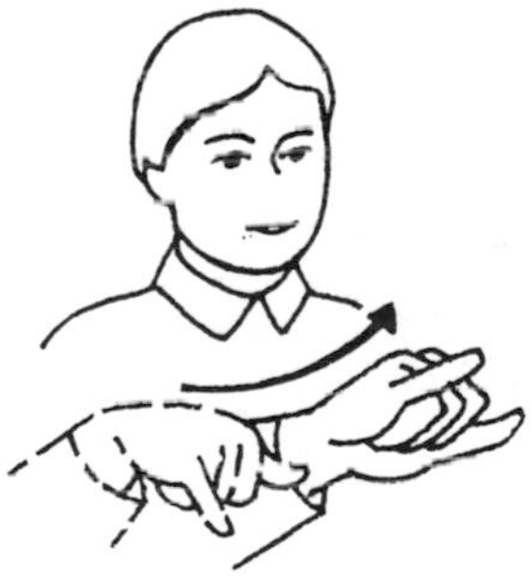

still 5

confused with

1

mule 2

rabbit 2

horse 1

FACE

Handshape: 1
Location: Face
Movement: Large circle

confused with | understandable

handsome **1**

6

confused with

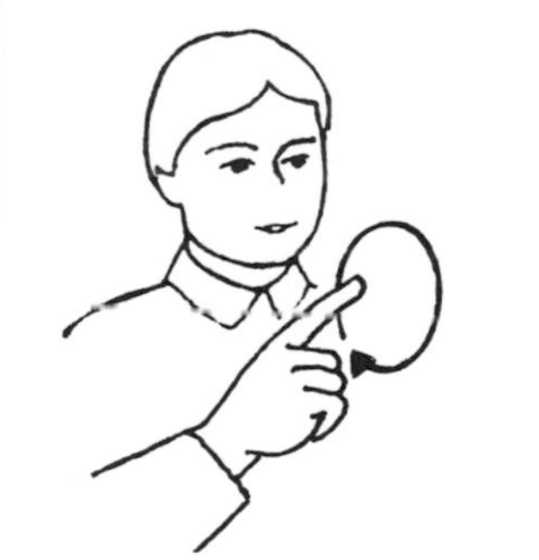

round **2**

who **6**

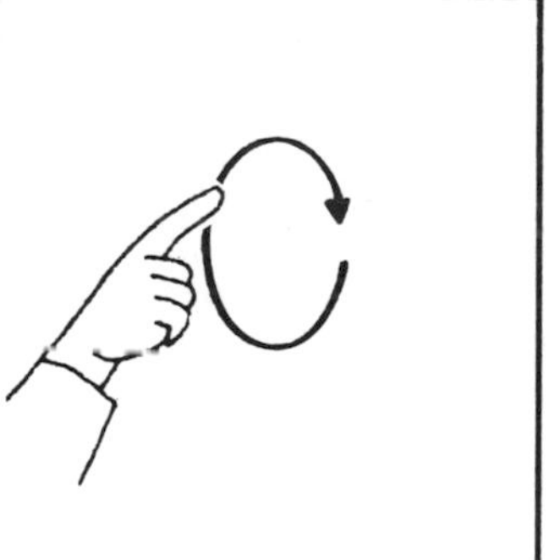

0

confused with

smile **1**

say **1**

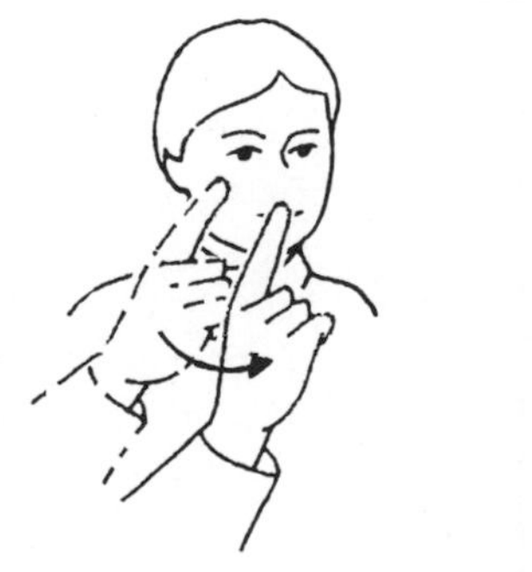

deaf **1**

0

SANDWICH

Handshape: C; open B
Location: In front of body
Movement: Slide in and out twice

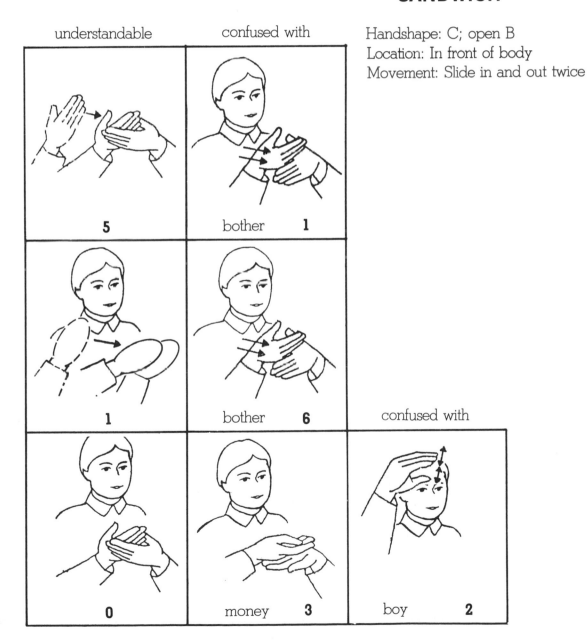

understandable

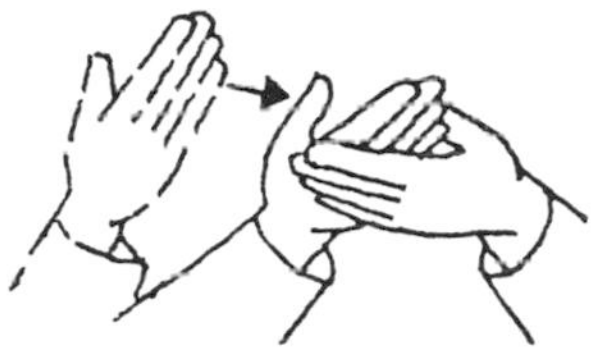

5

confused with

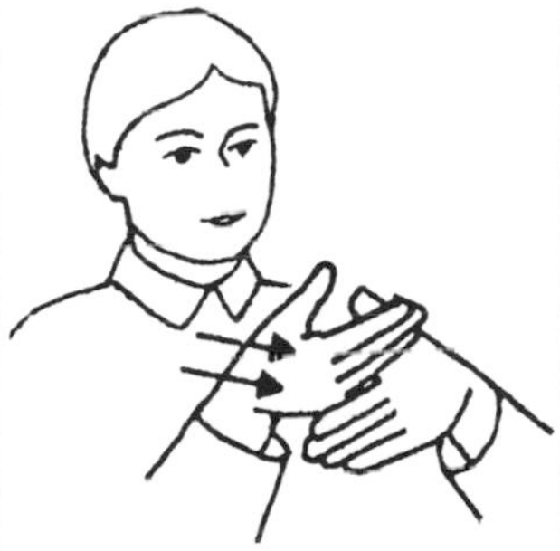

bother **1**

1

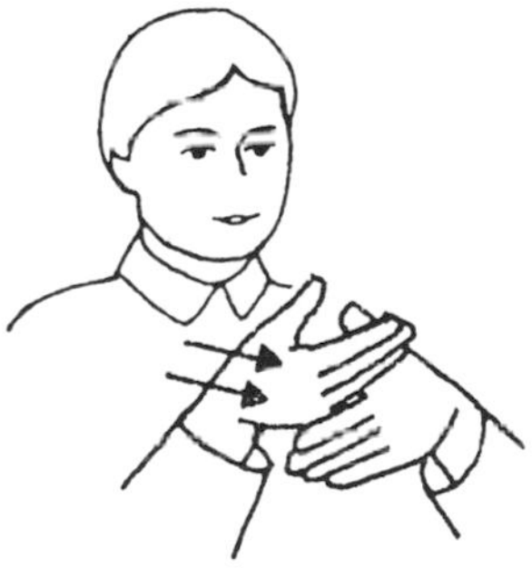

bother **6**

0

money **3**

confused with

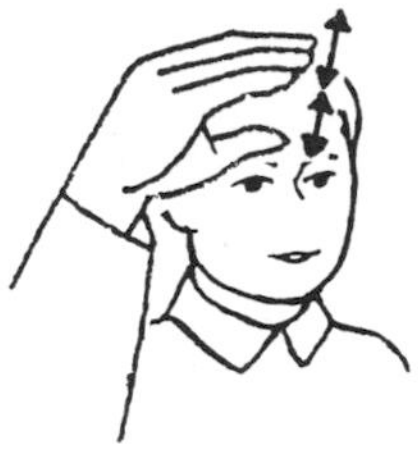

boy **2**

GIRL

Handshape: A, thumb extended
Location: Cheek
Movement: Forward on cheek

	confused with	understandable
	 any **2**	 **4**
confused with	 aunt **1**	 **1**
 brown **4**	 beer **5**	 **1**

EAR

Handshape: G
Location: Earlobe
Movement: Pinch

understandable

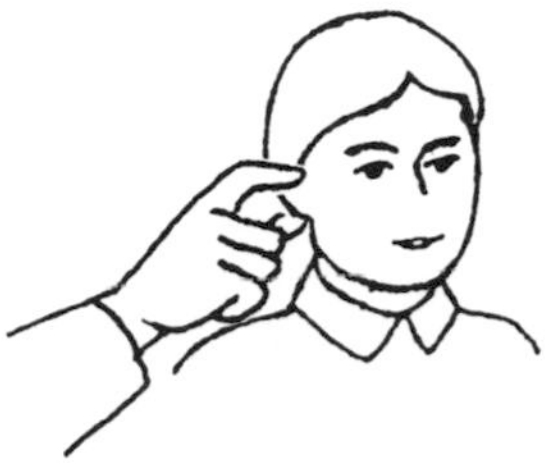

5

confused with

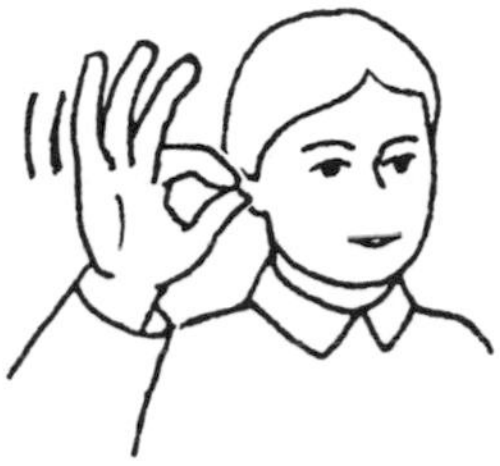

earring 2

1

hearing 7

0

"G" 1

confused with

curious 1

DENTIST

Handshape: D
Location: Teeth
Movement: Tap twice

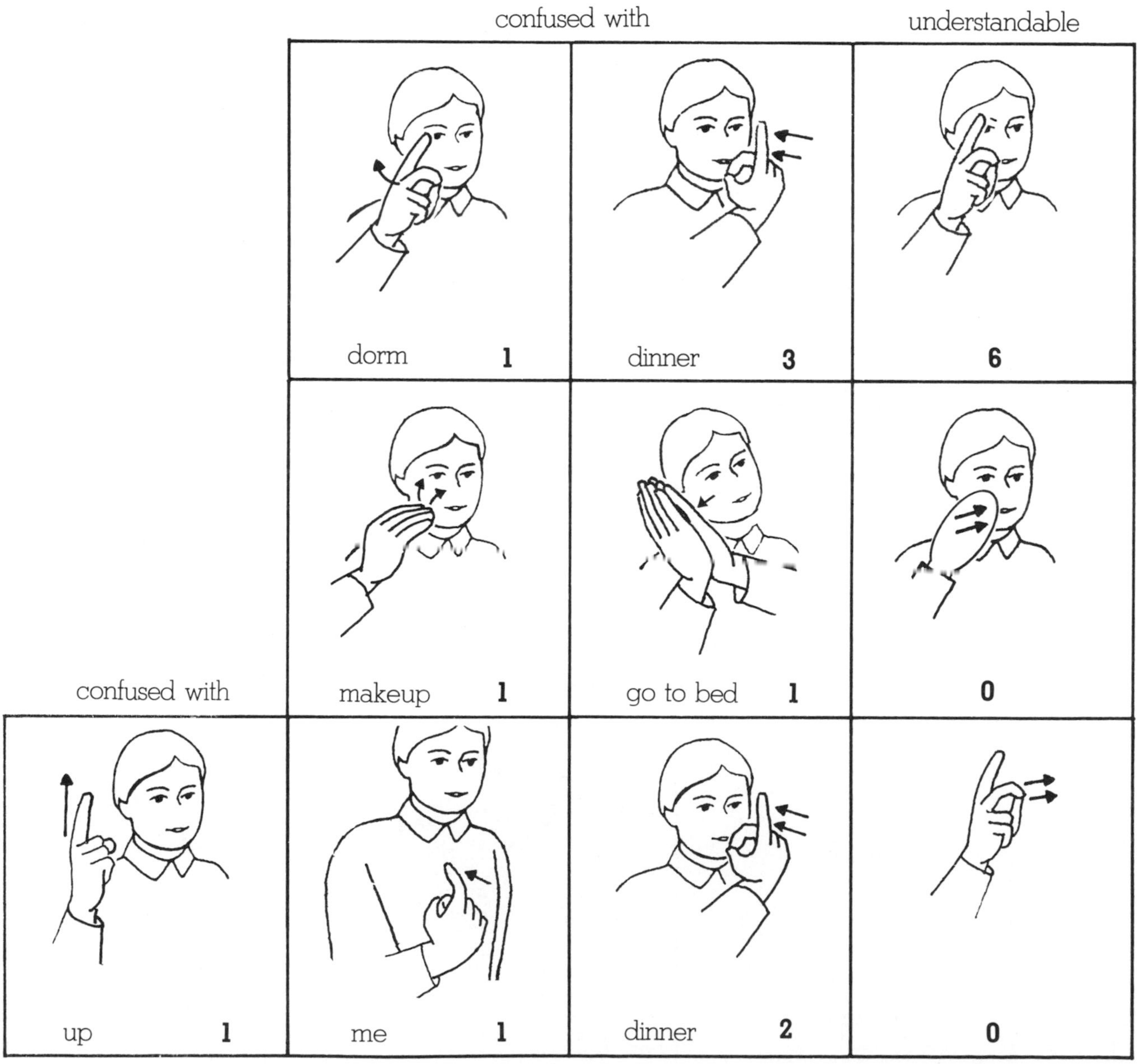

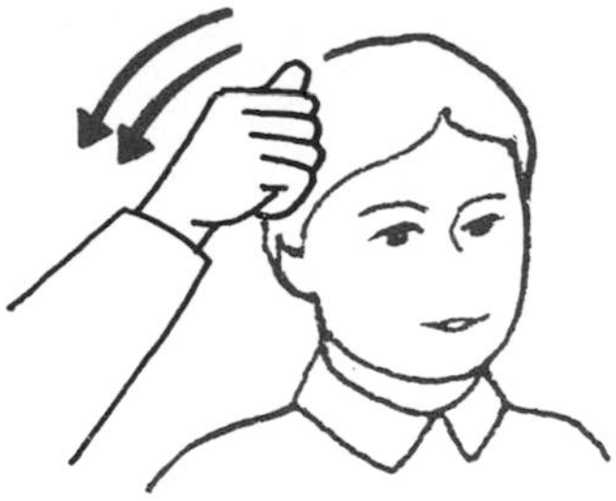

BRUSH

Handshape: A
Location: Head
Movement: Down Twice

understandable

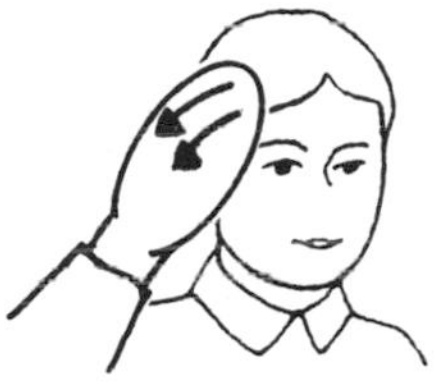

4

confused with

2

phone **4**

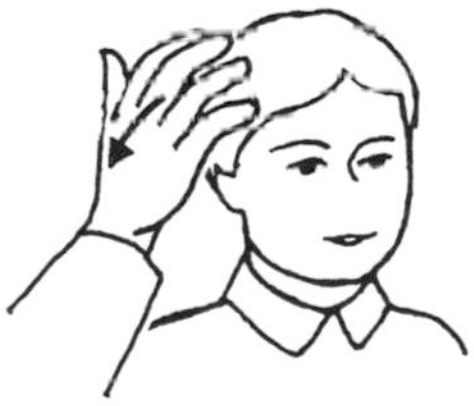

comb **1**

0

ice cream **2**

coffee **1**

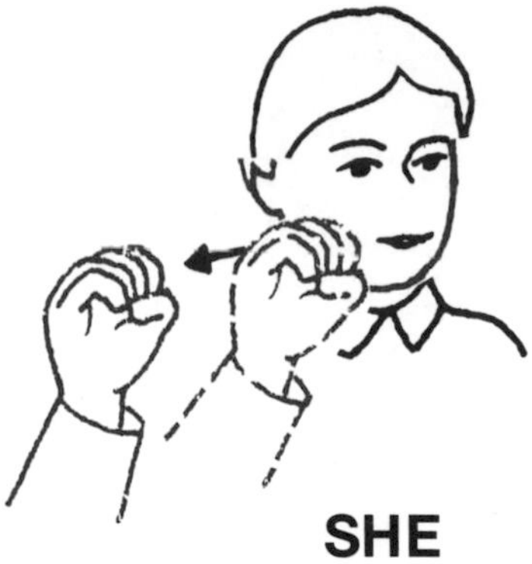

SHE

Handshape: E
Location: Cheek
Movement: Forward

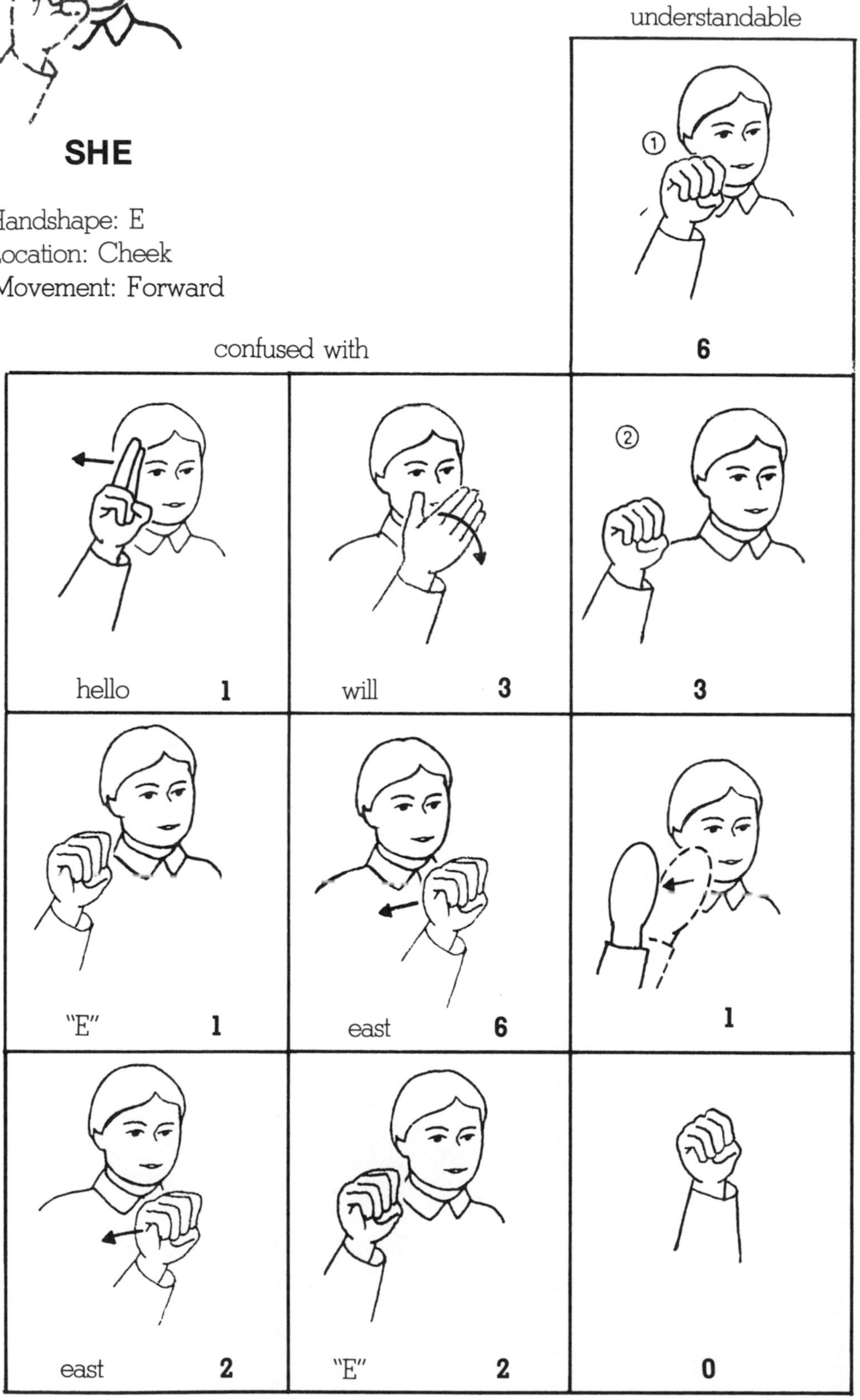

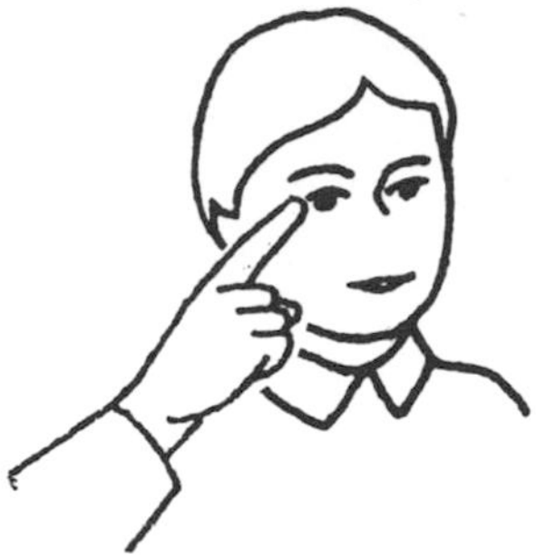

EYE

Handshape: 1
Location: Eye
Movement: Point

understandable	confused with		confused with
 4	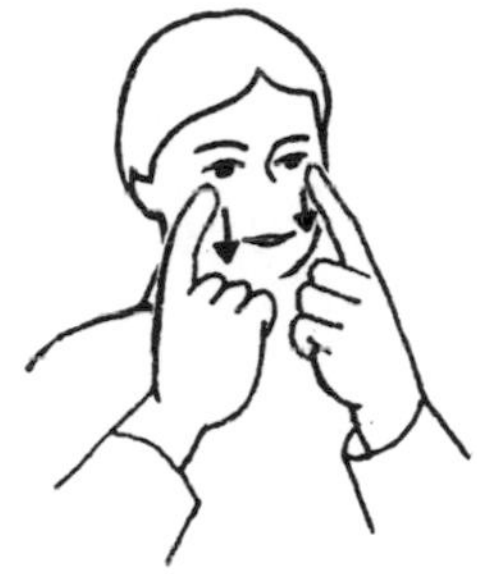 cry 2	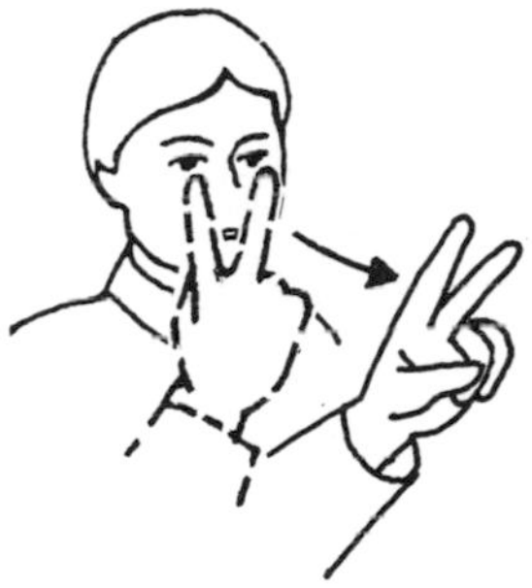 witness 2	 let me see 1
 2	 see 5	 know 3	
 0	 me 8		

MOUTH

Handshape: 1
Location: Mouth
Movement: Circle

confused with	confused with		understandable
 who **1**	 lips **1**	 napkin **1**	 **4**
	 think **1**	 who **4**	 **1**
	 shhh / quiet **3**	 red **3**	 **1**

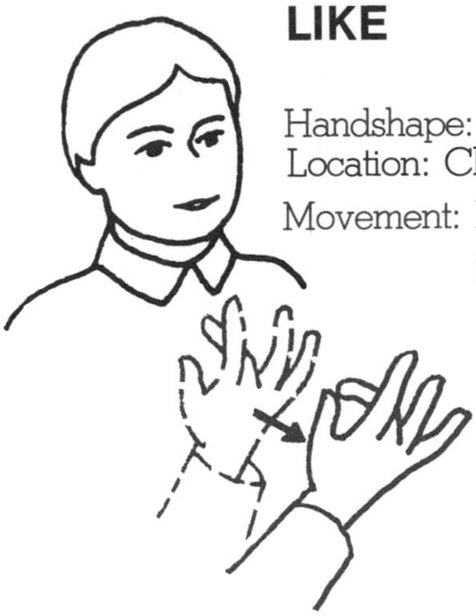

LIKE

Handshape: 5
Location: Chest
Movement: Forward, close middle finger and thumb

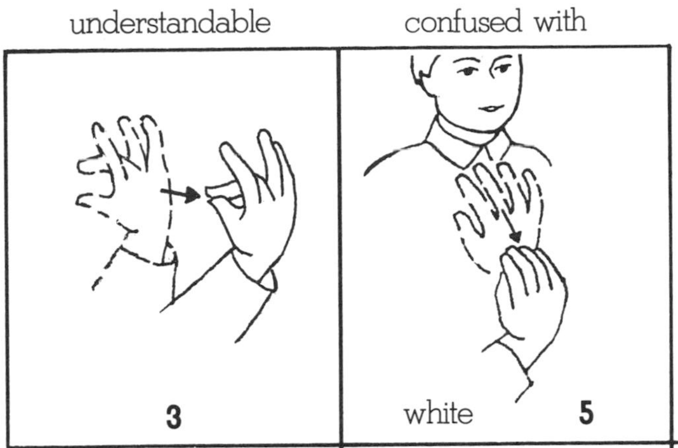

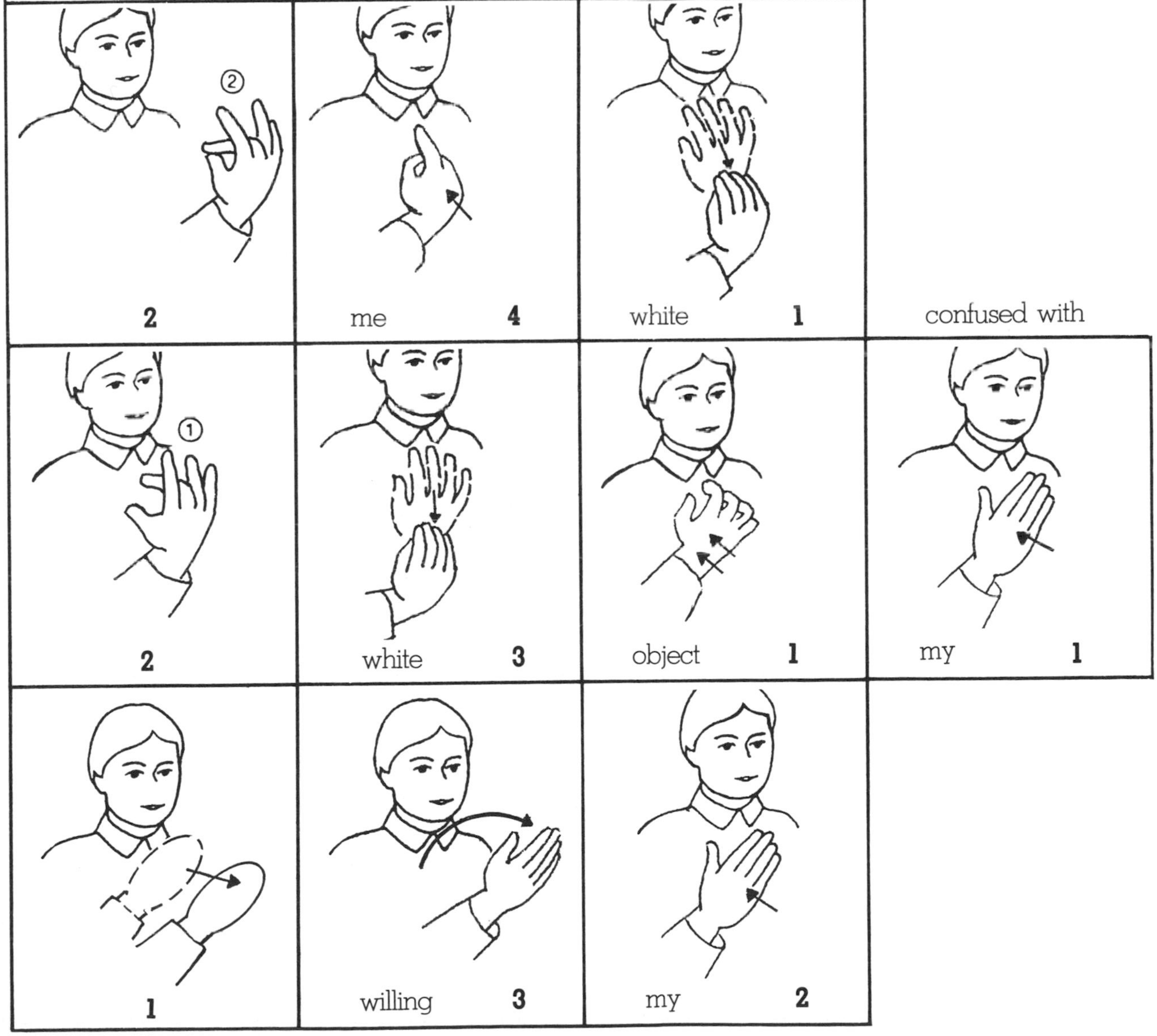

SHEET

Handshape: S, both
Location: Chest
Movement: Up to shoulders

understandable

7

confused with

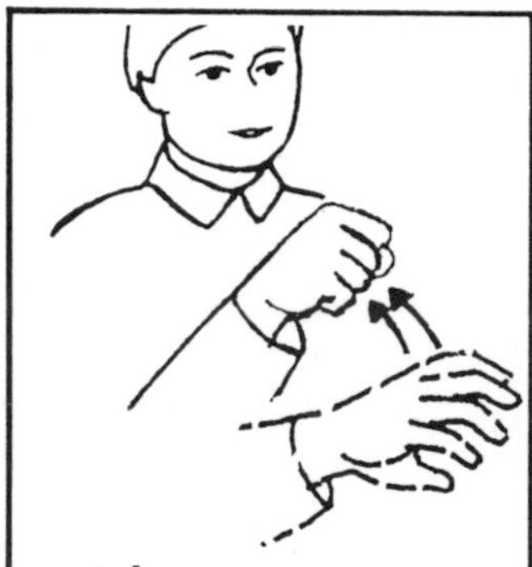

take over **1**

take **7**

0

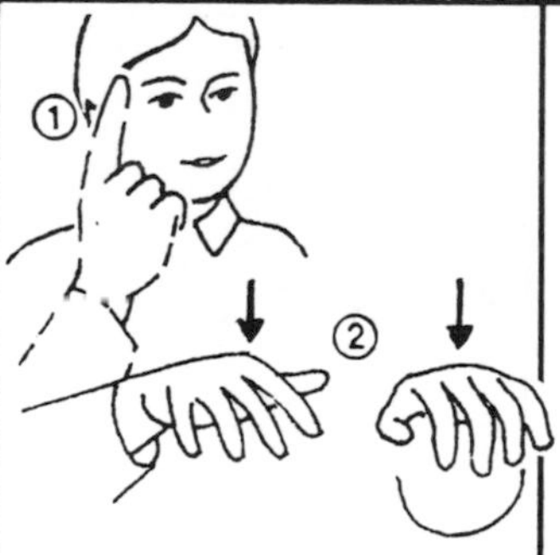

shock **1**

take **3**

0

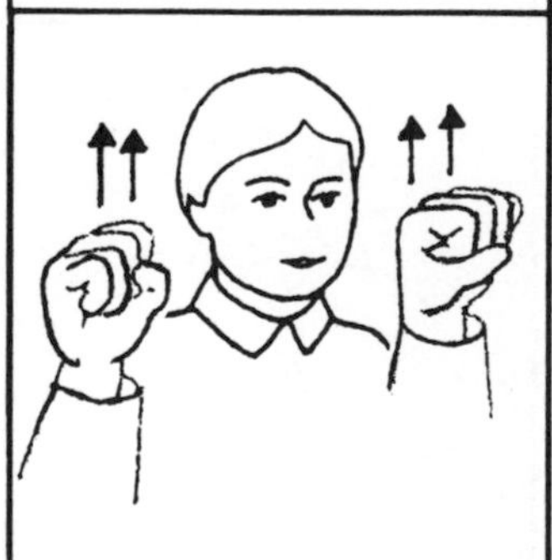

exercise **1**

able / can **5**

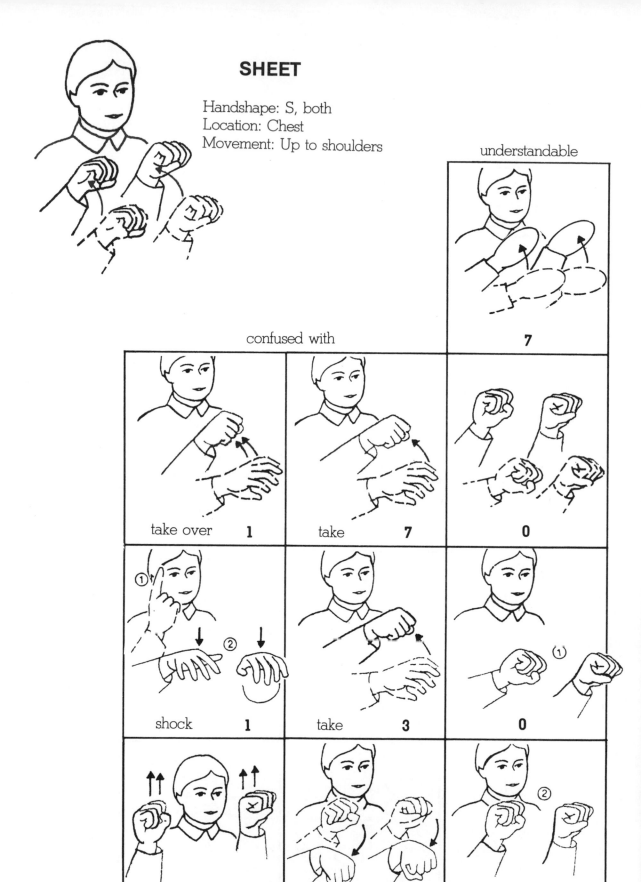

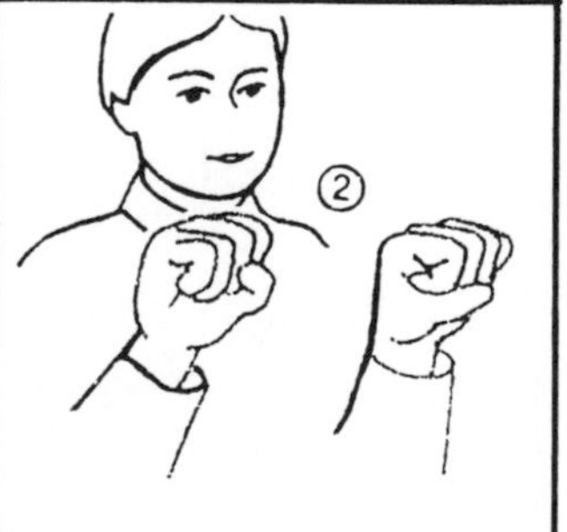

0

SUN

Handshape: C
Location: Eye
Movement: None

understandable

4

confused with

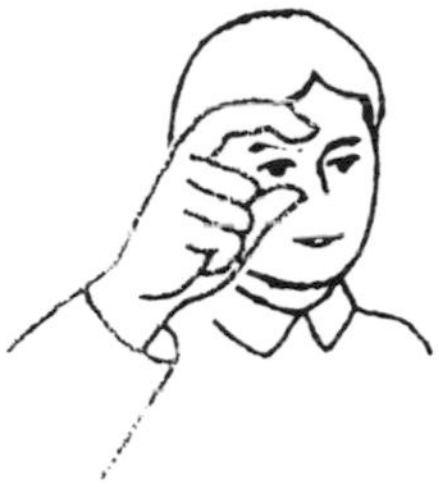

moon 2

wake 1

1

1

confused with

drunk 2

police 1

confused with

"C" 1

DOLL

Handshape: X
Location: Nose
Movement: Brush nose twice

confused with	understandable
 fool **3**	 **3**
 shave **4**	 **2**
 mouse **4**	 **0**

understandable confused with

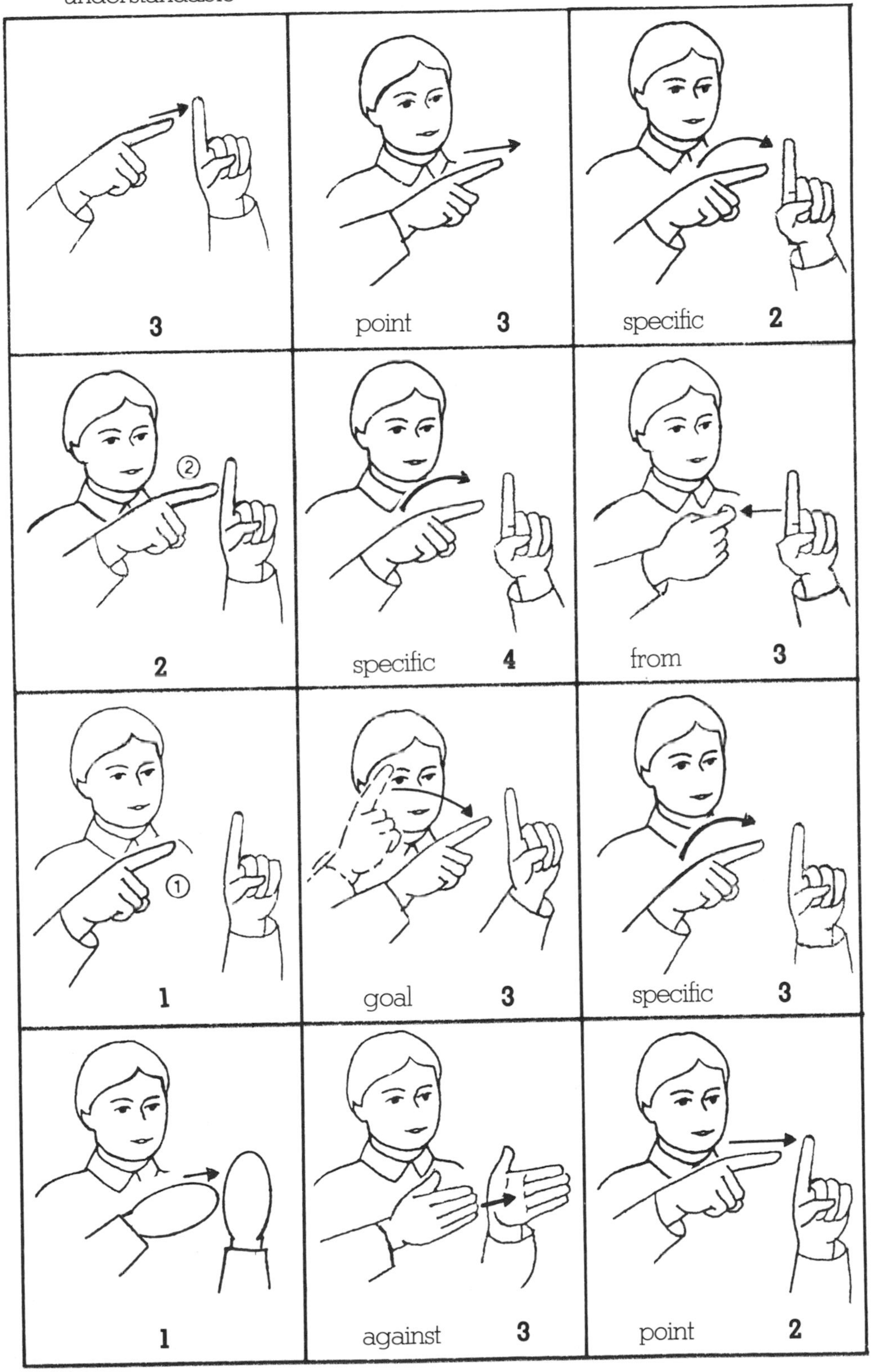

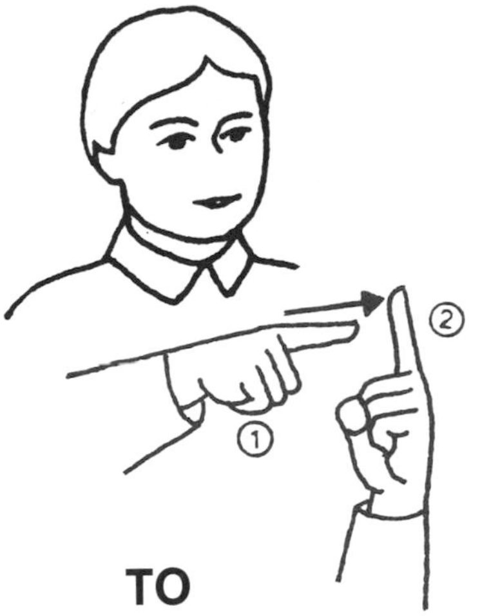

TO

Handshape: 1, both
Location: In front of body
Movement: Touch fingertips

ORANGE

Handshape: C
Location: Mouth
Movement: Open and close fist

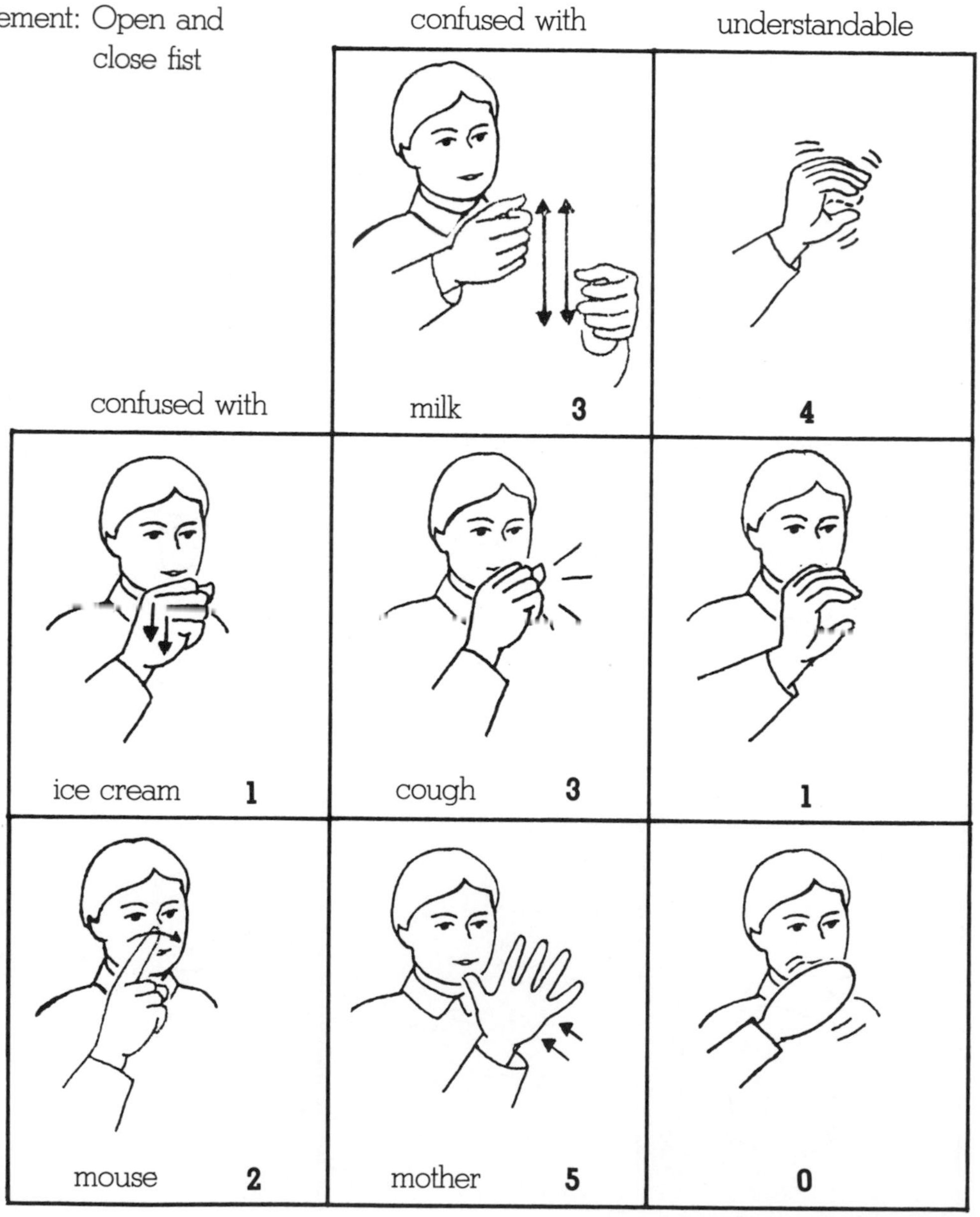

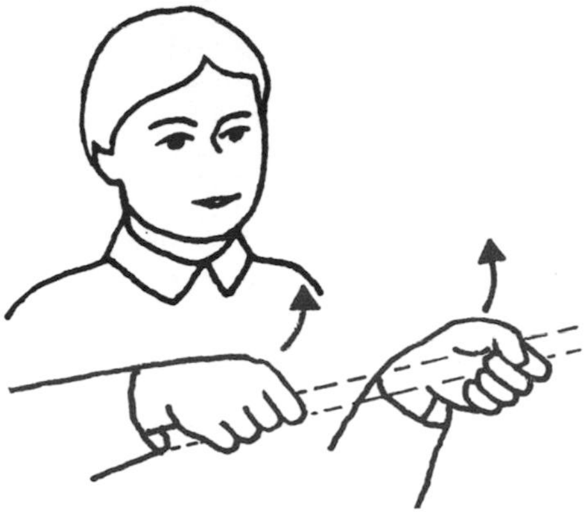

DIG

Handshape: S, both
Location: Waist
Movement: Mimic shovelling

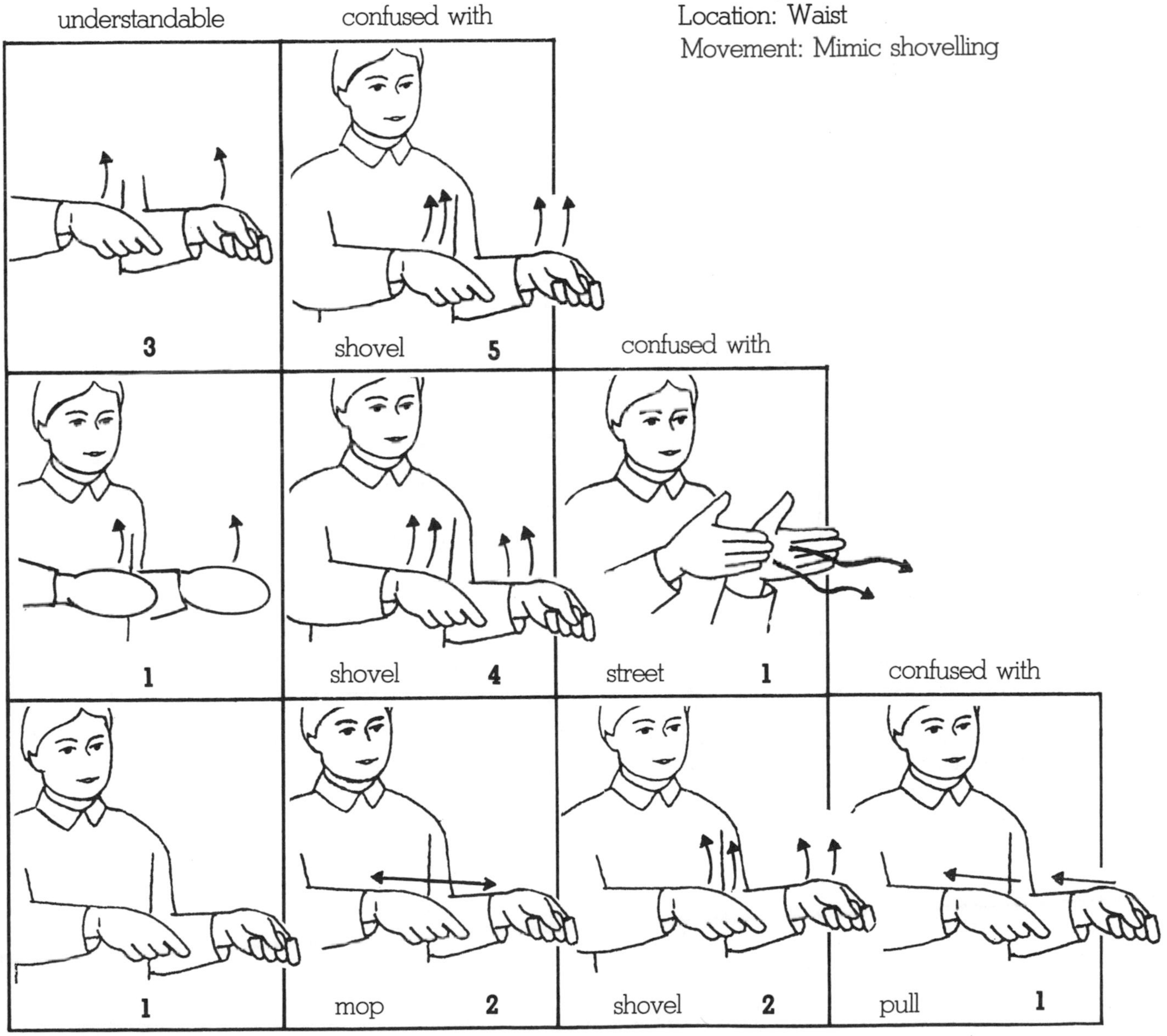

COLOR

Handshape: 5
Location: Chin
Movement: Wiggle fingers

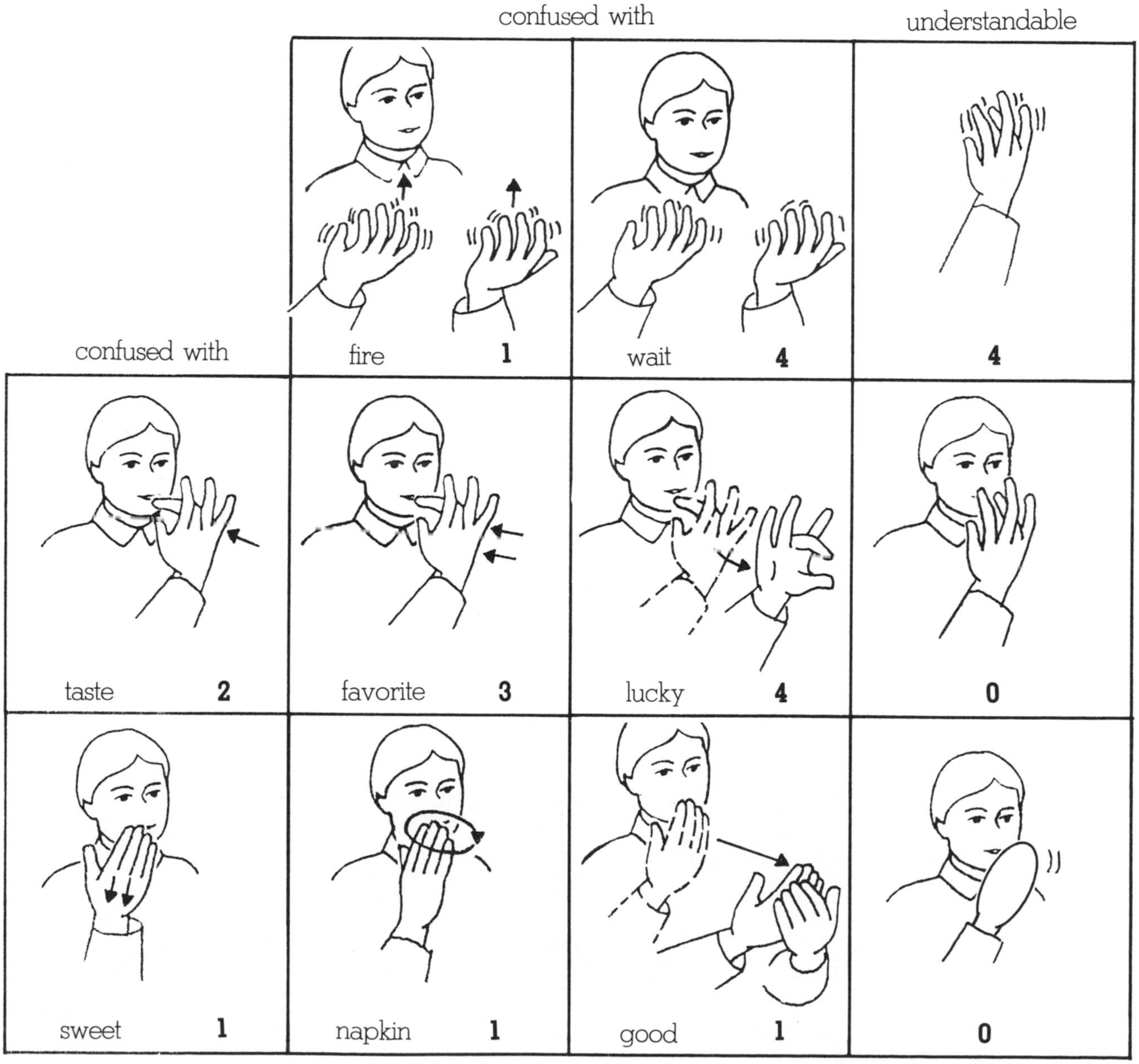

TASTE

Handshape: 5, bent middle finger
Location: Chin
Movement: Tap

understandable | confused with

3

sick 4

prefer 1

2

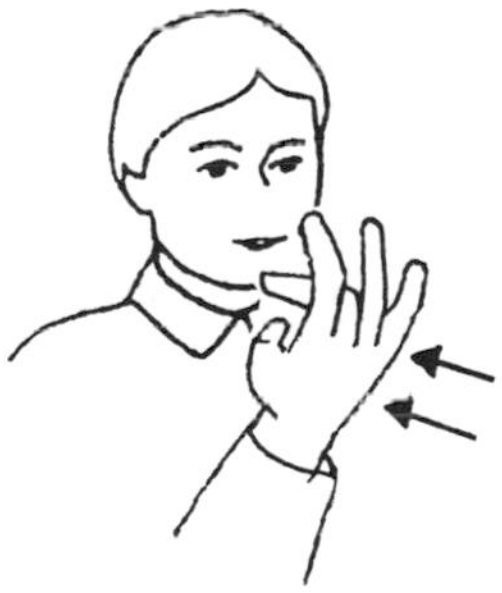

prefer 6

confused with

0

eat 3

good 2

food 1

HER

Handshape: R
Location: Cheek
Movement: Forward

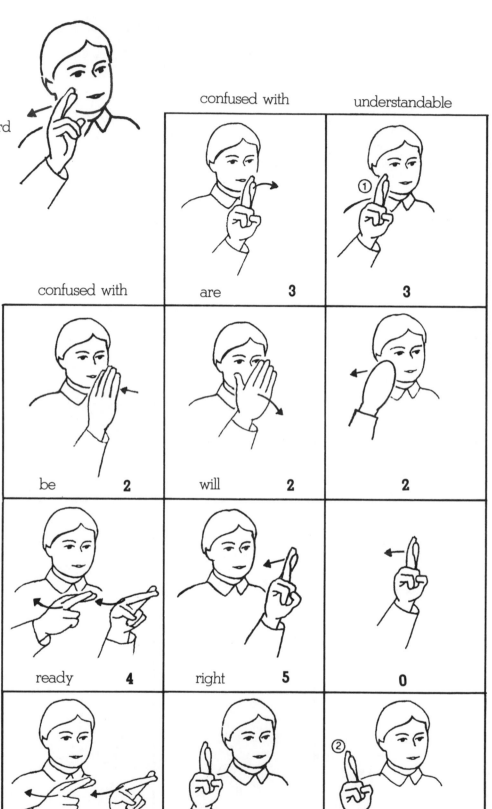

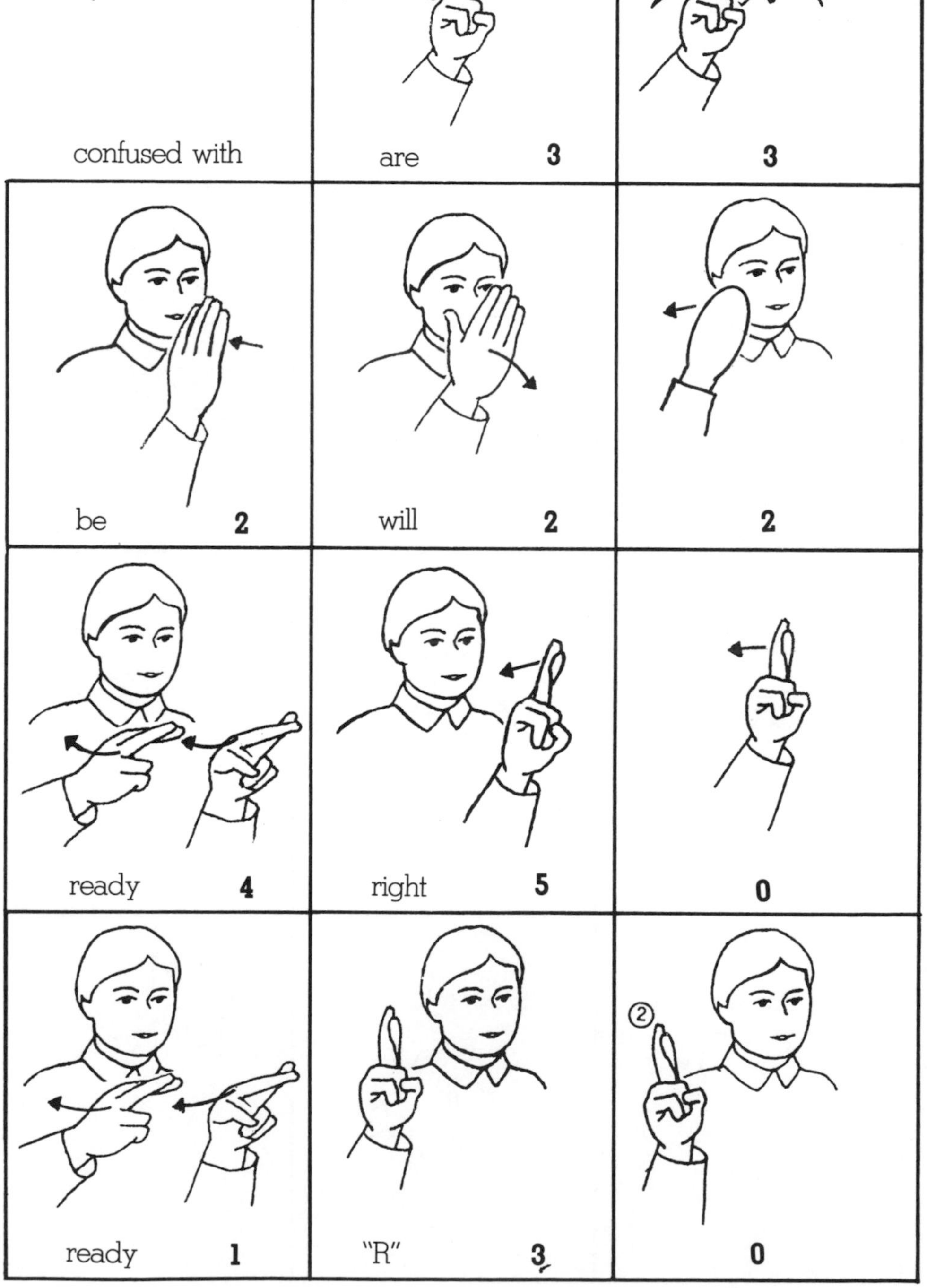

KNIFE

Handshape: H, both
Location: In front of body
Movement: Scrape twice

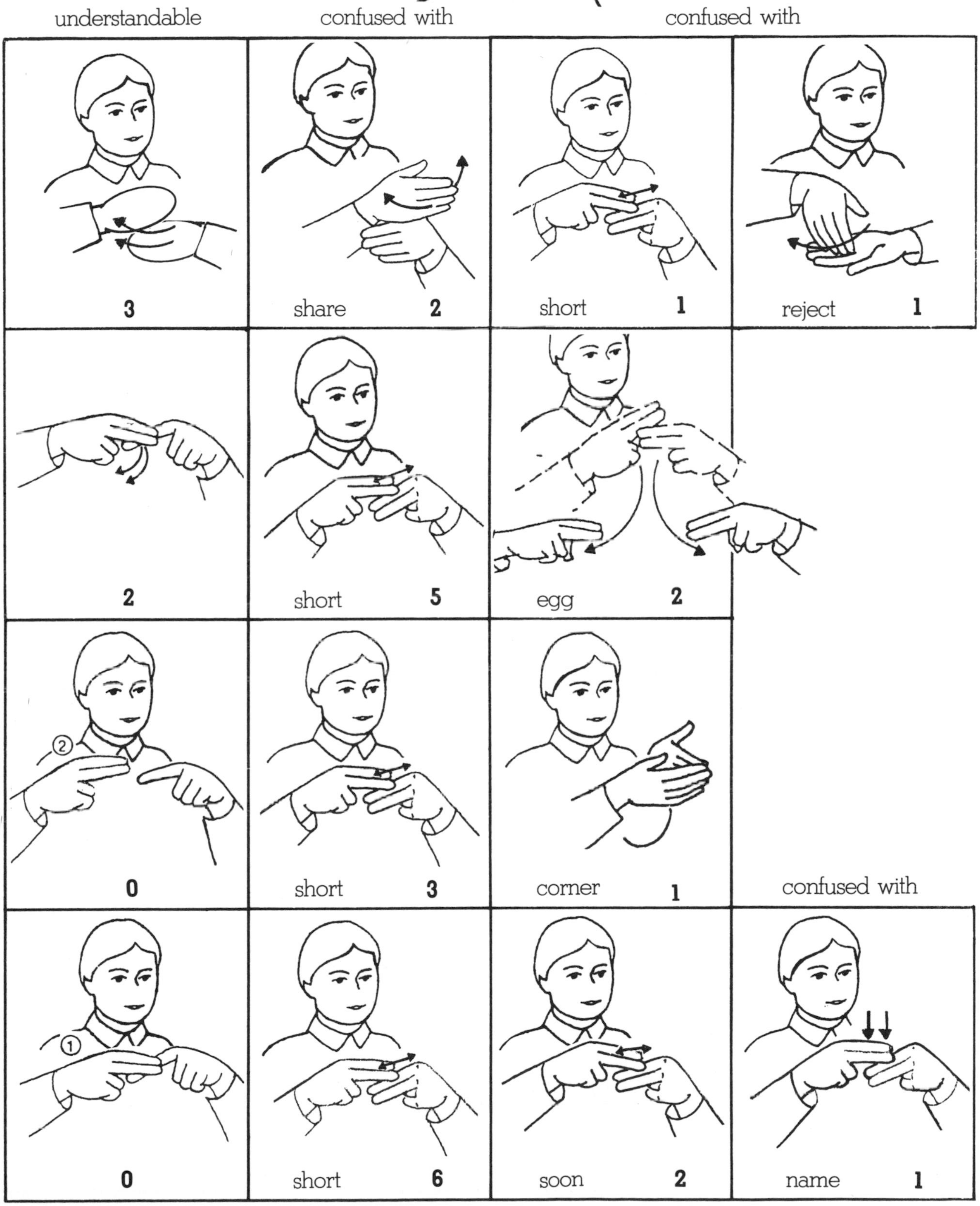

CHAIR

Handshape: H, both
Location: In front of body
Movement: Tap twice

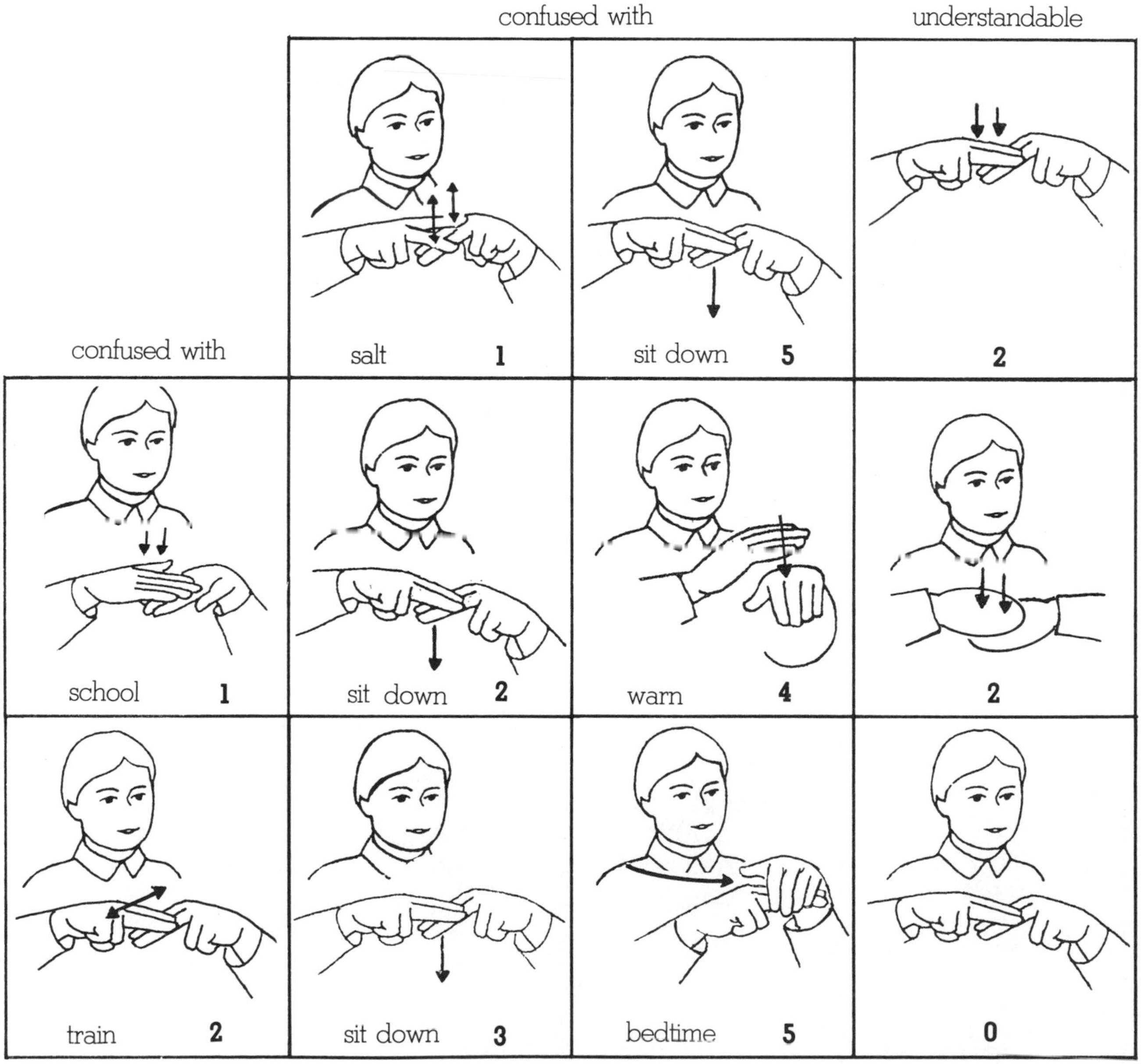

NAPKIN

Handshape: Open B
Location: Mouth
Movement: Small circle

understandable — confused with — confused with

GRANDFATHER

Handshape: 5
Location: Forehead
Movement: Two short arcs forward

confused with

understandable

invent **5**

3

confused with

smart **3**

concept **3**

0

smart **2**

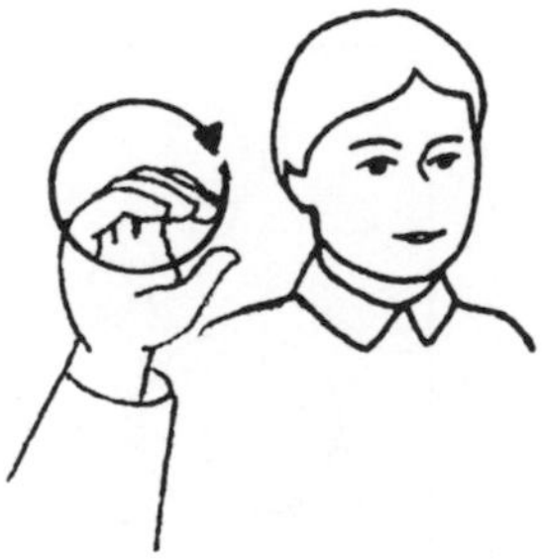

Christmas **3**

0

TRIPLE FRAGILE

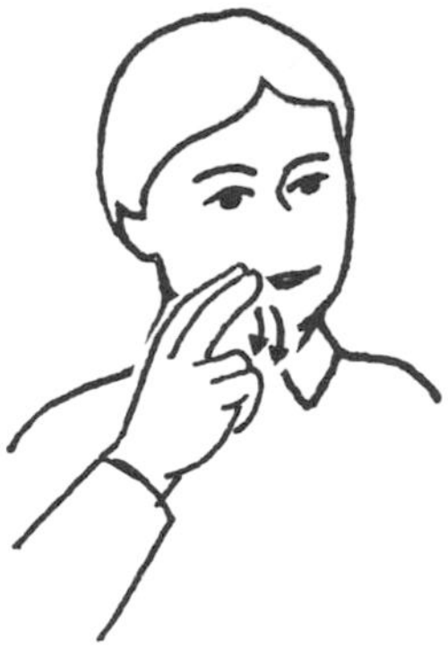

CANDY

Handshape: H, thumb extended
Location: Side of mouth
Movement: Brush down twice

understandable	confused with	
 **2**	 sweet **4**	 cute **2**
 **2**	 cute **7**	 sugar **1**
 **1**	 cute **8**	 sugar **1**

LETTER

Handshape: A, thumb extended
Location: Mouth
Movement: To palm

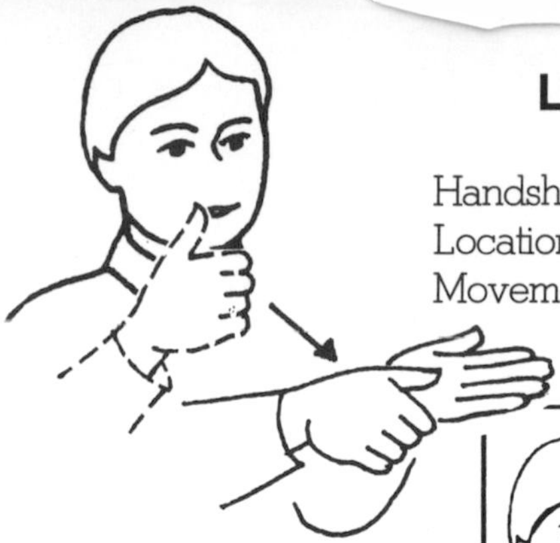

confused with

understandable

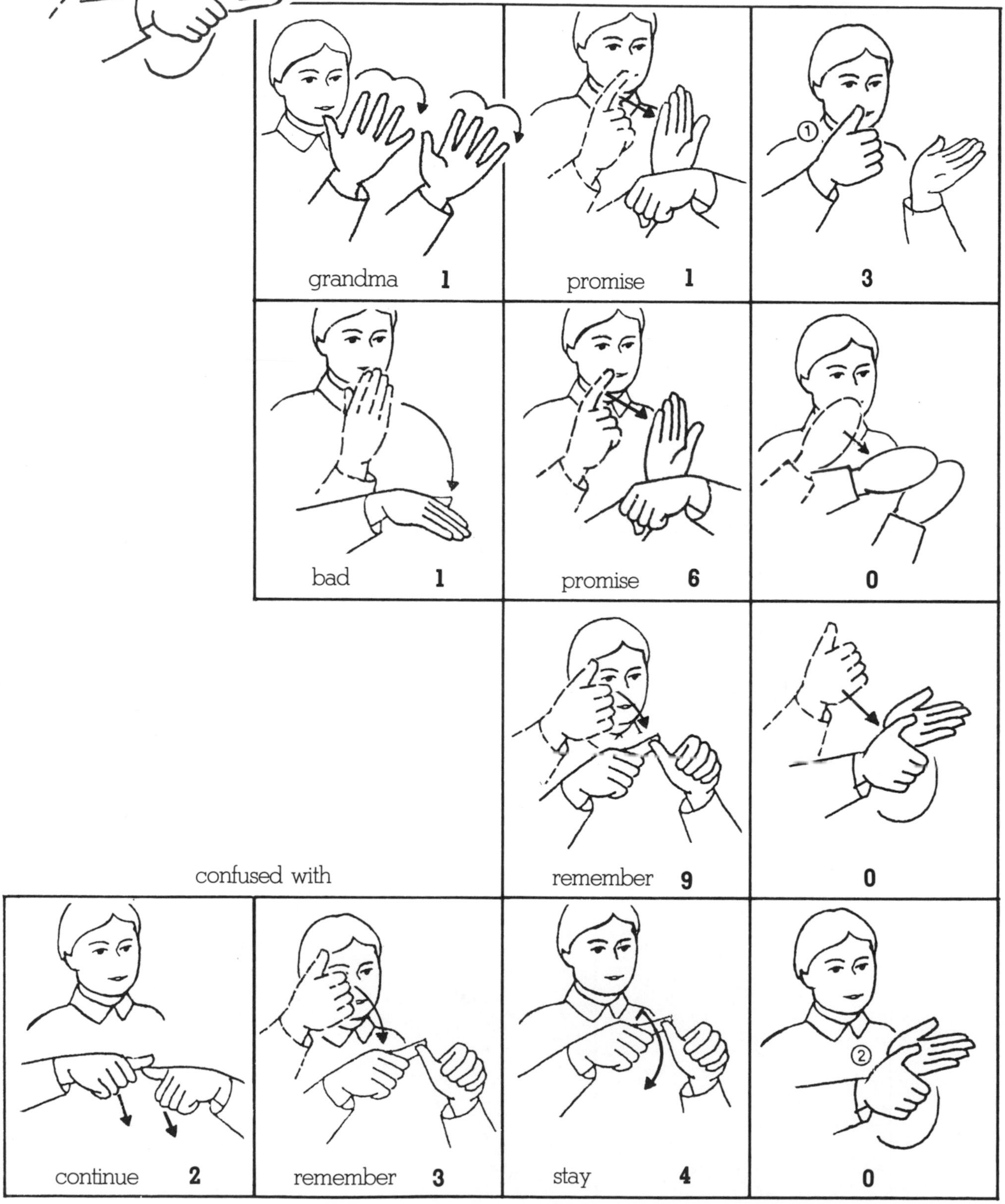

HIS

Handshape: S
Location: Temple
Movement: Forward

understandable

2

confused with

she 1

0

south 3

0

0

confused with

hello 6

CAKE

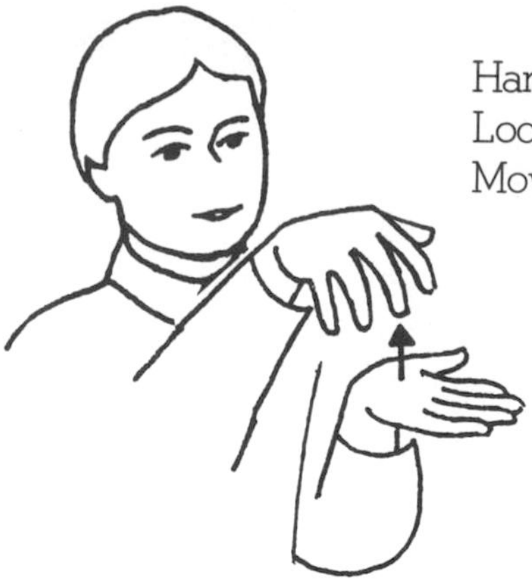

Handshape: C; open B
Location: In front of body
Movement: Up and spread fingers

confused with | understandable

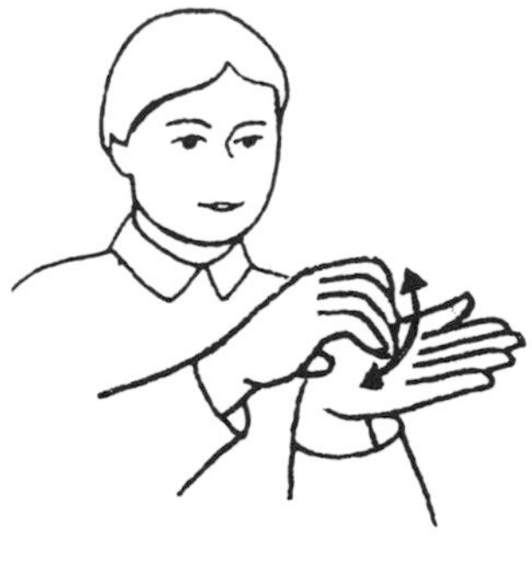

cookies **2**

rich **4**

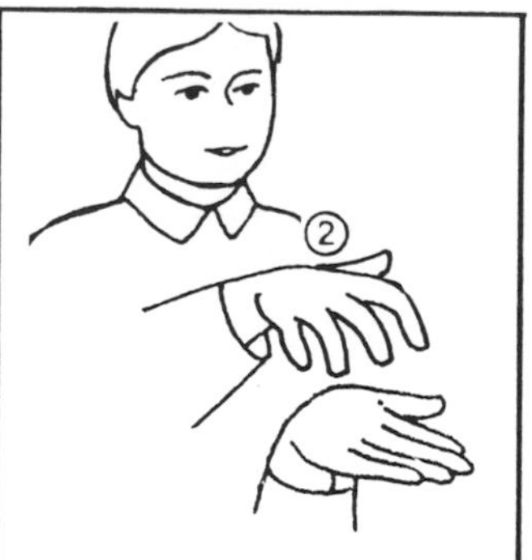

1

confused with

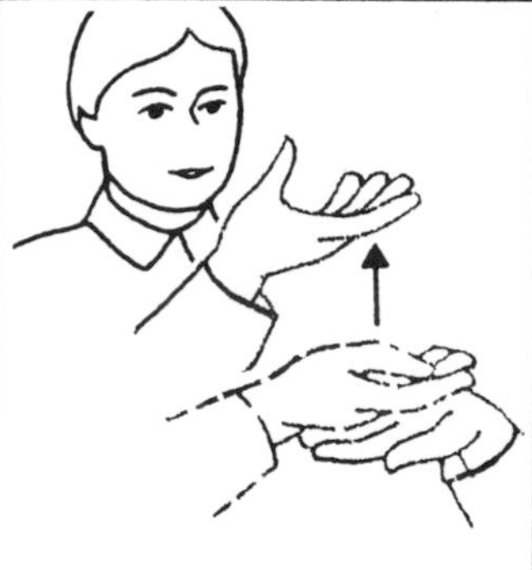

rich **1**

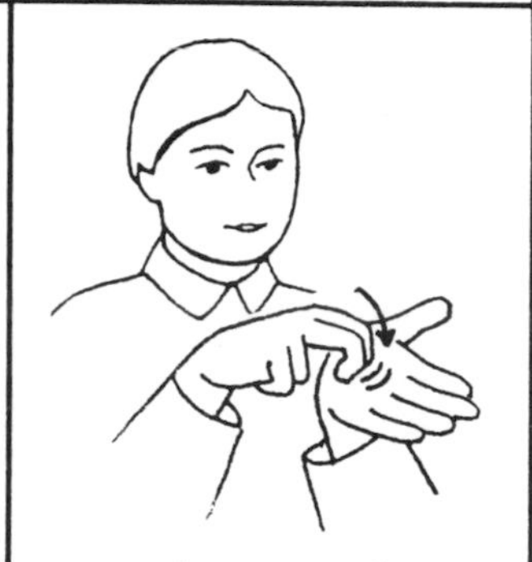

weak **2**

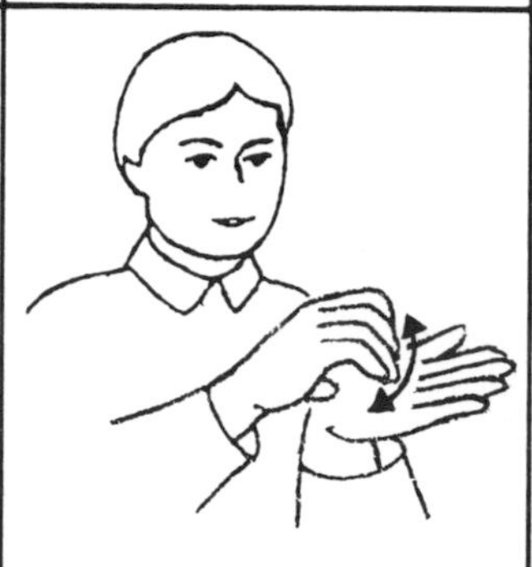

cookies **5**

1

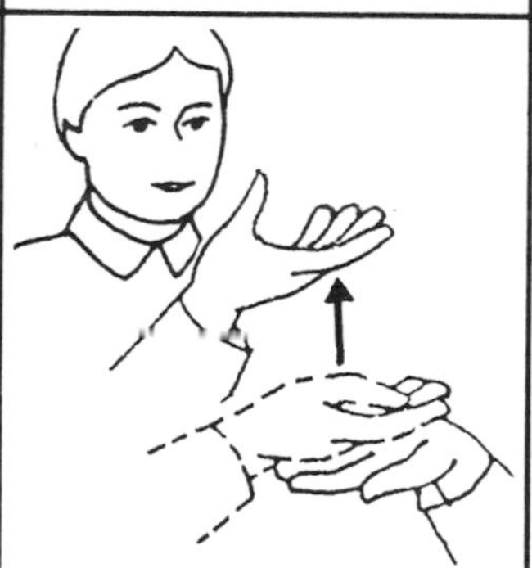

rich **6**

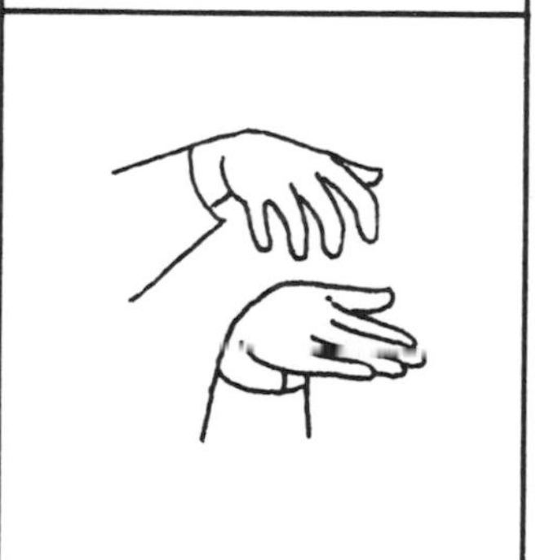

1

confused with

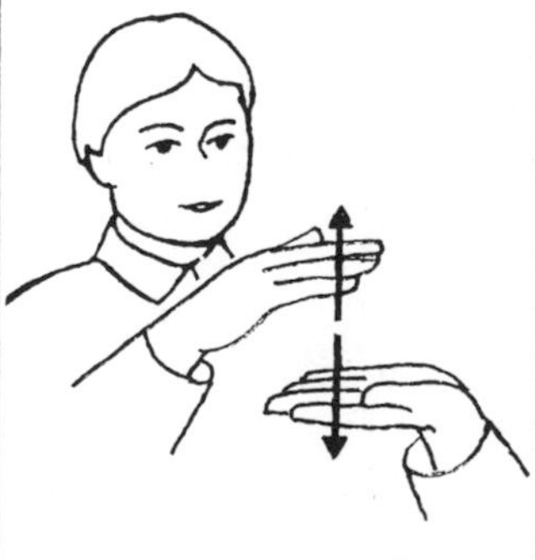

big **1**

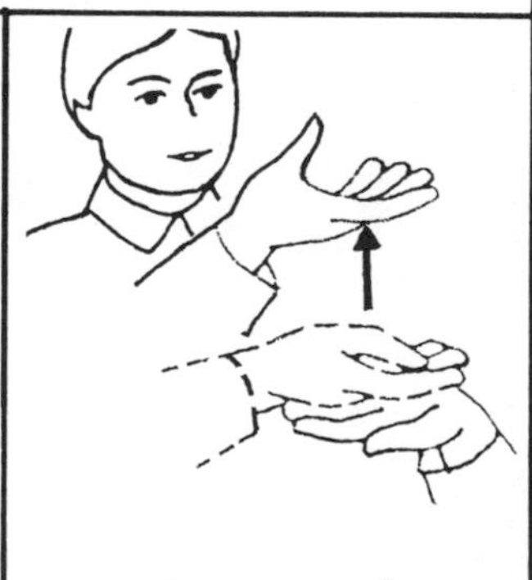

rich **1**

college **2**

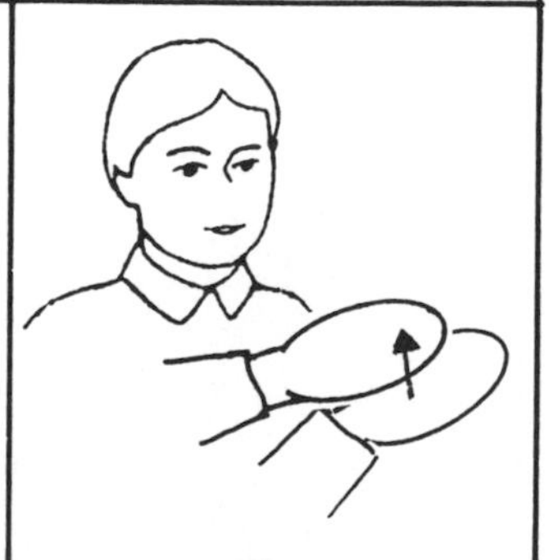

0

KISS

Handshape: Flat O
Location: Side of mouth
Movement: To cheek

understandable

2

confused with

0

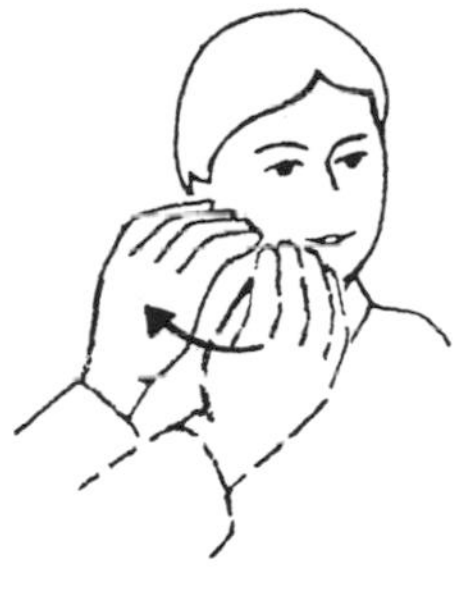

home **5**

bed **1**

0

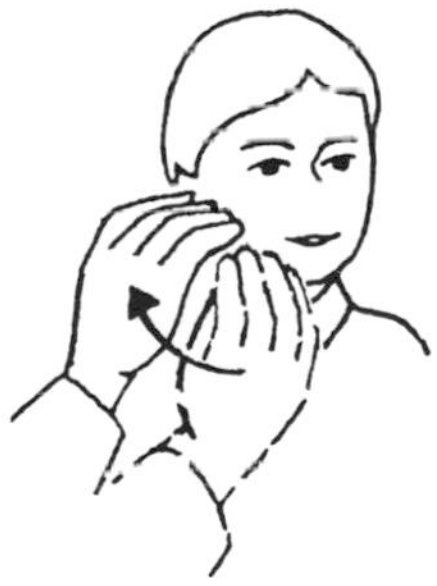

home **2**

white **2**

confused with

0

eat **7**

food **2**

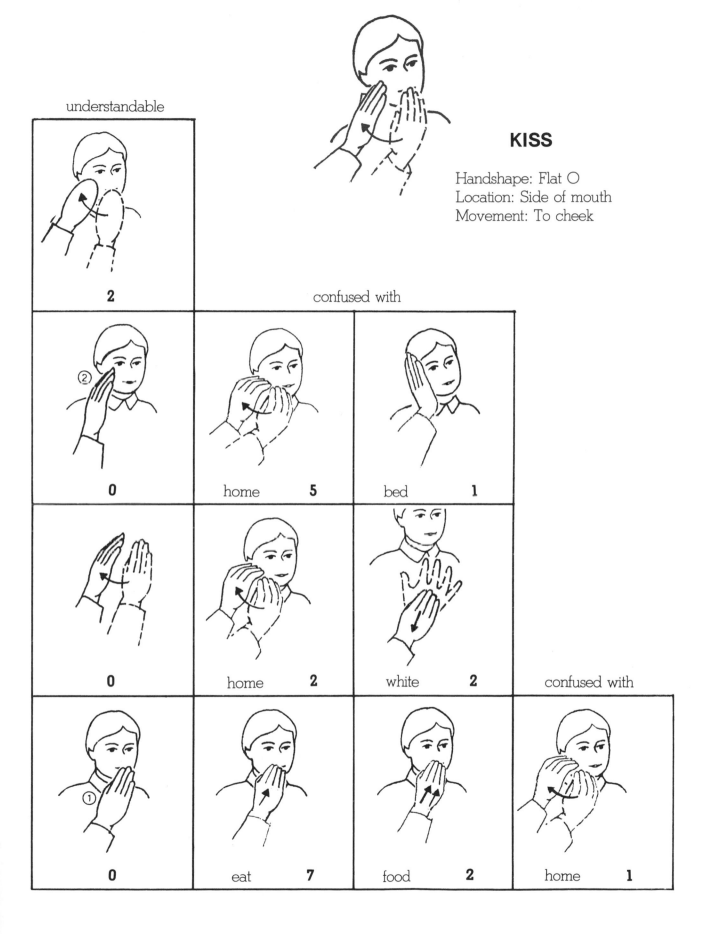

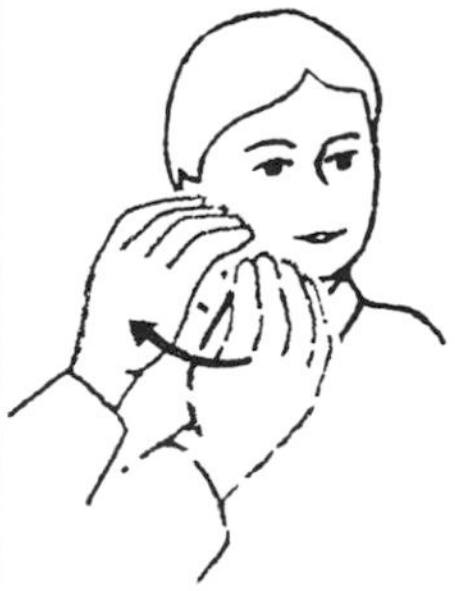

home **1**

JUICE

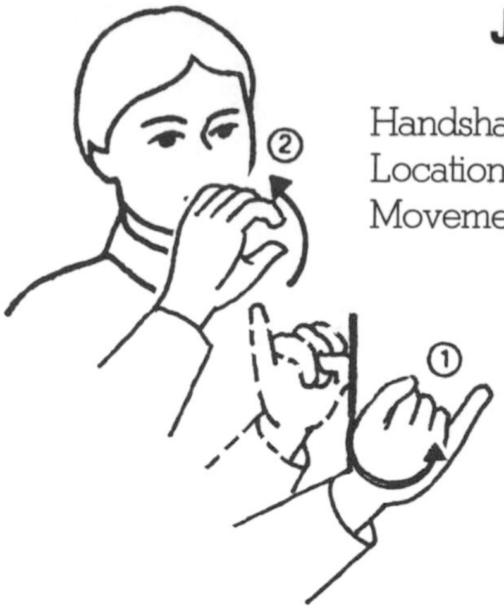

Handshape: J to C
Location: Mouth
Movement: Draw J, mimic drinking

confused with

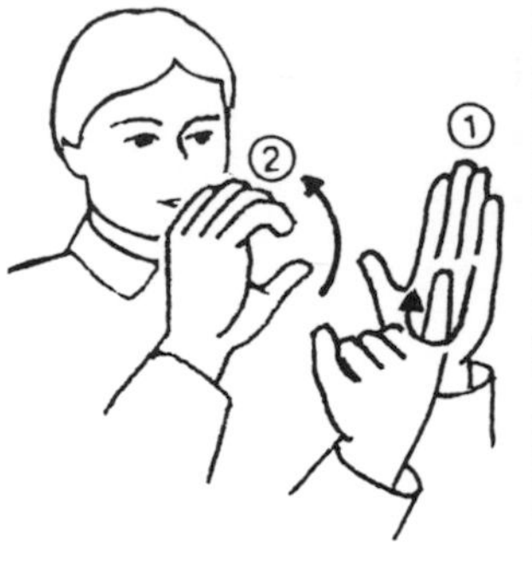

just drink 2

understandable

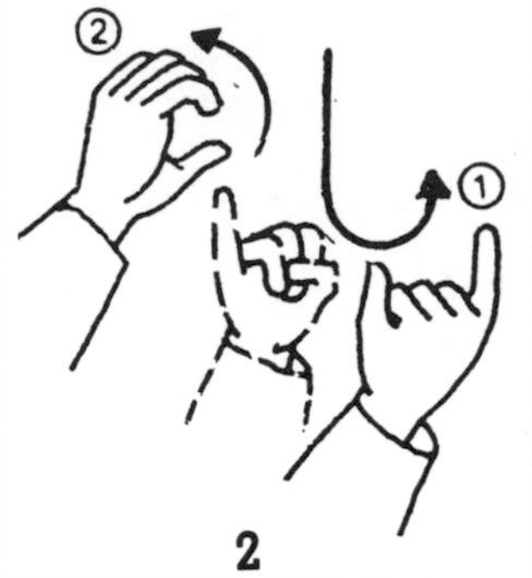

2

confused with

drink 1

just drink 3

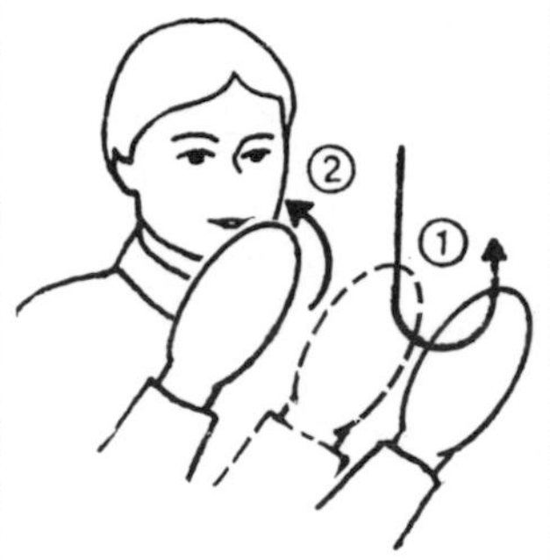

0

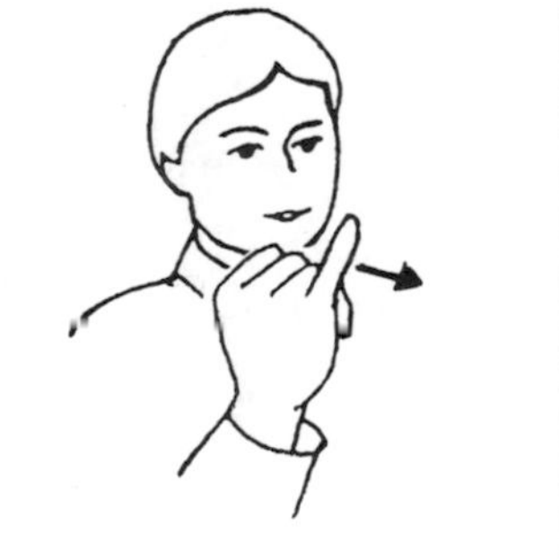

is 10

0

drink 10

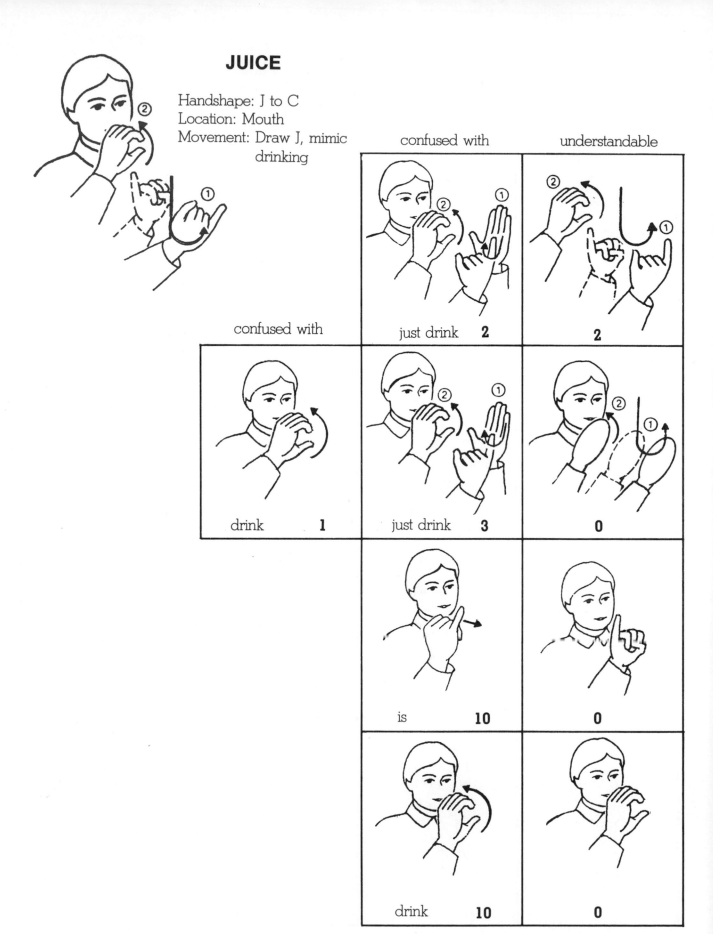

0

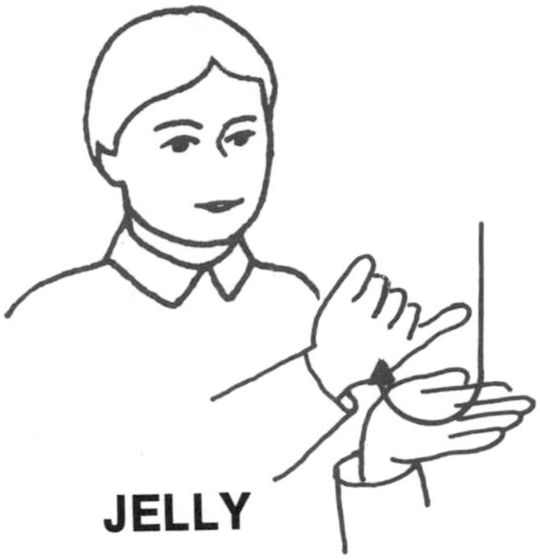

JELLY

Handshape: J; open B
Location: In front of body
Movement: Brush finger on palm

understandable

1

confused with

just 9

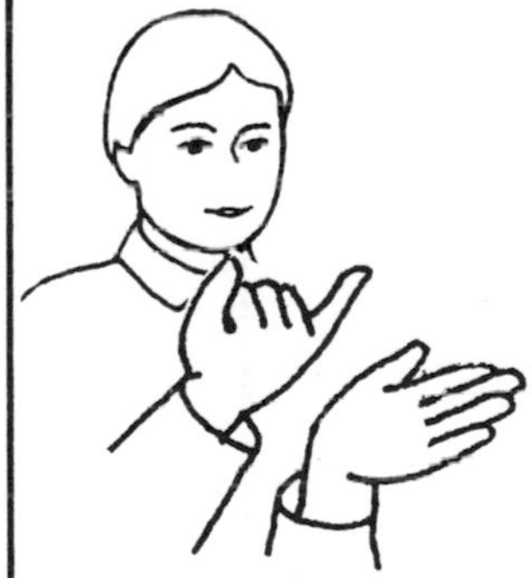

1

just 4

confused with

artist 3

0

broom 2

butter 1

FORK

Handshape: V; open B
Location: In front of body
Movement: Tap twice

confused with | understandable

good night **1**

bad **3**

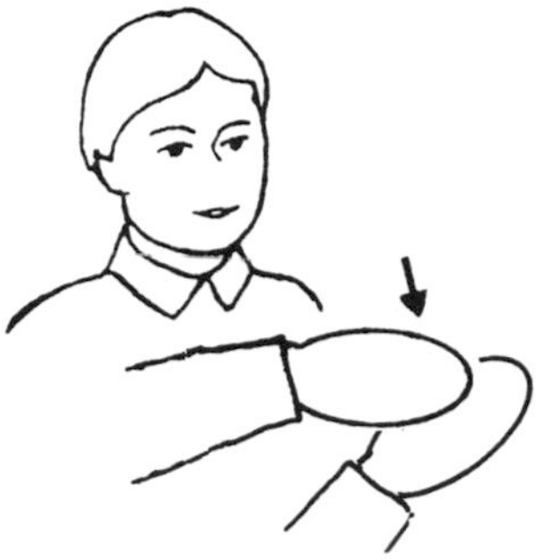

1

slip **1**

double **6**

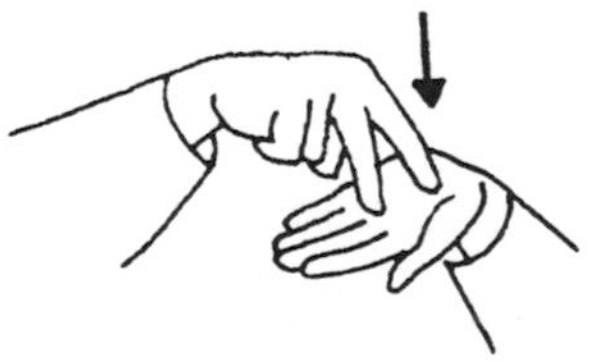

1

stand **9**

0

PANTS

Handshape: Open B, both
Location: Hips
Movement: Brush up twice

understandable confused with

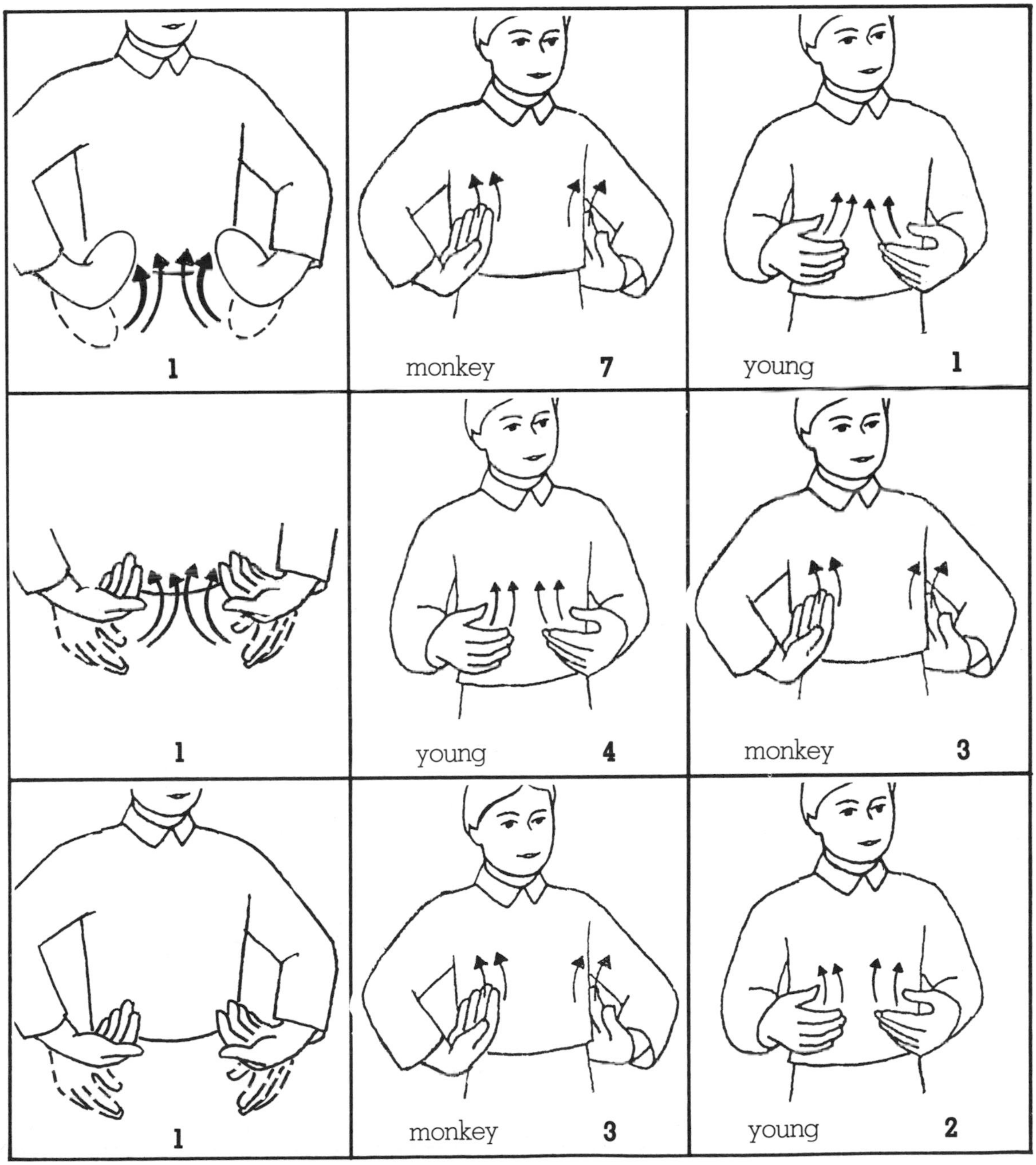

THURSDAY

Handshape: H
Location: In front of body
Movement: Small circle,

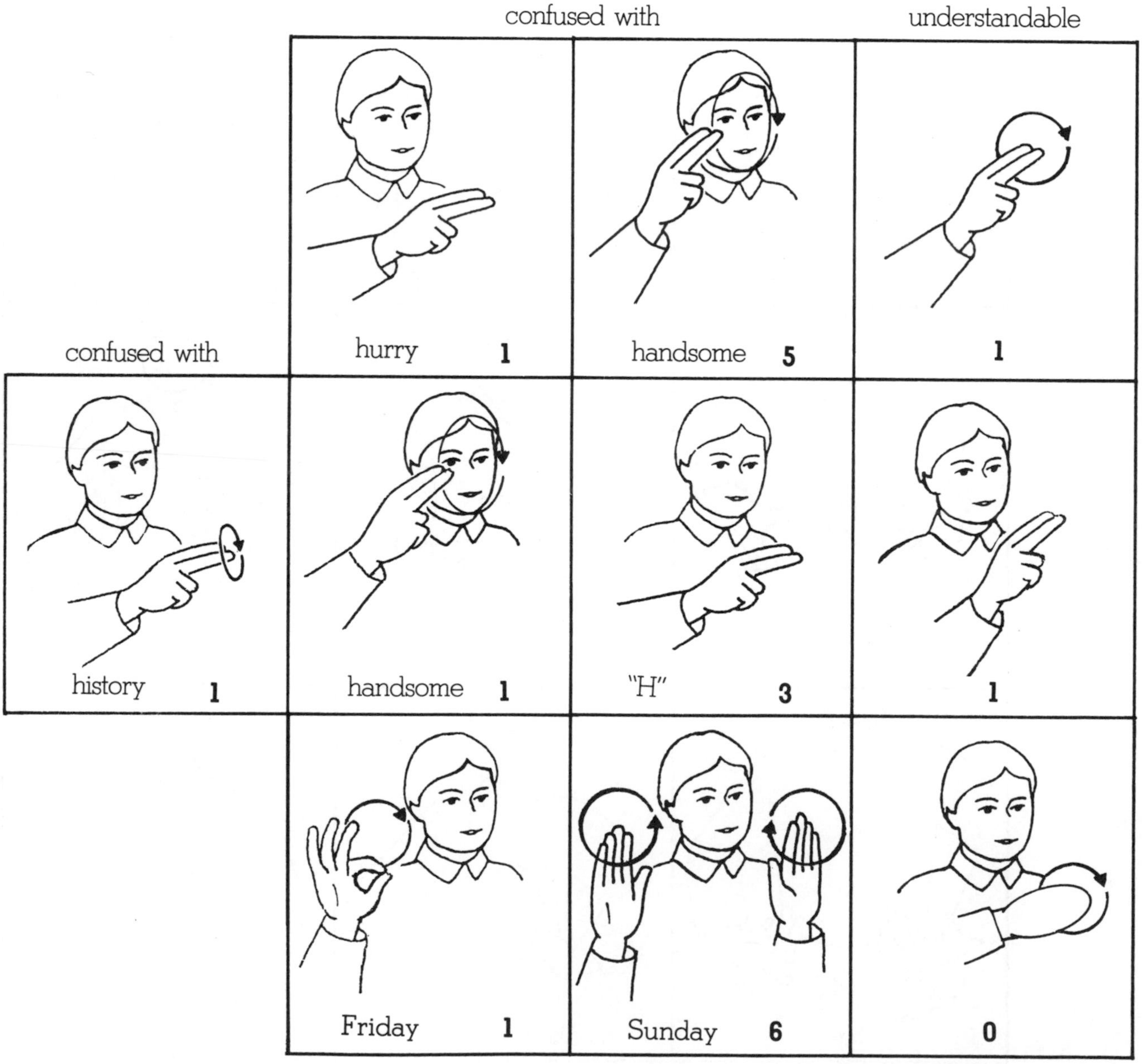

AM

Handshape: A
Location: Mouth
Movement: Forward

understandable	confused with		confused with
 1	 patienace 4	 secret 2	 private 1
 0	 yourself 5	 patient 1	
 0	 be 4		

PEPPER

Handshape: F
Location: In front of body
Movement: Shake

		confused with	understandable
		 preach **7**	 **1**
		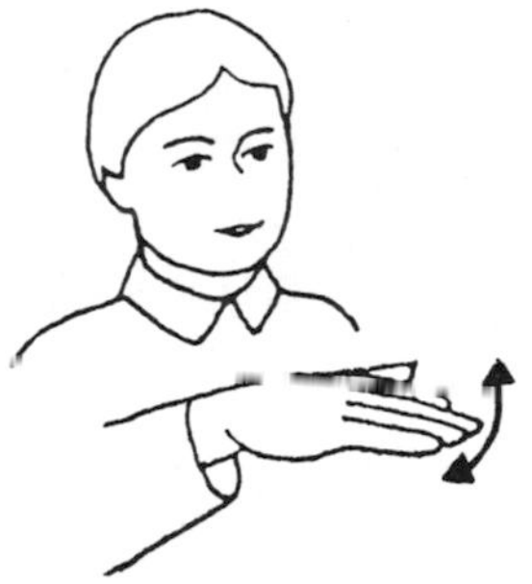 good-bye **9**	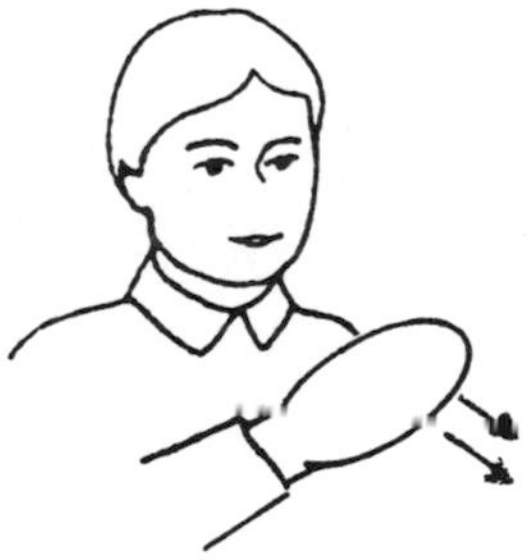 **0**

confused with			
 nothing **1**	 Friday **2**	 nine **2**	 **0**

THOSE

Handshape: Y; open B
Location: In front of body
Movement: Tap palm twice

understandable | confused with | confused with

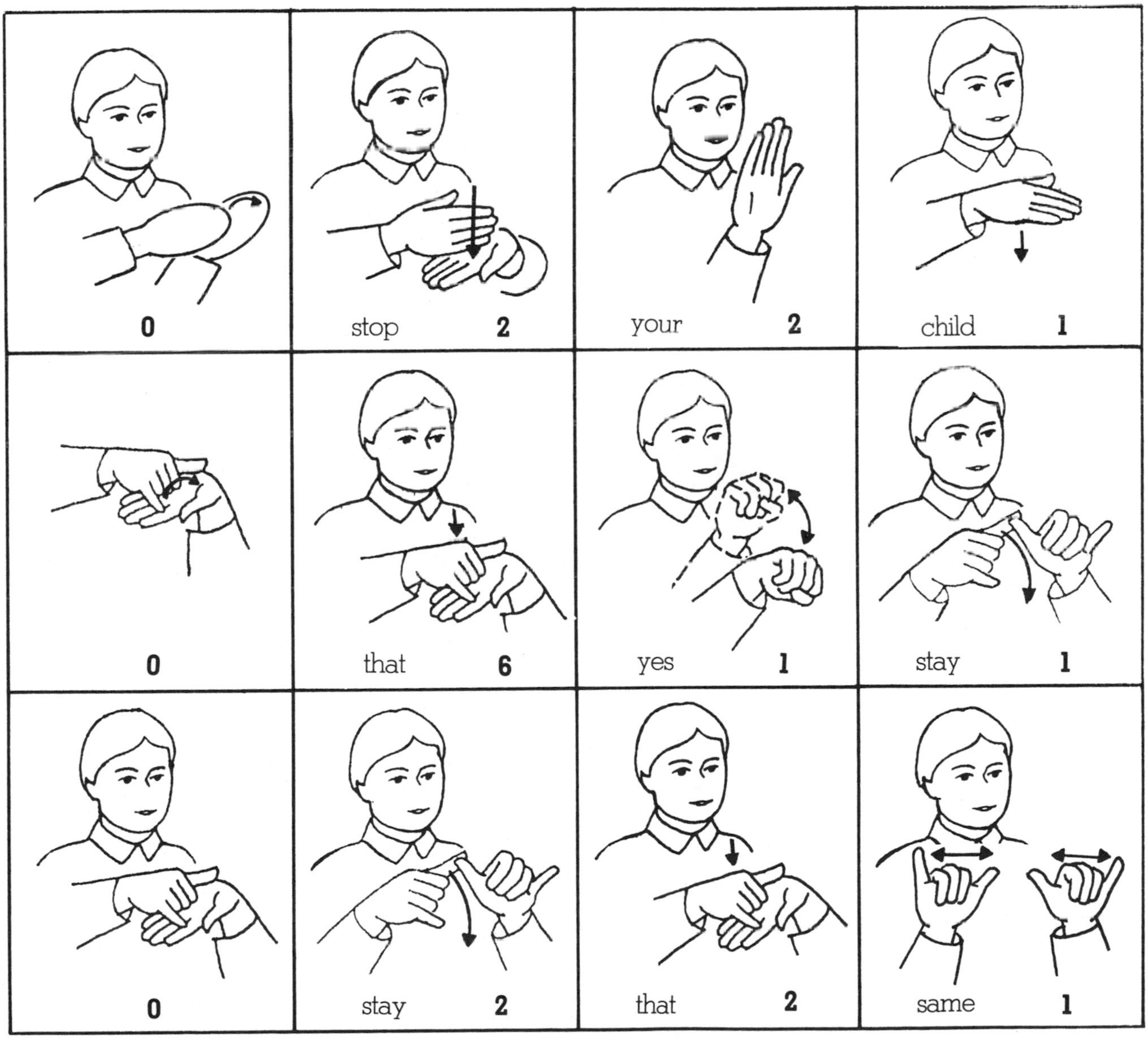

TOY

Handshape: T, both
Location: In front of body
Movement: Back and forth

SHIRT

Handshape: Open B, both
Location: Chest
Movement: Tap, lower, tap again

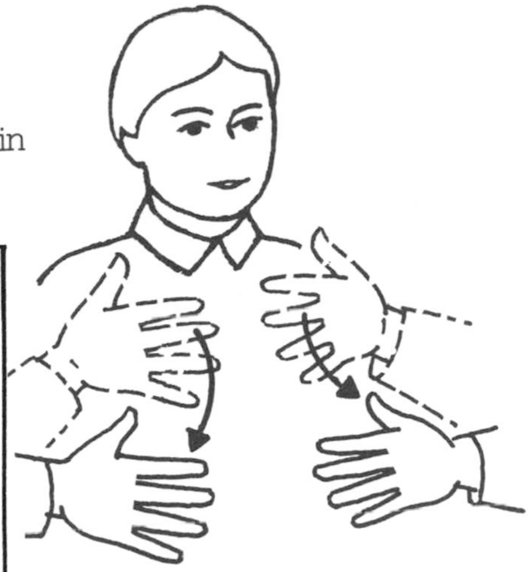

understandable

confused with

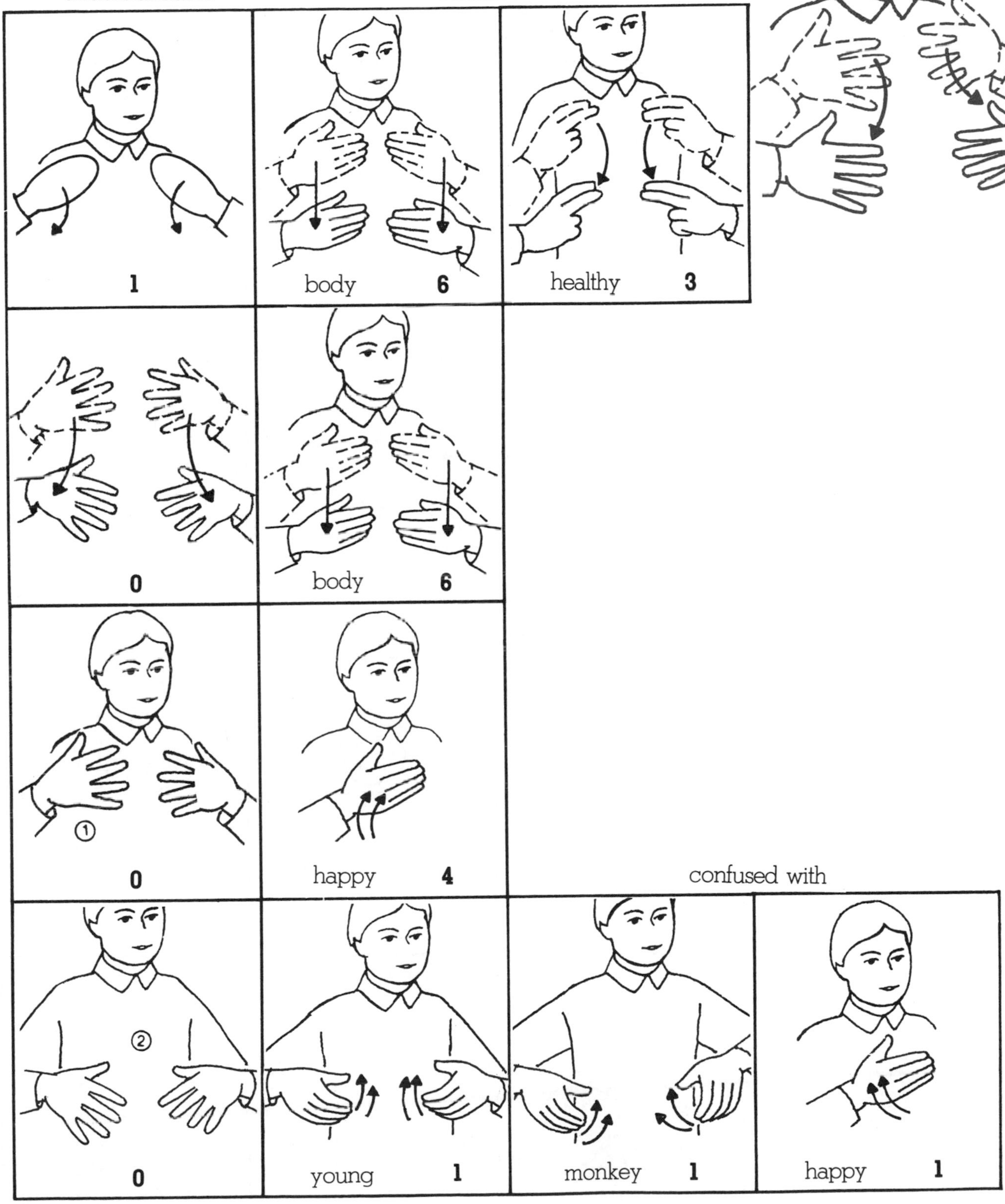

LIGHT

Handshape: Flat O, both
Location: In front of body
Movement: Open fingers

SHAVE

Handshape: X
Location: Cheek
Movement: Scrape twice

understandable | confused with

0

girl 5

confused with

0

should 5

need 5

confused with

0

tomorrow 3

yesterday 3

girl 2

HANDKERCHIEF

Handshape: Bent G
Location: Nose
Movement: Down and close fingers

confused with

understandable

funny **1**

cold **7**

0

confused with

and **1**

white **1**

soft **3**

0

stink **2**

cold **7**

0

pill **1**

eat **4**

0

Appendix

Table A-1. Robustness percentages for *handshape* eliminated condition

Sign	% Recognized	Page	Sign	% Recognized	Page
All	100	87	Drink	90	15
Good morning	100	111	Soda pop	90	79
Morning	99	3	Bread	89	10
Boat	99	36	Brown	89	134
Today	99	110	Time	89	5
Bring	99	62	Money	88	66
Day	99	7	Climb	88	32
Clothing	98	132	Movie	88	53
Talk	98	325	Hand	88	138
Baby	98	6	Man	87	210
Close	97	30	Glass	87	11
House	97	75	Make	87	61
Fish	97	42	Bear	87	13
Why	97	69	Please	86	71
How	97	122	See	86	60
Fall	97	68	Will	86	52
Clean	96	37	Butterfly	86	9
Afraid	96	46	Comb	85	154
Happy	96	27	Head	85	208
Hospital	95	63	Lion	85	47
Window	95	139	Over	84	91
Good	95	77	Ugly	84	149
Dance	95	12	Coffee	83	100
Paint	94	39	Show	83	57
Elephant	94	188	Help	82	18
Cook	94	108	Flower	82	143
Sick	94	38	Swing	82	107
Stop	94	55	Have	81	19
Swim	94	190	Skirt	81	191
Kick	93	23	Bowl	81	80
Out	93	4	Jump	81	78
Before	93	76	Blue	81	50
Paper	93	98	Stand	81	14
Open	93	90	Home	80	130
Bad	93	112	Pig	80	173
In	93	28	Coat	79	265
Here	92	64	Slow	79	119
Table	92	44	Noon	78	120
Tree	91	16	Thirsty	78	213
Our	91	21	Down	78	216
Carry	91	128	School	78	243
Your	91	131	Sweep	77	43
Goodbye	91	17	On	77	141
Afternoon	91	31	Toothbrush	76	253
Book	91	74	Woman	76	271
Sad	91	20	Put	76	29
Buy	91	73	Later	75	33
Quiet	90	22	Pain	75	115
Cry	90	102	Grandmother	74	301

Sign	% Recognized	Page	Sign	% Recognized	Page
Hurt	74	35	Bathe	50	151
Sheet	72	320	Come	50	179
Key	72	41	Up	50	219
Want	71	34	Santa Claus	49	235
Monkey	70	263	Rain	47	148
Box	69	153	Snow	45	268
Laugh	69	220	Angry	45	123
Hungry	69	244	Mouth	44	318
Sunday	69	250	Dog	44	260
Music	68	124	Play	44	144
Now	68	118	Door	43	237
Apple	68	164	Sock	43	228
My	68	117	Bed	43	285
And	68	121	Airplane	43	150
Can't	67	48	Take	43	269
Wash	67	215	Train	43	144
Night	66	40	Get	42	267
Bus	65	221	Rabbit	41	127
Nurse	64	116	Brush	39	315
Love	64	45	But	39	225
Boy	64	249	You	38	126
Give	63	160	Eat	38	137
Go	63	142	Horse	38	222
Yesterday	62	140	Bicycle	38	226
Face	62	310	Move	38	136
Banana	61	183	Cup	37	58
Me	61	56	Sit	37	180
Sign	61	168	Father	36	189
Hat	60	266	Look	36	147
Under	59	125	Breakfast	36	289
Brother	59	290	Pretty	36	224
Butter	59	54	Ride	34	135
Tea	59	164	Draw	33	152
Some	57	217	Store	33	233
Person	56	146	White	33	286
Won't	56	133	Sleep	33	238
Birthday	56	280	Nose	33	195
Black	56	231	Dirty	32	270
When	55	49	Turkey	31	273
Walk	55	51	Squirrel	31	233
Egg	54	172	Sister	31	297
Dress	53	230	Pillow	31	292
Throw	53	259	Mother	31	192
Glove	53	171	Grandfather	29	332
Duck	52	262	Cat	29	187
Fire	51	178	Which	29	59
There	51	256	Party	29	129
Off	51	274	Mop	28	293
We	51	229	Soap	28	232

Sign	% Recognized	Page	Sign	% Recognized	Page
Heavy	27	278	Fork	12	342
Truck	26	218	Hot	12	109
Child	26	291	Shoe	12	72
Hurry	26	65	Ear	11	313
Turtle	26	214	Salt	11	103
I	25	70	Is	10	176
Knife	25	329	Church	10	86
Many	25	236	Cow	10	309
Number	25	67	Name	10	162
After	24	298	Medicine	9	106
Where	24	156	Shirt	8	349
Hair	24	196	All-gone	8	305
Think	23	282	Purple	8	186
Napkin	23	331	Red	8	261
Police	21	227	Like	8	319
Her	21	328	Pencil	8	254
Spoon	21	294	Lunch	7	185
Car	21	239	Girl	7	312
Potato	21	234	Sun	7	321
For	20	303	Ice Cream	7	207
Teeth	20	283	Sandwich	7	311
Thank You	20	296	Big	6	206
Fast	20	170	Milk	6	202
Eye	19	317	Cold	6	99
Kiss	19	339	Teach	6	248
Stay	18	247	Shower	6	258
Do	17	84	Tomorrow	6	276
Candy	17	335	To	6	323
Bird	17	155	Pants	6	343
What	16	200	Four	5	96
This	16	246	Friend	5	95
Chair	15	330	Us	5	181
Doctor	15	275	Right	5	97
Sorry	15	159	Cookie	5	169
Room	14	199	Saturday	5	194
Telephone	14	167	Tuesday	4	272
From	14	165	Green	4	101
Friday	13	203	No	4	161
Same	13	88	Shave	4	351
She	13	316	Letter	3	336
Different	13	288	Light	3	350
Meat	13	85	Yes	3	201
Yellow	13	255	These	3	281
Who	13	287	Water	3	104
Work	13	89	Wednesday	3	306
Can	13	175	Wrong	3	177
Dig	13	325	Not	3	205
Run	13	83	Those	3	347
With	13	245	More	2	163
Towel	12	304	Juice	2	340

Sign	% Recognized	Page	Sign	% Recognized	Page
Glasses	2	307	Scissors	0	284
Dentist	1	314	One	0	252
Monday	1	166	His	0	337
Doll	1	322	Taste	0	327
Break	1	209	Ten	0	257
Old	1	308	That	0	182
Jelly	1	341	Are	0	277
Five	1	174	Am	0	345
Dinner	0	193	Three	0	94
Fight	0	204	Toilet	0	279
Cake	0	338	Thursday	0	344
Handkerchief	0	352	Touch	0	108
Color	0	326	Toy	0	348
He	0	302	Two	0	105
Orange	0	324	Pink	0	295
Ball	0	93	Pepper	0	346
Little	0	251	Write	0	92

Table A-2. Robustness percentages for *location* eliminated condition

Sign	% Recognized	Page	Sign	% Recognized	Page
Fire	100	178	Boat	97	36
Happy	100	27	Today	97	110
Cup	100	58	Tea	97	164
Bring	100	62	Money	97	66
Name	100	162	Walk	97	51
Ball	100	93	Milk	96	202
Baby	100	6	Butter	96	54
No	100	161	Green	96	101
Break	100	209	Hurt	96	35
Which	100	59	Butterfly	96	9
Play	100	144	Climb	96	32
Morning	99	3	Fish	96	42
Party	99	71	Now	96	118
Do	99	84	Number	96	67
Cold	99	99	Out	96	4
Work	99	89	Cook	95	108
Shoe	99	72	Can't	95	48
Make	99	61	Egg	95	172
Church	99	86	And	95	121
Look	99	147	Night	95	40
Same	99	88	Saturday	95	194
Write	99	92	Have	95	19
Three	99	94	In	95	28
Want	99	34	Slow	95	119
Coffee	99	100	Why	95	69
Paint	99	39	Blue	95	50
Later	98	33	Draw	95	152
Friend	98	95	Meat	95	85
Time	98	5	Soda pop	95	79
Yes	98	201	Bread	95	10
Us	98	181	Afternoon	95	31
Angry	98	123	Good-bye	95	17
More	98	163	Buy	95	73
Fall	98	68	Close	95	30
Put	97	29	Stand	94	14
Person	97	146	From	94	165
Dance	97	12	Glass	94	11
Afraid	97	46	What	94	200
Clean	97	37	Paper	94	98
Turtle	97	214	Sign	94	108
Sit	97	184	Glove	94	171
When	97	49	Banana	94	183
Sorry	97	159	Hurry	94	65
Help	97	18	Love	93	45
Run	97	83	Fast	93	170
Rabbit	97	127	Salt	93	103
All	97	87	Monday	93	166
Quiet	97	22	Here	93	64
Nurse	97	116	Over	93	91

Sign	% Recognized	Page	Sign	% Recognized	Page
Move	93	136	Bowl	83	80
Stop	92	55	Hot	82	109
Right	92	97	Train	82	144
Bear	92	13	Bad	82	112
Kick	92	23	On	81	141
Four	92	96	Sad	81	20
Please	92	71	Ice Cream	80	207
My	92	117	Water	80	104
Fight	92	204	This	79	246
Me	91	56	Stay	79	247
Can	91	175	Dog	78	260
Key	91	41	Squirrel	77	223
House	90	75	Pencil	76	254
Open	90	90	Eat	76	137
Talk	90	8	Truck	76	218
Pain	90	115	Many	75	236
Jump	90	78	Five	75	174
Tree	89	16	You	74	126
Show	89	57	Potato	74	234
Good morning	88	111	One	73	252
Won't	88	133	Hungry	73	244
Day	88	7	We	72	104
Two	88	105	Toilet	72	279
Before	88	76	Home	72	130
Ride	88	135	Take	72	269
Will	87	52	Pretty	71	224
That	87	182	All-gone	71	305
Touch	87	180	Wash	71	215
Friday	87	203	Up	71	219
Sweep	87	43	Yellow	71	255
Come	86	179	For	70	303
Go	86	142	But	70	225
Cookie	86	69	Lion	69	47
Under	86	125	Tuesday	68	272
Bathe	86	151	Birthday	68	280
Drink	86	15	Some	66	217
Our	85	21	Ten	66	257
Lunch	85	185	Get	65	267
Purple	84	186	Sock	65	228
Rain	84	148	Woman	65	271
Movie	84	53	Tomorrow	65	276
Box	83	153	Throw	64	259
Table	83	44	School	64	243
Where	83	156	Noon	64	120
Medicine	83	106	Hospital	63	63
Music	83	124	Teach	63	248
Swing	83	107	Heavy	63	278
I	83	70	Hand	63	138
Book	83	74	Bus	63	221

Sign	% Recognized	Page	Sign	% Recognized	Page
Elephant	63	188	Mop	34	293
With	63	245	Door	33	237
Wednesday	63	306	Thank you	33	296
Scissors	63	284	Sister	33	297
Sick	63	38	Monkey	33	263
Horse	62	222	Brother	33	290
Soap	61	232	Like	33	319
Bicycle	61	226	Car	32	239
Little	61	251	To	30	323
Yesterday	58	140	These	28	281
Red	58	261	Pink	28	295
Window	57	139	Taste	27	327
Carry	57	128	Dig	25	325
How	57	122	After	24	298
Snow	57	268	Comb	24	154
Down	56	216	Coat	24	265
Sunday	56	250	Breakfast	21	289
Different	56	288	Dress	21	230
Clothing	56	132	Spoon	21	294
Telephone	55	167	Chair	21	330
Brown	55	134	Knife	20	329
Santa Claus	55	235	Cat	20	187
Wrong	55	177	Grandmother	20	301
Sleep	54	238	Doll	19	322
Sandwich	52	311	Towel	18	304
Thirsty	52	213	Juice	17	340
Old	52	308	He	16	302
See	51	60	Not	15	205
Ugly	49	149	Candy	15	335
White	49	286	Room	15	199
Off	49	274	Turkey	15	273
Shower	48	258	Think	14	282
Is	48	176	There	13	256
Your	47	131	Bed	12	285
Store	46	233	Child	11	291
Flower	45	143	Pepper	11	346
Airplane	45	150	Head	10	208
Dinner	45	193	Cow	10	309
Doctor	44	275	Jelly	9	341
Police	44	227	Give	9	160
Color	44	326	Mother	8	192
Who	44	287	Cake	8	338
Orange	43	324	Thursday	8	344
Girl	42	312	Glasses	8	307
Bird	42	155	Sun	7	321
Good	41	77	Fork	6	342
Black	40	231	Hat	6	266
Swim	38	190	Mouth	6	318
Laugh	37	220	Cry	6	102
Are	35	277	Pants	6	343

Sign	% Recognized	Page	Sign	% Recognized	Page
Toy	5	348	Pig	1	173
She	4	316	Her	1	328
Face	4	310	Nose	1	195
Napkin	3	331	Teeth	1	283
Hair	3	196	Toothbrush	1	253
Am	3	345	Father	1	189
Duck	3	262	Letter	1	336
Skirt	3	191	Handkerchief	0	352
His	3	337	Boy	0	249
Those	3	347	Grandfather	0	332
Kiss	2	339	Big	0	206
Man	2	210	Brush	0	315
Light	2	350	Eye	0	317
Ear	2	313	Apple	0	264
Pillow	2	292	Shave	0	351
Dirty	1	270	Sheet	0	320
Dentist	1	314	Shirt	0	349

Table A-3. Robustness percentages for *movement* eliminated condition (center location)

Sign	% Recognized	Page	Sign	% Recognized	Page
Which	100	59	Same	88	88
Telephone	100	167	Hair	87	196
Sick	99	38	Have	87	19
Morning	99	3	Help	86	18
Write	99	92	Green	86	101
Water	99	104	Sweep	85	43
Run	99	83	Baby	85	6
Father	98	189	Turtle	85	214
You	98	126	Airplane	84	150
Meat	98	85	Five	84	174
Four	97	96	Nose	84	195
Three	97	94	Number	84	67
Stand	97	14	Shoe	83	72
Is	97	176	Noon	83	120
Time	97	5	Medicine	83	106
Ball	96	93	Sad	83	20
Drink	96	15	Not	82	205
Wrong	96	177	Bird	82	155
Out	96	4	Eat	82	137
I	95	207	Cold	81	99
Night	94	40	Me	81	56
Right	94	97	Dance	81	12
Hurry	94	65	Butter	81	54
Lion	94	47	Dinner	80	193
Talk	93	8	Good-bye	80	17
Butterfly	93	9	Salt	80	103
Bear	93	13	With	80	245
Church	92	86	Party	79	129
Cup	92	58	In	79	28
Glass	92	11	Climb	79	32
Want	92	34	Pig	79	173
Friend	92	95	Afternoon	78	31
Key	91	41	Little	78	251
Hurt	91	35	Thirsty	77	213
Do	91	84	Happy	76	27
Love	90	45	Dirty	76	270
Tree	90	16	Police	76	227
See	90	60	Monday	76	166
Big	90	206	Ride	76	135
Bread	89	10	Shower	76	258
Mother	89	192	Cookie	76	169
Cry	89	102	Boy	75	249
Later	89	33	Close	75	30
Cat	88	187	Nurse	74	116
Walk	88	51	Pain	74	115
Work	88	89	Table	74	44
Two	88	105	My	73	117
When	88	49	Sorry	73	159

Sign	% Recognized	Page	Sign	% Recognized	Page
More	73	163	Won't	54	133
Teach	72	248	Think	53	282
Rabbit	72	127	And	52	121
No	72	161	Bicycle	51	226
Now	69	118	Apple	50	164
Wash	69	215	Clothing	49	132
Teeth	69	283	Squirrel	49	223
Angry	69	123	Hat	49	266
There	67	256	Yellow	49	255
Duck	67	262	Stop	49	55
Movie	66	53	Pretty	49	224
Laugh	66	220	Skirt	48	191
Name	66	162	Lunch	48	185
Will	66	52	Cow	48	309
Your	66	131	Purple	47	186
Clean	65	37	Stay	46	247
Under	65	125	Ear	46	313
One	65	252	Rain	45	148
Boat	65	36	Hand	44	138
Ten	64	257	Look	41	147
Some	64	217	Play	40	144
Blue	64	50	Store	40	233
Paint	62	39	Ugly	40	149
Up	62	219	Doctor	40	275
Red	62	261	Sun	40	321
Touch	62	180	Comb	40	154
Glasses	61	307	Make	38	61
Are	61	277	Go	38	142
These	61	281	Eye	38	317
Dress	61	230	Soap	35	232
Carry	60	128	Here	35	64
Music	59	124	Draw	33	152
Horse	59	222	Tuesday	33	272
Show	59	57	Pillow	33	292
How	59	122	Doll	33	322
This	59	246	Bed	31	285
Toothbrush	59	253	On	31	141
That	58	182	Sock	31	226
Flower	58	143	Bathe	29	151
Down	58	216	Child	29	291
Fish	56	42	Black	29	231
Slow	56	119	Who	28	287
Dentist	56	314	Car	28	239
Train	56	144	Pink	27	295
Brown	55	134	Scissors	26	284
Can	55	175	Money	24	66
Towel	54	304	Where	24	156
Afraid	54	46	Potato	24	234

Sign	% Recognized	Page	Sign	% Recognized	Page
Person	24	146	Am	8	345
Please	24	71	Jelly	8	341
Saturday	23	194	Fight	8	204
Toilet	21	279	Heavy	6	278
Over	20	91	Dog	6	260
Brush	20	315	Mouth	6	318
Hungry	19	244	Pants	6	343
School	19	243	Sit	6	184
Monkey	18	263	Thursday	5	344
Yes	18	201	Get	5	267
Spoon	17	294	Orange	5	324
Old	17	308	Coffee	4	100
Tea	17	164	All	3	87
Taste	15	327	Ice cream	3	207
Glove	15	117	Sandwich	3	311
Wednesday	15	306	Mop	3	293
Milk	14	202	Chair	3	330
What	14	200	Grandfather	3	332
Come	14	179	White	3	286
Sunday	14	250	Those	3	347
Breakfast	13	289	Color	2	326
Friday	13	203	Fire	2	178
Grandmother	13	301	Toy	1	348
Throw	13	259	Swim	1	190
Girl	13	312	All-gone	1	305
Sign	13	168	Face	0	310
Egg	11	172	Fork	0	342
Napkin	11	331	Off	0	274
Swing	11	107	Juice	0	340
Snow	9	268	Shave	0	351
Candy	9	335	Paper	0	98
Dig	8	325	Pepper	0	346

Table A-4. Robustness percentages for *movement* eliminated condition (initial/final locations)

Sign	% Recognized	Page	Sign	% Recognized	Page
Day	95	7	Jump	25	78
Put	92	29	Box	24	153
Can't	82	48	Many	22	236
Give	82	160	Different	20	288
Our	78	21	Sleep	18	238
Room	75	199	To	17	323
Truck	71	218	Like	17	319
Hospital	68	63	Soda pop	17	79
Yesterday	67	140	Her	16	328
Move	66	136	Banana	16	183
Quiet	65	22	Take	16	269
Home	61	130	Letter	14	336
From	60	165	For	14	303
Hot	55	109	Elephant	13	188
Fast	53	170	His	11	337
Window	52	139	Head	10	208
Good	50	77	Cake	10	338
Pencil	48	254	Open	8	90
Bring	48	62	Coat	8	265
Kick	48	23	Thank you	7	296
Us	48	181	Cook	7	108
But	47	225	After	6	298
Bus	46	221	Break	6	209
Turkey	44	273	Man	5	210
Book	41	74	Knife	3	329
Why	39	69	Kiss	3	339
Fall	38	68	Bad	2	112
He	36	302	Today	2	110
Buy	35	73	Handkerchief	1	352
Before	32	76	Woman	1	271
Tomorrow	31	276	Light	1	350
House	31	75	Good morning	0	111
She	30	316	Brother	0	290
We	29	229	Sheet	0	320
Door	27	237	Birthday	0	280
Bowl	27	80	Sister	0	297
Santa Claus	25	235	Shirt	0	349

Index

All words and numbers in **boldface** represent the target functional vocabulary; all others represent signs that were confused with the simpler forms.